Chemistry at Oxford
A History from 1600 to 2005

Chemistry at Oxford
A History from 1600 to 2005

Edited by

Robert J.P. Williams, John S. Rowlinson and Allan Chapman
University of Oxford, Oxford, UK

RSCPublishing

The picture on the cover shows the Chemistry Laboratory of 1860, the Abbot's Kitchen, adjacent to the Research Laboratory of 2004. The pictures appear separately in the text.

LEEDS LIBRARIES AND INFORMATION SERVICE	
LD 4060725 9	
HJ	07-Sep-2009
507.1	£54.95
S035179	

ISBN: 978-0-85404-139-8

A catalogue record for this book is available from the British Library

© Royal Society of Chemistry 2009

All rights reserved

Apart from fair dealing for the purposes of research for non-commercial purposes or for private study, criticism or review, as permitted under the Copyright, Designs and Patents Act 1988 and the Copyright and Related Rights Regulations 2003, this publication may not be reproduced, stored or transmitted, in any form or by any means, without the prior permission in writing of The Royal Society of Chemistry or the copyright owner, or in the case of reproduction in accordance with the terms of licences issued by the Copyright Licensing Agency in the UK, or in accordance with the terms of the licences issued by the appropriate Reproduction Rights Organization outside the UK. Enquiries concerning reproduction outside the terms stated here should be sent to The Royal Society of Chemistry at the address printed on this page.

Published by The Royal Society of Chemistry,
Thomas Graham House, Science Park, Milton Road,
Cambridge CB4 0WF, UK

Registered Charity Number 207890

For further information see our web site at www.rsc.org

Preface

The University of Oxford is almost eight hundred years old. It is an unusual university in that it developed as a collection of many independent colleges loosely held together within it. It was only four hundred years after its foundation that chemistry was recognised as a separate science. At that time Oxford became one of the, if not the, first universities to have both a chemistry professor and a laboratory. Subsequently, that is after 1700, it lost its leading position and when Oxford chemistry was revived from 1850 to the First World War it was split between university and college interests and responsibilities. It has taken most of the next one hundred years to today to resolve the ensuing academic and organisational problems. During Oxford chemistry's idiosyncratic history the relationship between it and subjects such as geology (mineralogy) and medicine, which were its origins, have changed and we do not follow such subjects once they became located in separate departments. Both in the progression of college/university relations and of subject matter changes have not been made so much on a rational basis as by the influence of particular people. This of course is the case of the history of any local human endeavour. It is this that makes the history of chemistry in Oxford so fascinatingly different from that of anywhere else and from the rational development and growth of the subject everywhere much though we shall draw attention to Oxford's considerable contribution to that growth.

The index of names covers only those who have contributed to the teaching, research and administration of chemistry at Oxford and occasionally elsewhere. Authors of secondary sources are not indexed. Biographical information is given in full only for the leading figures and is minimal for those who are still alive.

We thank here the following for general advice and criticism while acknowledgment of more specific assistance will be given in the chapters, the figure legends and the tables.

Chemistry at Oxford: A History from 1600 to 2005
Edited by Robert J.P. Williams, John S. Rowlinson and Allan Chapman
© Royal Society of Chemistry 2009
Published by the Royal Society of Chemistry, www.rsc.org

John (Michael) Brown, Cliff Davies, Bob Denning, Robert Fox, John Freeman, Susan Green, Peter Grout, Karl Harrison, Annie Jamieson, John H. Jones, Paul Kent, Jeremy Knowles, Matthew Lodge, Graham Richards, Rex Richards, John Shorter, Tony Simcock, Keith Waters.

Thanks are also due to the Library staff of the Royal Society of Chemistry for their help in finding biographies in the Society's publications, to Joyce Thompson and Colin George of the University Estates Directorate, to Simon Bailey, University Archivist, and Robin Darwall-Smith, Archivist of Magdalen College, Oxford.

Susie Compton acted as our general secretary and deserves our special gratitude.

Abbreviations of commonly used sources,

DSB	*Dictionary of Scientific Biography*, ed. C. C. Gillispie, 18v., New York, 1970–1981
ODNB	*Oxford Dictionary of National Biography*, Oxford University Press, 60v., 2004
HCP	*Hebdomadal Council Papers, OUA*
OUA	*Oxford University Archives*
OUG	*Oxford University Gazette*

Oxford, April 2008
R.J.P.W.
J.S.R.
A.C.

Contents

Chapter 1   **An Outline of the History of Oxford University with Reference to its Chemistry School**
Robert J.P. Williams

1.1	An Introduction to the University	1
1.2	The Beginnings of Chemistry within the University	4
1.3	The Creation of Chemistry Departments	8
1.4	The Teaching of Chemistry	12
1.5	A Summary of Chemistry's Development	13
	References	15

Chapter 2   **From Alchemy to Airpumps: The Foundations of Oxford Chemistry to 1700**
Allan Chapman

2.1	Late Medieval English Alchemy	17
2.2	The Hon. Robert Boyle and his Chemical World	20
2.3	Where were the Laboratories?	25
2.4	Oxford's 'Invisible' Chemists: The City Apothecaries and their Laboratories	26
2.5	The Oxford Airpump Discoveries	30
2.6	John Mayow	34
2.7	Thomas Willis	36
2.8	The Revd John Ward: Amateur Chemist and Physician	38
2.9	The Ashmolean Laboratory, 1683	40
	Acknowledgements	45
	Notes and References	46

Chemistry at Oxford: A History from 1600 to 2005
Edited by Robert J.P. Williams, John S. Rowlinson and Allan Chapman
© Royal Society of Chemistry 2009
Published by the Royal Society of Chemistry, www.rsc.org

Chapter 3 The Eighteenth Century: Chemistry Allied to Anatomy
Peter J. T. Morris

3.1	Introduction	52
3.2	Chemistry in the Eighteenth Century	53
3.3	Oxford in the Eighteenth Century	56
3.4	The Teaching of Chemistry in Eighteenth-Century Oxford	58
3.5	The Revival of Chemistry after 1775	65
3.6	Conclusion	71
	References and Notes	73

Chapter 4 Chemistry Comes of Age: The 19th Century
John S. Rowlinson

4.1	The Aldrichian Chair	79
4.2	Charles Daubeny and Reform	83
4.3	The Museum	93
4.4	Benjamin Brodie	96
4.5	William Odling and his Demonstrators	103
4.6	The College Laboratories and the Growth of Physical Chemistry	113
	References	123

Chapter 5 Research as the Thing: Oxford Chemistry 1912–1939
Jack Morrell

5.1	Introduction	131
5.2	The Impact of Perkin	132
5.3	The Contributions of the Colleges	141
5.4	The Mancunian Inheritance	157
5.5	The Dr Lee's Chair and Old Chemistry	164
5.6	The Chemical Synthesiser	170
5.7	X-Ray Crystallography	173
5.8	Careers: The Lure of Industry	175
5.9	Conclusion	178
	Notes and References	179

Chapter 6 Interlude: Chemists at War
John S. Rowlinson

References	192

Chapter 7 Recent Times, 1945–2005: A School of World Renown
Robert J.P. Williams

7.1	General Introduction to the Period: The Three Centres of Influence		195
	7.1.1	The Three Periods 1945 to 1965, 1965 to 1980, 1980 to Today	199
	7.1.2	Summary	204
7.2	Recruitment and the Nature of Professorships and Fellowship/Lectureships		206
	7.2.1	A Note on Women Fellows in Chemistry	213
7.3	The Undergraduate Entry into Oxford and the Chemistry Course		213
	7.3.1	The Butler Education Act 1944	213
	7.3.2	The Structure of the Chemistry Course	216
	7.3.3	The Content of the Undergraduate Course	218
	7.3.4	The Graduate School	224
7.4	The Three Professors and the Three Departments of 1945		225
	7.4.1	Hinshelwood and Physical Chemistry	226
	7.4.2	Robinson and Organic Chemistry	231
	7.4.3	The Third Professor: Sidgwick	236
	7.4.4	The Acting Heads and Nature of the Third Laboratory of Inorganic Chemistry (1945–1963)	238
7.5	Research 1945 to 1965		241
	7.5.1	Theory and Mathematical Research	244
	7.5.2	Summary	246
7.6	Research 1965 to 1980		247
	7.6.1	The Revival of Inorganic Chemistry	247
	7.6.2	Organo-Metallic Chemistry	251
	7.6.3	Traditional Organic Chemistry	253
	7.6.4	Physical Chemistry	254
	7.6.5	Chemical Crystallography and Biophysics	256
	7.6.6	Theoretical Chemistry and its Short-Lived Department	258
	7.6.7	The Enzyme Group	258
	7.6.8	Life in Oxford, 1945–1980	261
7.7	Research: 1980 to 2005		262
	7.7.1	Introduction	262
	7.7.2	Physical Chemistry	263
	7.7.3	Theoretical Chemistry Department	266
	7.7.4	Organic Chemistry	267
	7.7.5	Inorganic Chemistry	270
	7.7.6	Oxford Chemistry Today, 2008	273

Appendix 1   The Laboratories   275
Acknowledgement   282
Appendix 2   The Chemistry School Finances   282
Acknowledgement   283
Notes on Oxford University   283
References   286

Index of Names   292
Index of Subjects   301

CHAPTER 1

An Outline of the History of Oxford University with Reference to its Chemistry School

ROBERT J.P. WILLIAMS

Inorganic Chemistry Laboratory, Oxford University, South Parks Road, Oxford OX1 3QR

1.1 AN INTRODUCTION TO THE UNIVERSITY

We give this brief historical introduction to Oxford University to show how it has become so particularly constituted. Without such an introduction readers might well be puzzled as to how and why the university has become as it is today and why its history has greatly affected its chemistry school.[1] The usual structure of universities and of their departments elsewhere is that they are more or less top-down in governance. There is a governing body or council including a Vice-Chancellor and it, with an administration, effectively makes policy much though a democratic assembly of professors and some lecturers, the Senate, can vet proposals. This assembly is mainly a safeguard against mal-administration but it has little real control over finances and major policy decisions.[2] Within this structure, a chemistry department is constrained by its head, a professor, who manages, largely, the distribution of its finances and staffing. The lecturers have a well-defined set of duties beneath the professor but with a fair degree of freedom in research objectives. It would be stretching a point to say that the department is democratically organised. In fact it could well be that a genuinely democratic system is not to the advantage of a department or a university. It must be remembered that the English universities, to which we refer with the

Chemistry at Oxford: A History from 1600 to 2005
Edited by Robert J.P. Williams, John S. Rowlinson and Allan Chapman
© Royal Society of Chemistry 2009
Published by the Royal Society of Chemistry, www.rsc.org

1

exception of Oxford and Cambridge, were established around or after 1850 to serve their local communities, were set up with a model in mind, and with a bias toward sciences linked to industry. Their birth and whole subsequent background is so different from that of Oxford that they serve only to show how very different Oxford and its chemistry department are due to their history.

In marked contrast then to the other British universities, Oxford University is similar only to Cambridge University in that both organisations, including their chemistry schools, are largely bottom-up, Cambridge to a lesser degree than Oxford.[1,3-5] Here, bottom-up implies that ultimate power over all matters rests largely with the collective will of individuals, professors and fellow/lecturers, not with central authority. Unlike that in other universities, appointment at Oxford of a lecturer today is joint, with a fellowship, between the university and a college, hence the description fellow/lecturer. It is the dual position with two sets of duties together with the separate creation of professorships that has led to the peculiarities of Oxford chemistry.

Oxford University is associated in the first instance with different colleges, 39 today, listed in Table 1.1, each with an independent governing body of fellows (tutorial teachers), statutes and finances. For several centuries the colleges as they grew in numbers paid little regard to any central university authority, which in any event was managed by a committee of their heads. A fellow was

Table 1.1 List of colleges with chemistry fellows today.

'Men's' Colleges	'Women's' Colleges
University (1249)	Lady Margaret Hall (1878)
Balliol (1263)	Somerville (1879)
Merton (1264)	St Hugh's (1886)
Exeter (1314)	St Hilda's (1893)
Oriel (1326)	St Anne's (1952)
Queens (1340)	
New (1379)	
Lincoln (1427)	
Magdalen (1458)	
Brasenose (1509)	
Corpus Christi (1517)	
Christ Church (1546)	
Trinity (1554)	
St John's (1555)	
Jesus (1571)	
Wadham (1610)	
Pembroke (1624)	
Worcester (1714)	
Hertford (1740)	
Keble (1868)	
St Edmund Hall (1938)	
St Peter's (1947)	
St Catherine's (1963)	

There are eleven additional graduate colleges not necessarily with chemistry fellows. The colleges today are all mixed, women and men, and the above listing is by origin.

active in a college but took little part in the university. However, in the nineteenth and especially the twentieth century, with the expansion in numbers and academic topics, increased common centralised powers and actions were obviously necessary in addition to the previously accepted minor common need of the university for a body with control over regulations and examinations leading to degrees.[1] The new central requirements especially after 1850[6] were increasingly for an administrative centre, linked to better lecturing and examining facilities, inter-college exchange of view and cooperation, laboratories and more extensive libraries, and to meet the demands from outside authority for the university to show that it, like other universities, served the community at large. The role of a college fellow then increased in the university for he was required to assist in lecturing in courses for all undergraduates. The number of faculties and sub-faculties quickly increased, all linked to the university. The resultant augmented power of central administration was in fair part due also to its increasing financial strength given to it for these purposes. It became financed by government grants, not given to colleges, and it had to respond step by step, as far as possible, to the conditions attached. By 1965 it had appointed three professors of chemistry with independence from colleges in newly built laboratories in which the professors and fellow/lecturers had their own research units. It is easy to see how this change in financial strength and responsibilities with the need to organise led to a clash between the bottom-up system, with its highly independent internally democratic colleges, and the requirement to make top-down overall decisions. However, any top-down proposal was and is still open to a vote by Congregation, effectively all the fellows. The problem became more noticeable in chemistry especially after 1912 and is there today for all to see[7] and its effects upon the chemistry school are also still very apparent. As the university took over the running of laboratories from the colleges the very strong link between a chemistry fellow and a college has slowly been replaced by a stronger link to the laboratory, effectively the university.

We give next a somewhat more detailed historical account since it will serve to make clear the way the chemistry school at Oxford has developed within the collegiate and dominantly an arts university.[1] In this account we shall stress the marked contrast between any top-down governance and largely bottom-up Oxford in which, as we have indicated, individuals or groups of them, such as heads of colleges, differently minded professors or even fellow/lecturers, can exert considerable influence from their power-bases upon the overall organisation, the administrative centre, defining the degree to which the centre can act. This is the background against which the changes in teaching and research have evolved, and is essential for an understanding of the unique nature of the Oxford Chemistry School. We shall see that in fact individuals have coloured the School very considerably. There has been little attempt at central management of it until relatively recently. The very essence of the chemistry school even today arose from the roles played by single colleges and, quite separately, the university in appointments, financing, teaching and laboratory construction sometimes due to individual initiatives.

1.2 THE BEGINNINGS OF CHEMISTRY WITHIN THE UNIVERSITY

The University at Oxford began to take shape in the form of a loosely organised body of scholars who had taken up residence in Oxford in and before 1200. Certainly by 1230 this loose corporation of scholars or masters could be called a university with a Chancellor but without a written constitution.[1] The Chancellor could refer matters to the congregation of masters for judgement. The teaching, in somewhat confusingly called 'schools', was based on masters who came to Oxford and had most frequently obtained a qualification from an earlier 'school' or university, such as Paris. They taught young men from grammar schools who had knowledge of Latin and who were to be trained in modes of argument based on logic or religious texts. In this way they became, after obtaining a qualification given by the university, able to be lawyers or clergymen for example. The Oxford 'schools' before and after becoming somewhat organised were financed in part by monies in the form of income from well-wishers and in part by fees. Of considerable interest to the subject matter of this book is that the students were made to read and discuss the works of Aristotle amongst others as probably the only basis of any taught science, at that time a part of philosophy. There is in these works only an attempt to describe the environment we experience by way of very general concepts that were applied to all the subject matters we now divide in sciences such as physics, botany, zoology, chemistry, geology and so on. Chemistry as a separate discipline did not exist.[3] By 1355 a charter, backed by royal authority, raised the status of the university to be the dominant power in the City of Oxford. Notice that a powerful external body was then concerned about the university but it is only after five more centuries that it began to exert a strong influence on it except concerning religious matters.

In the earliest times scholars resided in various protected large private houses or halls but these halls had but a small income from endowments given by persons of little stature. The teaching was done in rooms associated with these halls all under general university regulations[1] but the university had no premises of its own though it used the Congregation House, built in 1320, in St Mary's Church. By bringing together a small number of halls, now with endowments from particular rich and often influential persons, some of these collective units became known as colleges, born as Christian Catholic societies. At first these colleges, see Table 1.1, were still little more than organised places in which to live and to be taught, protected by isolation from the Town. Their scholars, from what we now call students or undergraduates all the way up to the Masters, now called fellows, were still embraced within the university and at this stage the colleges had little independence in matters of teaching. A point of great interest for this book and the particular development of its subject matter, chemistry at Oxford, in the whole context of the colleges and the university is that this dominance of the university was lost to the colleges. Before 1600 the colleges had increased in number to fifteen and had gained greatly relative to the university through endowments and fees. They had growing privileges associated with their increasing wealth and each had its own statutes and a "democratic" governing body composed of fellows, which elected any addition to its numbers. They now used their resources to turn what had been merely places of board and lodgings into independent centres of learning with a nascent tutorial system. What a fellow of a given college did in 'research' was a matter for him. The colleges still resembled religious establishments in that their resident tutors, fellows of colleges, with a head, all appointed internally, were looked after (board and lodgings) by the individual colleges. Fellows were not permitted to marry (until after 1870) as was the case in foundations based on monastic traditions. Each college came to have it own library, dining rooms and chapel. Any visitor to Oxford today can see that the earliest foundations such as University, Balliol, Merton, and Exeter Colleges, quickly had impressive buildings quite separate from the university that had virtually none until after 1500. The governance of the university itself had also largely fallen into the hands of heads of colleges leaving the university with the management of lectures, examining, the giving of degrees and a general voice in the moral and religious behaviour and practices of college fellows. Individual fellows of colleges clearly did not need to bother much about the university. It is the creation of this loose organisation that has led to a bottom-up system in the university.

At this time the university had not changed its governance greatly. It had a Chancellor and a Vice-Chancellor who worked with the Congregation of Masters (fellows) to consider the programmes of work that had to be completed for a degree. However, by 1600 there had been a development of academic staff separate from the colleges in that the University had received some endowments for professorships,[1] which could be thought of as maintaining a stronger line in academic work from the main body of college fellows and they were not teaching (tutorial) fellows. Remember that fellows were not appointed or paid by the university, as the professors were, but by the colleges. The professors gave public university lectures but the distinction between lectures given by professors and by fellows of colleges was not closely made at first. Very generally speaking, by accepted statutes a professor came to be an authority in a subject and was paid by funds, deliberately set aside by endowment or later by the university. He was usually made a fellow of a specific college and allowed to dine but not necessarily to live there, certainly not if he was married. The upshot of this well-meant intention of broadening teaching and allowing research separate from an internal college base had its merits and mattered little in Arts subjects but much later it did create an unforeseeable problem within chemistry and sciences generally. Once professorships in these subjects (here chemistry in particular) were associated with laboratories they became new centres of power separate from the colleges, as we shall see. Sciences such as chemistry could and did grow separately in the university and in colleges but did so noticeably only in the nineteenth century.

Somewhat sadly the initial intention of the original masters of the university that students should reach a greater general level of understanding by disputation was slowly lost. As a consequence, by 1650 or even earlier, the colleges became more or less places of leisurely education for gentlemen. The teaching of logic had been largely replaced by a humanistic approach based on classical literature and containing little, if any, science of any kind. It was the persistence of this style that left Oxford resistant to the changes in the outside world for the best part of two more centuries.

It is against this background that we must see another important development of the university from when it began to have buildings of its own.[8] The earliest university buildings, of little consequence to later divisions of power, were associated largely with the Arts, ceremony and administration – the Divinity School, (1440–1490), the Bodleian Library (1630), the Sheldonian Theatre (1667), the Clarendon Building (1713), and the Radcliffe Camera (1749). They did not introduce novelty to the lives of most fellows even to today, but were very useful, acceptable additions to general facilities. It was the creation of the (old) Ashmolean Museum with a *chemistry laboratory* in its basement (1683) that was different in that it was here that a new discipline, the teaching of and experiments in science, could be carried out, see Chapter 2. The laboratory had no parallel except in a few college rooms and private houses. Its functions were part of a *university* activity under a Museum director, a university appointment, who was also the first university professor of chemistry and he was not necessarily very college-minded. (A professorship at that time was of very little note or power, however.) The laboratory had a purpose, the study by the then novel logical empirical, experimental, methods, which had been uncovered in the first part of the seventeenth century but were held in little esteem by many fellows of colleges in so far as they were then concerned about logic at all.

Now, as will be outlined in Chapter 2, the creation of a laboratory in the Museum arose through a breakaway from traditional Aristotelian science to empirical science in Europe that was introduced in Oxford by a group, not just fellows of colleges and not associated with a particular college led by Warden Wilkins of Wadham College. They worked together more in the style of a club promoting experimental philosophy, devoted both to academic and practical matters. Some of their work, often in groups, testing and confirming observations, gave one strand to the beginnings of chemistry, for example the studies by Boyle, often referred to as the first modern chemist. The other strand, which aided practice of this science only, was the efforts from earlier centuries, recorded as alchemy, to reach material "purity" often thought of as turning everything into gold. This club of experimental philosophy played a large part in the foundation of The Royal Society in 1660. Note again the impact was by a group of individuals, some associated with, even fellows of, colleges, but not by a central university organisation or by a college. Unfortunately, shortly after 1700 the impetus generated in chemical studies due to this group and the standing of the Ashmolean laboratory was largely lost. Whether this was due to lack of novel instrumentation or of a general disinterest in the colleges due to the switch away from logic to *literae humaniores*, or due to the distraction of the advances in astronomy is unknowable. Much of the so-called Scientific Revolution of 1640–1780[9] appears to have left Oxford untouched. For more than one hundred years little chemistry seems to have been done, except at Christ Church, the first college to have a laboratory, and we see the initiatives of Wilkins and

An Outline of the History of Oxford University

Boyle leading elsewhere to the work of Priestley and Dalton in Britain and of figures such as Scheele, Berzelius and Lavoisier abroad, as described in Chapter 3. It is only after 1800 that Oxford chemistry was to have a professor again but even then the subject was taught merely as something a gentleman ought to have heard about and no research of note was done. During all the period from 1700 even to 1850 most Oxford colleges and much of the activities of the university kept themselves almost isolated from the money-making industrial outside world that became assisted by those in the new Universities despisingly called "tradesmen" by some in Oxford. This was contrary to the objectives of Wilkins and his group in the middle of the seventeenth century. A good deal of this College wish to avoid especially industry and its financial objectives lasted long into the twentieth century and remains one cause of some tension within the university even today.[7] The activity in the university laboratory, in the Ashmolean, such as it was, was moved in the mid-19th century to a Magdalen College site.

In the second half of the 19th century the study of chemistry was more genuinely revived. We see this in two different developments in addition to the continuation of a professorship in chemistry from 1803; first in the setting up by the *university* of a separate Honours School of Chemistry, 1850, and then in the building of a chemistry laboratory, the Abbot's Kitchen as part of the new Museum for natural science (1860) with a considerable extension in 1879. Much of the change was due to pressure from a Government Commission.[6] The second development was of more intense research activities, largely by fellows, in a few independent *college* laboratories, which had begun in a somewhat relaxed fashion almost one hundred years previously in Christ Church.[10] In fact, approaching 1900, it was not the professor, or the workers in the university laboratory such as it was, who had the major impact upon the study of chemistry but a few individuals in three or four college laboratories one of which (Balliol-Trinity), it so happens, became the focal point of one part of the future Oxford Chemistry School in the next century.[10] They, and a few others of two or three more colleges, were also in part the chemistry teachers for all the colleges though much teaching was done by outside tutors not college fellows. Note again that college innovation was in a particular college by a few fellows, not top-down but bottom-up. This period is the subject of Chapter 4. Further developments took place after 1850 in that colleges became forced to help finance the university through an Act of 1881.[6] With greater financial strength the university on the advice of the professor appointed one or two and a slowly increasing number of demonstrators, as they were called, to assist in laboratories, mainly as teachers but they were only fellows of colleges by agreement. As a result a professor newly appointed after 1910 found several demonstrators in his laboratory, some of whom were fellows. The professor could take charge of those who had no fellowship but faced a difficult task with those who were fellows for they had by then an independence in teaching and research. We shall see that the relationships from 1920 on were far from comfortable for many years. An interesting change around the end of the nineteenth century was the dissociation from medicine, especially so-called anatomy, and from mineralogy.

1.3 THE CREATION OF CHEMISTRY DEPARTMENTS

From the beginnings of the 20th century almost to the end of the Second World War in 1945 the college laboratories continued on their own idiosyncratic path, as analysed in Chapter 5. The dominant group were the fellows of Balliol and Trinity Colleges concerned largely with a very successful, internationally regarded, study of chemical kinetics and not with preparative or analytical chemistry, that is with a special academic discipline aiming to explain why and how chemical reactions occurred and certainly not concerned with producing new substances. It is difficult today for us to see how unusual it was to engage in this kind of study and in depth. From sometime late in the 19th century this research fell under the name of *physical chemistry*. As described above, before this period for fifty years the University Chemistry Laboratory, from 1860 to 1910, and its successive professors, had achieved very little in any branch of chemistry. Meanwhile, the university was again under pressure from Government Commissions to show its willingness to modernise not only but especially in sciences such as chemistry.[6] A battle developed between those fellows who wished to reform the university and a die-hard group that wanted to keep their traditional way of life.[1] By 1910 the reforming party had effectively won. A very significant step for chemistry and the university then came with the opportunity to replace the ineffective Odling, the then professor of chemistry. In 1912 William Perkin, a man of very different persuasion and style from the Oxford academics, was elected. He was from Manchester, had great energy and ambition, and industrial connection and finance, a combination previously unknown in Oxford. Moreover, the Council of the university had promised or bribed him with a large sum of money and offered a loan as well as a site on which to build a new laboratory under him. Now Perkin was a chemist concerned with the analysis and synthesis of new organic compounds and he was not particularly interested in the colleges or perhaps even academic attitudes including teaching. There were very few college fellowships devoted to chemistry coupled or not to demonstratorships associated with university laboratories at this time. He immediately set about making the *organic chemistry* laboratory[11] the home of a group of research workers not connected to the fellows of colleges, some of "his" demonstrators paid by the university, as well as arranging the chemistry course to meet his demands for research students. He did this by persuading the university to add a fourth year to the course in which students did only research, (presumably his research!). The research remained with connections to industry. To put it crudely the university had elected a "tradesman", with a completely different style and attitude (and not just to chemistry), those of top-down Manchester University, from those held in college laboratories with the individualist, enquiring mind approach. Following Perkin, Robinson was elected (1930) and maintained this style. The status of organic chemistry quickly reached that of the college-based specialised physical chemistry. Increasingly organic and later other chemical research came to depend on outside money that was siphoned through university channels not via colleges, thus helping to increase the strength of the central administration especially when more general funding went through that channel too. The obvious problem was the division of this funding between colleges, administration and laboratories. As we shall see, as more science professors were elected and new laboratories built the university allowed more and more appointments of demonstrators and assistants by professors, so creating an under-privileged group relative to college fellows and decreasing the strength of the colleges in the sciences relative to departments under a professor. Each professor in effect had an independent department and we can follow this development to today.

In the years between the two world wars the separation in the two styles of chemistry remained although the majority of colleges were hardly involved. The university side of chemistry clearly needed balance since Perkin only interested himself in organic chemistry and even in 1912 the need for a second chair had been aired. To this end in 1919 a second chair of chemistry was inaugurated and filled by a professor who considered himself to be a physical/inorganic chemist. The selection of Frederick Soddy[12] as the Dr Lee's professor proved to be unfortunate for the combination of his difficult nature and the unwillingness of the university and certain powerful college fellows, to support him. They were also demonstrators including Sidgwick in the Dyson Perrins Laboratory and Lambert appointed before Soddy's arrival in the old laboratory. This opposition, especially to the building of a further new laboratory, finally planned in 1935, led to almost twenty years of no development of this side of university chemistry. Meanwhile research in the Balliol-Trinity college laboratory, now in part supported by university finance, and that of organic chemistry almost raced ahead. In the end, 1937, Soddy resigned and the university promptly elected the very successful leader of Balliol-Trinity College style of academic physical-chemistry, C.N. Hinshelwood, to his chair. In no time at all it was agreed that, again with the help of outside money, a new laboratory should be built for Hinshelwood, seen to be a college man, and that he should also be placed in charge of the Old Chemistry Laboratory, the first to be called the *Physical* and the second the *Inorganic Chemistry* Laboratory. In 1944 the newly formed Chemical Crystallography Department was also put under him but note that as a result there was no pressure to do inorganic chemistry research and the planned large extension of the old, now inorganic, chemistry laboratory of 1935 was dropped.

Thus by 1945, that is after a break in most activities due to the Second World War, see Chapter 6, there were three independent chemical departments in Oxford with different limited personal interests. They had arisen not from planning chemistry but from the internal confusion of interests between different groups of college chemistry fellows, professors, and the members of university governing bodies. One was for organic chemistry for many years under an organising chemistry professor, Robinson, with interests in research and largely not in teaching, linked to industry, and that has its historical origins in large part in the creation of buildings and a professorship independent from colleges. It had grown from the Old Ashmolean laboratory (1683) to the Museum of Natural History general chemical laboratories (1860 and 1879) but only as organic chemistry alone when the separate Dyson Perrins Laboratory was built for Perkin and organic chemistry. This was a direct university development placed under a professor in which several demonstrators worked and not all were connected to colleges. It introduced a sense of local top-down power. The second, the physical chemistry laboratory, new in 1941, and its staff, even its professor, had attitudes that evolved from colleges, more or less intellectual thinking and teaching of individuals, into a part of physical chemistry. The professor, Hinshelwood, used the old Chemistry laboratory, now nominally the third department and called the Inorganic Chemistry Laboratory, for more physical chemistry. These two laboratories under Hinshelwood allowed individual demonstrators, almost invariably fellows of colleges, much freedom. They were somewhat closer in organisation to bottom-up than to top-down, typical of a college, as we describe in Chapter 7. In some ways the activity in Robinson's organic laboratory could have been said to deny the whole nature of a collegiate university but for one fact. This was that the laboratory and its services were at least in part financed by the university (as were the other two but to a greater extent) so that colleges that appointed chemistry tutorial fellows who had an interest in organic chemistry could expect to place these fellows in the organic chemistry laboratory, a procedure which was natural to the other two laboratories. It is the partial resolution of this confusion of two styles with the widening of interest in the study of chemistry, against the background of the resistance of some of the colleges, especially by their arts fellows, to involvement with the outside world or to control by the university, which is being slowly reached.

Between the two world wars the financing of parts of the university's activities by the government increased considerably.[1] In particular an agreement was reached by which the position of a college fellow was amalgamated with functioning as a university lecturer or as a university demonstrator in the case of a scientist. Unfortunately in many ways, the agreement did not lead to a uniform scheme for all fellows. For the most part an Arts fellow became a CUF, Common University Fund lecturer, while a Science fellow became a university demonstrator. The salary structures were different in that two thirds of the salary of the CUF lecturer came from a college, while only one third of that of a university demonstrator came from that source. The old adage that he who pays the piper calls the tune has slowly been realised in that power of appointment of an arts fellow/CUF lecturer still rests quite strongly in the hands of a college but that of a science fellow/demonstrator (renamed lecturer in 1964) has moved much more in favour of the university. In fair part the Arts fellow has kept a strong loyalty to a college, opposed to central governance, while the Science fellow has moved increasingly toward the university/laboratory organisation. We shall see this movement in chemistry particular in the last chapter. All of this confused fellowship structure has to be seen against the continued background from after 1850 to 1964 of the appointment of demonstrators (or research officers) by the professors of chemistry entirely financed by the university and without a college connection. Only in 1964 was the situation rectified so that all university demonstrators (now lecturers) were appointed jointly as college fellows. It will be seen immediately that these changes in structure affected not only the way chemistry is financed, by colleges and the university and industry, but the content of the course and research in the subject since they are influenced by the degree to which finance comes with demands. The fight for control over the central administration is ongoing, while more and more financial power moves into the university, which faces greater and greater pressure to act as the government wishes.[7] (As an example the distribution of teaching fees is now in the hands of the university, whereas earlier colleges received fees directly.)

It went seemingly unnoticed in 1945 that the two special interests in the three laboratories left out both the deep earlier connections of chemistry with the mineral (inorganic) elements, and the preparation of new chemicals from them, and with a large part of biological chemistry and especially medicine, the very basic initial interests in the whole subject. (Note that Perkin and his successor, Robinson, did study small-molecule natural products and were aware of the growing pharmaceutical industry but did not develop interest in large biological molecules.) It has taken the many years since 1945 to come to grips with these omissions, especially through the separate chemical activities in other departments; Earth Sciences, Material Science and Biological Sciences as well as many changes in chemistry itself. In particular, in 1963 a separate chair in *Inorganic Chemistry* was created and the new professor inherited the now much-extended (1957) old chemistry laboratory. There, mineral-related chemistry and surprisingly biological chemistry were partly revived. In reality there were now three separate chemistry departments with little exchange between them and for some time little coordination of their activities. Note again that they had arisen in part from the nature of the subject but in part from the haphazard growth of the University, the colleges and the professorships. It is in the history of Oxford's collegiate university that we can see how on the one hand colleges have come to pick and choose which part of chemistry they will teach, often limited by financial consideration, while on the other no guidance came from the university, which was also more or less opportunistic in its selection of the first professors without giving the subject's divisions much thought. These professors were allowed to create departmental laboratories, linked together casually, if at all, in an academic sense. They and their teams then occupied the laboratories with fellows from colleges with similar labels, organic, physical and inorganic and later theoretical and biological but with individual ambitions. As if to establish this departmental structure laboratories were built on well-separate parts of land, Figure 1.1. As we shall see in Appendix 1 to Chapter 7 the consequence is an almost impossible task of today for the integration of them all. The fault lies clearly with the collegiate structure with its belief in individual freedoms as opposed to cooperative university activity with the resultant weak central university authority which finds it difficult to give chemistry a genuine organisation. The history of the developments to 2005 while the school increased greatly in size and prestige will be difficult to describe fully here just because of its extent based on individual endeavour but an outline will be given in Chapter 7. It is very noteworthy that in recent times, Oxford Chemistry has been larger than and at least the equal of all similar departments in the UK.

Figure 1.1 An aerial view of the chemistry laboratories and those of several other departments. The line of trees from the dome on the right runs west to east along South Parks Road. In the immediate foreground (right) is the Museum with the small tower of the Abbot's Kitchen, the earliest purpose-built laboratory alongside to the north, sandwiched between it and the library. The museum is on Parks Road, which runs due north to the left. The old laboratory is connected to the Inorganic Chemistry Laboratory in a quadrangle facing South Parks Road. Running east along the road we come to the Pharmacology Laboratory (in the shade) used as an overflow for chemistry more recently, and then the Dyson Perrins (Organic) Chemistry Laboratory, the modern extensions of which reach from South Parks Road almost to the tall sunlit building of Biochemistry. The next building on South Parks Road, partly lit by the setting sun is the Physical Chemistry Laboratory. The large sunlit structure towards the top centre of the photograph is the new Chemistry Research Laboratory. In the top left-hand corner is the Zoology/Psychology block and in the bottom left-hand corner are Earth Sciences, Physics and Physiology buildings. A plan of the site is given in Chapter 7. This photograph is the property of Mr K. N. Harrison and Mr M. Lodge of the Inorganic Chemistry Laboratory, Oxford and is published with their permission. See also Figure 7.4.

1.4 THE TEACHING OF CHEMISTRY

We now return to the nature of teaching that had evolved in the Arts from the very beginning of the university and that like research had fallen into two parts by 1800, that in colleges in a tutorial style and that by the university professors, lectures only. It is likely that the structure of college teaching had developed from a cosy social pattern following the establishment of richly endowed colleges. The college fellows made independent selection of young male students, few other than from the well-off at first, whom they taught in a person-to-person relationship and this remains so today except that selection has no deliberate social bias and some 50% are not so clearly related to very well-heeled parents. A big change is that women students are closely equal in numbers to men today. The style and division of teaching in chemistry was taken over from the Arts toward the end of the nineteenth century, when chemistry was a matter of learning facts and experimental methods, much though there was no obvious reason for chemistry to be taught in tutorials. Generally speaking chemistry was not taught in this way in other universities but just by lectures and demonstrations. At Oxford, however, the establishment of a newly taught subject such as chemistry could only come about through college controlled entry of students and, as existing fellows, mainly not interested in chemistry, had established an intimate fellow/student relationship, it was natural that they should impose the style on the new science disciplines never considering if it were appropriate. It was an inevitable consequence of a conservative 'bottom-up' college approach and has proved to be self-sustaining as education developed. This is not to deny its virtue in general but only to pose the question of its applicability to chemistry teaching in particular. It can also lead to a very conservative attitude if pupils succeed their tutors as fellows though not necessarily in the same college, a not unusual event. However, chemistry teaching from 1930 to today has changed very rapidly. Together with the demands on research and administration the college teaching requirements and duties certainly cause considerable stress for the fellow/lecturer in chemistry and for the professors today.

1.5 A SUMMARY OF CHEMISTRY'S DEVELOPMENT

To summarise the particular nature of the chemistry school at Oxford we note the following leading pointers to its historical development, which we expand in the following six chapters. After its birth as a loosely organised single academic school there grew up in Oxford a university of many colleges, 39 today, Table 1.1, within which individuals, fellows, say 30 to 50 per college today, can act with a fair degree of independence in research and teaching. The independence of these fellows in democratic colleges was and is backed by considerable financial strength and statutes, limiting outside interference, as well as giving close contact with students and their education. The colleges have always elected their own fellows, but with direct university involvement more recently. As first they developed roughly parallel ways of teaching conservative classical attitudes to academic subjects and did their best to ignore external changes as they selected their teachers, (fellows) and students. Against this background the collegiate university took little note of the introduction of science and its increasing practical application from say 1700. After 1850 and increasingly this position was not tenable, the university in particular became pressurised by government and general outside concern.[6] At this time the response could not be coordinated. A few individual colleges set up science fellowships and the university created professorships respectively and separately. In the case of chemistry this required the development of expensive specialised laboratories for teaching and research both in a few colleges and separately in the university. The colleges proceeded in an arbitrary way at first through individual fellow's leadership to concentrate on a few selected aspects of chemistry.[10] The university from 1912 acted very differently and elected first one professor and then another who were given facilities quantitatively better financed than anything a college could afford. These professors could ignore the colleges and college modes of teaching. Unfortunately, the university did not manage the resources given to the first two professors in an even-handed way so that within it one branch of chemistry, organic, thrived while all else, inorganic and physical, remained dormant. The balance of the subject was partly saved by the work in one of the above groups of college fellows in a particular branch of physical chemistry. As the Second World War approached the maintenance of its laboratory and the wish to expand became increasingly difficult for a college and by 1941 it and most other lesser branches of college chemistry were transferred to a third university laboratory. Its main leader was made a professor with an interest in a particular part of (physical) chemistry. However, the selection of professors and indeed of fellow/lecturers was again in no way coordinated with the teaching of the subject or with the whole breadth of chemistry. Thus, parts of chemistry grew in an arbitrary and confused manner in both teaching and research with the colleges still electing teachers (fellows) at first and the university through the professors then providing research space for them to work besides the professor's group but not in a coordinated way. This led in 1965 to three chemistry departments in three separate laboratories. It is against this confusing background plus the changing manner of the selection of fellows by colleges with the university, of professors by the university, and of the way the university has slowly attempted to manage the organisation of chemistry that we shall view the history of the teaching and the advances of research chemistry in Oxford. Despite the muddling along approach (some may say just because of it) the school has undoubtedly made many great contributions to the subject.

Readers will be aware that chemistry, like physics, in the broadest sense underlies a wider subject area than that which is taught or could be taught in the discipline of the chemistry school at Oxford. The interactions between chemistry and biology, including medicine, earth sciences (geology) and physics are today very strong. The question of where to draw the dividing lines between subjects has not been faced directly so that the school has grown with its related sub-faculties through a sequence of decisions made against the background of a particular time and of the presence of certain individuals. A very unusual university structure has been built in an historical sequence, which we shall follow while noting which parts of science have been retained under the title chemistry and which have been lost. All these matters of teaching and research in chemistry fall under the general problem of the way in which Oxford

University may have to change.[13] For further details of the University and its Chemistry Departments today see reference 14. University publications are listed in notes 1 to 7 at the end of the book.

Note, in the book we shall not make extensive comparison between the overall history of chemistry and the contribution made to it by Oxford. The University was involved in the beginnings of the subject between 1600 and 1700 and, in a few colleges, with the introduction of physical chemistry at the end of the nineteenth century, and here we shall refer to work elsewhere. However it played very little part in the develpoment of inorganic and organic chemistry between 1700 and 1900, when these subjects were becoming well established. In the twentieth century Oxford played a major role in all parts of the subject and in this period we shall again refer to work elsewhere.

REFERENCES

1. *The History of the University of Oxford*, Volumes I to VIII (various editors), Oxford University Press, Oxford 1984.
2. U. Teichler, *Higher Education Systems: Conceptual Frameworks, Comparative Perspectives, Empirical Findings*, Science Publishers, New York, 2007.
3. F. M. Brewer, The Place of Chemistry – I. At Oxford, *Proc. Chem. Soc.*, 1957, 185.
4. F. G. Mann, The Place of Chemistry – II. At Cambridge, *Proc. Chem. Soc.*, 1957, 190.
5. M. D. Archer and C. D. Haley (eds.), *The 1702 Chair of Chemistry at Cambridge*, Cambridge University Press, Cambridge, 2005.
6. There have been several enquiries as to the nature of the university including The Royal Commission of 1850 and The Asquith Commission of 1919–1922 together with those on college and university statutes and finances 1871, 1873, 1877, and 1923 as well as some parts of the 'Robbins' Report on all universities in 1963. The enquiries have led to requests, even demands for change. In an effort to forestall further outside enquiry the university has attempted internal reform following the 'Franks' Commission 1966 and the North Report 2000. The degree of success of the "suggestions" in them has been patchy and recent attempts at reform have been rebuffed, see ref. 7.
7. Oxford Gazette (an official university publication) and Oxford Magazine (a publication of views, non-official, by members of the University) in the years 2006–2007, give a blow-by-blow account of recent conflict.
8. City of Oxford Royal Commission on Historical Monuments. His Majesty's Stationery Office, London, 1939.
9. *Thoughts on the Scientific Revolution* (various authors) *Eur. Rev.*, 2007, **15**, 438–512. Articles cover the effects of scientific thinking in the period 1550 to 1750.
10. K. J. Laidler, *Arch. Hist. Exact. Sci.*, 1988, **38**, 197.

11. R. Curtis, C. Leith, J. Nall and J. Jones, *The Dyson Perrins Laboratory and Oxford Organic Chemistry 1916–2004,* Published by John Jones, Balliol College, Oxford, 2008.
12. L. Merricks, *The World made New: Frederick Soddy, Science, Politics and Environment*, Oxford University Press, Oxford, 1996.
13. A. Kenny and R. Kenny, *Can Oxford Be Improved?* Oxford University Press, Oxford, 2007.
14. Oxford University website is www.ox.ac.uk. The Chemistry website gives much detail of the nature and activity of the department at www.chem.ox.ac.uk.

CHAPTER 2
From Alchemy to Airpumps: The Foundations of Oxford Chemistry to 1700

ALLAN CHAPMAN

Wadham College, Oxford, OX1 3PN

'As silver tried in a furnace of earth, purified seven times'
Psalm 12, v. 6

Unlike astronomy, which had an academic lineage in the University extending back to the twelfth century, chemistry was a relative latecomer. For not until the seventeenth century does one begin to find the science being pursued in a coherent way in Oxford. And even for most of that century, chemistry was cultivated not as a curriculum-related discipline, having no place in the academic *Quadrivium*, but as an independent or private pursuit for curiously minded gentlemen working beneath the broader umbrella of the University. Indeed, it was not until after Elias Ashmole's 'Musaeum' was opened in 1683 that public chemistry lectures came to be provided for the delectation of University gentlemen, and in the same year Dr Robert Plot was appointed Professor of Chemistry, though even then the subject had no formal curricular status.[1]

2.1 LATE MEDIEVAL ENGLISH ALCHEMY

On the other hand, chemistry and its older cousin, alchemy, had been actively pursued in England for several centuries before the seventeenth. Friar Roger Bacon had probably been the first Oxonian to conduct systematic experimental investigations in chemistry, as described in his *Opus Tertium* (*c.* 1268),[2] while it is apposite to note that one of the first historians of chemistry was an Oxonian, of Brasenose College, as well as a subsequent benefactor to the University. For Elias Ashmole's *Fasciculus Chemicus* (1650) and *Theatrum Chemicum Britannicum* (1652) are invaluable documents, containing extensive citations from English alchemical manuscripts to which Ashmole had access in the early seventeenth century; some of these now reside safely in the Bodleian Library, and some are preserved elsewhere or have since gone astray and survive only in Ashmole's published transcriptions.[3] For, as the *Fasciculus* and *Theatrum* make clear, late medieval England had a thriving community of 'chymists', many of them living in, or else members of, Benedictine or Augustinian religious houses, and often referred to generically as 'Canons'. What is more, Geoffrey Chaucer was quite familiar with this 'chymical' breed 300 years before Ashmole, and in his *Canon's Yeoman's Tale* (*c.* 1380), which Ashmole reprints in the *Theatrum*, Chaucer gives a detailed description of the apparatus, laboratory techniques, and failures of a fictional Canon whom one suspects may have been drawn from life, as narrated to the Canterbury pilgrims by his disillusioned 'Yeoman' or assistant.[4] It is quite possible, moreover, that Chaucer knew William Shuchirch, an alchemical Canon of St George's Chapel, Windsor, in the 1370s and 1380s, in his capacity as a technically minded Civil Servant and Clerk of the King's Buildings, with Chaucer's inevitable sojourns at Windsor with the Royal Court.[5] Further, Chaucer's presentation of his fictional Canon in verse is wholly authentic, for most of the English alchemists whose works are reprinted by Ashmole also wrote in verse form. And while we may now find it incongruous, and even tedious, to read of scientific procedures in verse form, one must bear in mind that in late medieval or Renaissance scholarly culture rhyming mnemonics formed an integral part of memory training; and to classically aware monks and clergy, for whom the recitation of the *Psalms* and poetic Latin Collects formed part of the daily liturgy (not to mention their familiarity with Virgil's *Georgics* on farming and Lucretius' *De Rerum Natura* on cosmology), it would have seemed quite natural.

Of course, there was nothing uniquely Oxonian or even English about alchemy in the late medieval period, though Ashmole's *Fasciculus* and *Theatrum* give a fascinating insight into the English scene as it was drawn from the manuscripts to which Ashmole had access. These English alchemical practitioners, moreover, were many in the 300 years or so running up to *c.* 1600, and Paul Antrobus has identified 129 individuals, though no one is claiming that this number is exhaustive. And one of the best-documented fifteenth-century figures was George Ripley (died 1490), a Canon of Bridlington Priory in Yorkshire, who not only seems to have enjoyed liberal funding for his researches, but also travelled (he appears to have studied at Louvain but not Oxford or Cambridge) and communicated with fellow-practitioners.[6] Ripley's *The Compound of Alchemie*, a work of some 86 printed pages and dated 1471, is included in Ashmole's *Theatrum*.

Indeed, we dismiss medieval alchemy at our peril, as modern scholars such as Lawrence Principe and William Newman have so abundantly shown. For not only did it develop many basic techniques, pieces of apparatus, and even substances such as mineral acids, used in modern chemistry, but it also possessed what was, within a post-classical frame of reference, an essentially creditable intellectual foundation. For the romantic image of the crazed gold-seeker owes more to nineteenth-century fantasies about pre-modern science than it does to the fruits of scholarly research conducted into medieval and early modern primary sources.[7]

For at the heart of medieval alchemy were three guiding principles. Firstly, there was the concept of isolating purity in what was perceived to be a fallen and contaminated world. And secondly, there was the Aristotelian idea of substance, based on the properties of the four elements: Earth (weighty), Water (moist), Air (airy), and Fire (hot). In the generally accepted world of Aristotelian science – which by 1400 formed an integral part of the curricula of Europe's universities – a physical substance was defined by its 'accidents', or combination of observable properties, which would hopefully lead us to its 'essence', or real nature. And thirdly, there was the agreed assumption that fire and intense heat were the essential forensic forces that could reduce all complex and compounded bodies to their fundamental parts. Now, if a skilful metallurgical chemist had produced a golden alloy in his crucible, how did he know whether or not it was true gold? He would test its 'accidents' by an assay process: if it looks like gold, feels like gold, weighs like gold, melts like gold, beats like gold, and resists mineral acids, then it must have the 'essence' of, and be, gold. Yet if it passes all these tests except, let us say, the test of acid resistance, then surely the alchemist must be almost there! For once he has rearranged the parameters of his experiment to make his alloy mineral-acid-resistant, then he knows he must have the 'essence' of true gold, for all of the Aristotelian related 'accidents' will then be present. It was often in pursuit of the last goal that alchemists bankrupted themselves, slid into criminal duplicity (as did Chaucer's fictional Canon), and sometimes went mad.[8] Yet their failure was not rooted in superstition or foolishness, but in a concept of substance and of purity that was, quite simply, fundamentally different from that of modern science. And in addition to the above, one might suggest that there was a fourth guiding principle: an assumed alchemical procreativeness, fertility, or power of multiplication. For just as plants and animals 'seeded' themselves and proliferated *ad infinitum* across the world, so it was believed that minerals and metals did the same, albeit much more slowly. Were metals such as silver and mercury simply immature forms of gold that, with a bit of informed human assistance, could speed up in their growth to become pure gold? And how exactly did one prepare that fabled – usually blood-red – powder, a pinch of which would turn a crucible full of boiling mercury into the purest gold or silver? Every alchemist knew someone who had mastered the secret, from Chaucer's Canon down to Johannes Baptista van Helmont in the early seventeenth century.

And then, transcending all the theories of substance, matter, and technique was the vision of alchemy as a spiritual quest. For only a man – and there seem to have been no female alchemists – who had put himself right with God could hope to plumb the depths to uncover the arcane knowledge whereby the

Creator had built His universe: to understand how metals 'seeded' in the earth, how elixirs imparted their power, the source of the force of Fire, or what astrological configurations were appropriate to make The Work succeed. Indeed, one of the reasons why medieval and Renaissance alchemical texts seem so impenetrable to many modern readers is because they are grounded within a mystical world view that draws upon Old Testament prophecies, angelic visitations, Jewish stories of Moses, Aaron, and Miriam, and the Golden Calf of the Book of *Exodus*, and upon New Testament analogies of Resurrection. Paracelsus' alchemical triad of Mercury, Sulphur, and Salt, encapsulating as it did the totality and unity of substance, mirrors the unity of the Holy Trinity; while alchemical Green Dragons, fiery flying Serpents, rampant Lions, noble Eagles, and conflicts between the fixed and the volatile mirror that Great Dragon, Leviathan, and other mystical beasts who appear in the *Psalms* and the Books of *Daniel* and *Revelation*, and who must be destroyed in the final conflict if peace, perfection, life, and alchemical success are to be won. Yet on one level, there was also something spiritually heterodox in alchemy: a gnostic belief that a sort of resurrection or eternity could be obtained by the exercise of the alchemist's own cleverness in the production of elixirs. And while individual alchemists no doubt covered a spectrum from heterodox gnostics to orthodox Christians, with hefty chunks of passed-down pagan mythology thrown in for good measure, none of them would have been *materialists* in the modern sense, for the whole nature of intellectual coherence itself was seen in spiritual terms.

2.2 THE HON. ROBERT BOYLE AND HIS CHEMICAL WORLD

The intellectual journey made by Robert Boyle in the 1640s and 1650s began not too far from this way of thinking. For Boyle's chemical mentors, such as Paracelsus who died nearly a century before Boyle was born, or van Helmont who passed away in 1644 when Boyle was seventeen, still shared in many respects this broader view of substance, purity, fire, multiplication, and the profound nobility and spiritual potency of the 'chymical' quest. And one could argue that, even down to his own death in 1691, Robert Boyle never really abandoned certain key aspects of it, for an awareness of the mystical, Divine causality of nature formed part of the very essence of his intellectual being. But what he did in a variety of laboratories – in continental Europe, in Pall Mall, London, at Stalbridge, Dorset, and most of all at his famous lodgings in Deep Hall on Oxford's High Street – raised serious doubts about the correctness of certain received chemical opinions and led him to prepare a variety of alternative models, some of which were to be of considerable use to his successors in Oxford and the wider world.[9] Yet to see Boyle as the self-conscious 'Father of Chemistry', who formulated Laws that drove out superstitions, misses the man and his creative world by many miles.

In many respects, indeed, seventeenth-century chemistry was more confused in its thinking than it had been two or three centuries before in the days of Chaucer or George Ripley. And this had largely come about as a result of a substantial new body of experimental and observational evidence that had started to emerge in the sixteenth century, only to grow and grow as the seventeenth century wore on. Of course, in addition to the fundamentally metallurgical chemistry of the medieval alchemists, there had been a veritable explosion in medical chemistry. Yes, it is true that the apothecary, or 'physician's cook', had fulfilled a time-honoured role in the compounding of medicines, but sixteenth- and seventeenth-century researches into anatomy and physiology had posed numerous questions about bodily substances and processes as well as opening up all manner of puzzles in pharmacology. Why, for instance, do some drugs, such as stupor-inducing opiates, have general effects upon the human body, whereas others such as the newly imported South American 'Jesuits' bark' or crude quinine act only on very specific clinical conditions, such as malarial fevers? And most dramatically, and controversially, there were the new metal-based drugs, promoted by Paracelsus and his radical disciples and denounced by conservative Galenic physicians, who tended to favour botanical or organic pharmaceutical preparations. Yet Paracelsus's loudly trumpeted mercury certainly provoked violent reactions within any human body into which its compounds were introduced. For mercury induced copious salivation, which contemporary aetiological theory claimed must be good for the new disease syphilis. Likewise, the newly discovered antimony was a powerful 'diaphoretic', capable of inducing abundant perspiration and vomiting, which were held to be potent weapons in the 'breaking' of debilitating and deadly fevers; while white arsenic preparations were used on all manner of diseases, from malarial fevers to cancerous tumours. These metallic drugs were seen as powerful purgatives in so far as they drove out of the body the excessive heat or superfluous moisture that physicians held lay at the heart of life-threatening fevers.

For quite simply, chemical thinking had in many ways become more diversified and labyrinthine as a result of all the new classes of phenomena – metallurgical, medical, botanical, and such – that were coming to light, and that were now begging a coherent explanation. And to this should be added the serious evidential challenges being posed by the broader undermining of Aristotelian science. Galileo's assault upon Aristotelian cosmology, and the wider fallout resulting from William Harvey's discovery of the circulation of the blood in 1628, are just two cases in point. For why, if Aristotle's science constituted an elegant intellectual unity, should we continue to invest trust in his ideas of substance if his cosmology and his physiology had been so disturbingly undermined by recent observations and experiments?

It had been Paracelsus in the early sixteenth century who led the first assault, as it were, on the idea of Aristotelian substance, suggesting (largely on philosophical grounds) that laboratory phenomena were best explained not in accordance with Aristotle's four elements, but with the above-mentioned Paracelsian *Tria Prima*, or three active principles: namely, Mercury (for fluidity and volatility), Sulphur (for inflammability), and Salt (for stability or fixity).[10] The real chemical art, argued Paracelsus, lay in unravelling the presence of these ingredients in all forms of chemical change. But more significant in practical terms were the 40 years' worth of intensive chemical research conducted just outside Brussels by J. B. van Helmont. For van Helmont largely dismissed Paracelsus' rather mystical chemical principles, and asserted that nature contained only two irreducibles: Air and Water. Earth, or at least earthy, heavy, matter, after all, could be formed from Water, as in the case of ice, crystals, and vegetation. (Van Helmont famously grew a willow tree in 200 lbs of furnace-dessicated soil, but irrigated with an abundance of water, thereby confirming his belief in the primacy of water; of course he completely missed the crucial role played by nitrogen and other atmospheric gases in the tree's growth.) And fire was really a sort of 'wild spirit', *Geist*, or *gas* that came off heated or decomposing matter. He saw flame as a species of excited glowing *gas*, while other gases, or spirits, had the power of suppressing flames. We now know that, from his carefully described experiments, van Helmont had most likely made – and sniffed – carbon and sulphur dioxides, carbon monoxide, chlorine, and a few more 'wild spirits' as well. On the other hand, he had no coherent concept of chemically specific gases as *elements* or fundamental building blocks, as came to be developed in the late eighteenth century. Rather, he had a notion of wild, pungent, and dangerous *airs* that were driven off certain substances during chemical reaction.[11] And by the early 1660s, Robert Boyle had all but abandoned his last remaining loyalties to any Aristotelian ideas of substance that he might ever have entertained, if we take his *The Sceptical Chymist* (1661) as a touchstone. For to Boyle, and in accordance with the writings of that other deeply influential anti-Aristotelian and spiritual father of the Royal Society, Sir Francis Bacon, Lord Verulam, heat seemed to be the product of mechanical motions rather than an innate characteristic of the irreducible element Fire, while his experiments with the newly invented airpump posed all manner of questions about the element Air and its relations with Water and Earth. Instead of the Aristotelian elements, Boyle began to develop a variety of atomic models, many of them, it is true, based on 95% philosophical speculation and 5% conjectural experimental data, in which the stuff of the natural world – sulphur, water, magnetism, metals, flame, flesh, trees, and so on – could be seen as analogous to a series of architectural structures. For at the heart of things must lie 'Elements . . . [or] Primitive and Simple, or perfectly unmingled bodies', or the fundamental components that were the basic ingredients of the 'mixt Bodies' into which they could be compounded, and from which they could be resolved and returned to their 'Primitive and Simple' state by chemical analysis.[12] These substances, moreover, existed as atoms or corpuscles that possessed the ability to combine with or stay aloof from each other; but in the same way that one could use the basic unit of, let us say, a common brick to construct a pigsty or a palace, so perhaps God had used fundamental units of substance, elements and atoms, to put together the complex world of nature. Of course, these corpuscles or atoms were not imagined as being identical in the way that bricks are, for some might be pointed or sharp-edged, such as those that formed acids that were sharp to the taste, while others might resemble 'buttons, others like loops, some like male others like female screws, so their parts may be so inter-woven', as Boyle thought of mechanical analogies whereby elements might combine to form mixtures and compounds.[13] Boyle came to see the chemical quest as working out and trying to demonstrate by experiments what might be imagined as this brick-to-building concept of the atomic theory, as advanced in *The Sceptical Chymist*. And one might suggest that there was a poetic irony in the fact that Sir Christopher Wren and Dr Robert Hooke (who were the architects of post-Great-Fire London) were not only eminent Fellows of the Royal Society in their own right, but also enduring friends of Boyle. Indeed, Hooke had been his assistant and pupil in Oxford, and went on to become his life-long friend, correspondent, dining companion, and beneficiary in his Will.

Yet the seriousness with which Boyle's atomism was taken, and the way in which the essential concept of an atomic basis for the structure of the natural world and of chemical reaction won intellectual credence across the board in late-seventeenth-century Europe, probably owed more to Boyle's wider reputation and personal character than it did to any particular set of experimental results. For before Boyle, atomism had a dubious status, primarily on theological grounds. Classical writers, such as Democritus, had associated atomism with blind chance, while the above-mentioned Epicurean philosopher Lucretius's *De Rerum Natura* (the *c*. 60 BC text of which was rediscovered in a monastery in 1417) had summed up all that had gone before by arguing that nature had no particular design, and all that we see around us in the world can be attributed to a blind concatenation of atomic melding and dissolving. The atoms themselves, however, were seen as eternal, indestructible, and capable of endless rearrangement.[14]

Boyle's atomism, to the contrary, was axiomatically associated with the Argument from Design, and atoms were the very bricks, metaphorically speaking, from which God had built His universe. Divine principles ran through physics, chemistry, medicine, and so on, and it was the scientist's job to elucidate that design and make it public, thereby making scientific research an act of worship in its own right. It was not for nothing that his friend and fellow F.R.S. John Aubrey (of Trinity College, Oxford) styled Boyle a 'Lay-Bishop', for in addition to his international standing as a scientist, Boyle was universally respected as a learned theologian, whose deep personal piety, simplicity of lifestyle, and numerous acts of charity were seen as the very model of how a first-rate Christian intellect should approach experimental science. By 'Christianizing' atomism, therefore, Robert Boyle paved the way for the wider acceptance and development of atomic concepts in science in general and in chemistry in particular.[15]

The first person to study the 'new' or post-Aristotelian chemistry in Oxford, however, seems to have been Dr Thomas Willis of Christ Church, who had, with his friends Ralph Bathurst and John Lydall, 'studied Chymistry in Peckewater Inne chamber' in the late 1640s.[16] Yet more of Willis anon. And then came Dr (later Sir) William Petty. Following the radical reorganisation of Oxford after the capture of King Charles I and the end of the first cycle of civil wars in the late 1640s, many new faces were intruded into the University by Parliament, to replace Royalist dons who would not forswear their loyalty to His Majesty and were thus ejected from their posts in 1648. To put it plainly, this brought about the prospect of a stern puritanical and republican management of the University, the effective abandonment of college autonomy and independence of Congregation, and a considerable amount of direct rule from Westminster. Yet it is one of the ironies of history that this potentially dire scenario gave rise to a flourishing of independent scientific research in Oxford. Not on a curricular level, of course, but as the unintended and incidental product of a group of gentlemen either intruded by Parliament or else living independently in Oxford. Dr William Petty was one of them. A Hampshire weaver's son, Petty had, after attending the local grammar school, gone off to sea as a lad. After various adventures, he later came to qualify as a physician at Leyden, where he learned the latest ideas in chemistry and anatomy, before coming to Buckley Hall and Brasenose, where he became a Fellow and Praelector in Anatomy after 1648.[17]

It was, however, after the Revd Dr John Wilkins entered into the Wardenship of Wadham College, in 1648, that a loose association of scientific friends began to meet in the Warden's Lodgings and to think of themselves as a 'Philosophical [or Scientific] Club'. The purpose of this Club – which had no more official status in the University than a private dining club or amateur singing group and was certainly *not* a directed research institute, as some writers have suggested – was to try experiments and discuss problematical topics in contemporary science.[18] On the whole, their interests tended to focus upon astronomy, optics, mechanics, and medicine, as only befitted the group's intellectual lineage back to Gresham College, London, where several of the men in the post-1650 Oxford group had first started to meet together around 1645, and may have formed part of what Robert Boyle referred to in his 1647 letter to Samuel Hartlib as the 'Invisible College'.[19] But at least four of its members had clear chemical interests. These included Robert Boyle, who at Wilkins's request came to live in Oxford around 1654–5, and stayed for over a decade. Then there was the young John Locke of Christ Church, who would qualify as a physician, and in his late 50s, and after publishing his *Essay Concerning Human Understanding* (1690), go on to win fame as the greatest philosopher of science of the age. We know that Locke spent considerable sums of money on chemicals, furnaces, and apparatus, though we are less sure of what he was trying to achieve.[20] Dr Thomas Willis also had an association with the Wadham group, though he seems to have been a resident Oxford victim of the same Parliamentary purge of 1648 that brought Petty, Wilkins, and others to Oxford. Willis's stalwart loyalty to the King, High Anglicanism, and dislike of the republican establishment had cost him what must have seemed at the time his career in Christ Church and the University, as his future father-in-law, the Very Reverend Dr Samuel Fell, Dean of Christ Church, had been evicted from his college – known in Oxford as 'the House' – and from his ecclesiastical Deanery. Yet now continuing to support himself by a private medical practice and probably by private medical teaching, Willis embarked upon a breathtaking series of researches, initiating the science of and coining the word 'neurology', as well as pioneering the science of organic chemistry. And the youngest member of this group was the twenty or so year-old Robert Hooke, also of

Christ Church, whose undergraduate brilliance, and possibly his prior reputation at Westminster School, had brought him into association with Dr Wilkins and his friends. And while Hooke's genius had first found expression in mechanics and astronomy, he cut his chemical teeth, as it were, when he became Willis's chemical assistant around 1657, before entering the employ of Boyle about a year later. Surviving correspondence tells us that Hooke was to work with Boyle in his Oxford and London laboratories down to 1662, when he became Curator of Experiments at the newly founded Royal Society, and even on many occasions thereafter. In 1663–4, for instance, he was still using Boyle's London address, 'Pall Mall', on his letters, when he wrote to Boyle asking him to pass on his, Hooke's, good wishes to Mr Crosse, the apothecary, and his wife.[21]

2.3 WHERE WERE THE LABORATORIES?

But where was all of this curiosity-driven chemistry being done in a University without any scientific premises, other than the Botanical Gardens? Quite simply, it seems to have focused on five, or possibly six, private laboratories, along with what the Oxford diarist Anthony Wood described as the 'privat elaboratories' – perhaps quite rudimentary ones – of the 'severall scholars' who attended Peter Stahl's [Sthael's] chemistry course between 1659 and 1665, in whose number Wood seems to have included himself.[22] The first and most famous of these laboratories was at Boyle's lodgings at Deep Hall in the High Street, in a house whose site is now marked by a plaque set into the wall of the early nineteenth-century extension of University College. Where Willis lived and performed his own researches and experiments before his marriage to Mary Fell and his purchase of Beam Hall in 1657, however, is unclear. He is referred to as having rooms in Peckwater Inn, a medieval property to the north of Tom Quad that was absorbed into Christ Church, although College records for the 1650s are imperfect, and it is not impossible that Willis, even if no longer a formal resident Member of Christ Church at the time, was still allowed to rent rooms in Peckwater, for the rental of College rooms by outsiders and resident alumni was an established seventeenth-century practice. This status, as perhaps a rent-paying tenant, could also explain the fact that Willis invited evicted Anglican clergymen to conduct illicit Prayer Book Services in his rooms, a practice *not* likely to have been tolerated for a resident member of the Foundation during the Puritan rule. Yet Willis came to own or lease two sites in Oxford, one of which was his house, Beam Hall, facing Merton College, and the other Bostar Hall and perhaps part of the Angel Inn, on the south side of the High Street on the corner of Logic Lane, only a few properties eastward of John Crosse's Deep Hall, where Boyle lived and worked. This latter site also seems to have been run as a private hospital by Willis and other physicians. Though no solid proof exists, it is probable that Dr Willis performed chemical experiments and possibly anatomical dissections in, or within the outbuildings adjacent to, each of these sites. A fourth chemical location was the Warden's Lodgings in Wadham, or more likely, outbuildings (some of which still stand) attached to the Lodgings, premises that around 1650 were recorded in one of Thomas Willis's notebooks as having running costs amounting to £36-17-8^d, a very substantial sum of money at that time.[23] And fifthly, because we know that several medically and chemically minded men met in Trinity College, just across the road from Wadham, during the presidency of Willis's friend Ralph Bathurst, it is not unlikely that some sort of laboratory was operational there as well, in private rooms or in an outbuilding that the Fellows could use for their own purposes. And sixthly, there were Peter Stahl's own laboratories, in which he used to teach chemistry to his many 'disciples' on a fee-paying basis, as he held no University appointment. Stahl taught successively in a 'tenement' near Boyle's Lodgings at Deep Hall, then, after several moves all around the eastern end of the High Street, 'at an antient hall called Ram Inn, in Allsaints parish in an old refectory of which he erected his elaboratorie and taught severall classes.'[24] Stahl, a native of Strasburg, Prussia, was brought over to England by Boyle in 1659 and seems to have settled, living and teaching privately in Oxford, as well as sometimes serving as 'Operator' to the Royal Society in London where, no doubt, he worked under Robert Hooke who was Curator of Experiments. It was on 23 April 1663 that Anthony Wood, along with around ten others, signed up as one of Stahl's new pupils, as did the young John Locke who, Wood recorded, made himself 'prating and troublesome' at the lectures. During his time in Oxford Stahl attracted a body of chemical disciples, who would later go on to a variety of careers, in which they would win eminence (including a future Bishop of Durham), which gives one some sense of the intellectual fascination that experimental chemistry exerted over the men of that age.

Indeed, practical chemistry flourished across Great Britain and Europe in the seventeenth century, as an expanding metallurgical industry, the production of dyestuffs, and the mass manufacture of saltpetre, gunpowder, glass, and other substances created a burgeoning demand for practical men who knew what to do in a laboratory.[25] It is hardly surprising that a body of scientifically minded young dons should have been curious about it, and so keen to learn from an expert.

2.4 OXFORD'S 'INVISIBLE' CHEMISTS: THE CITY APOTHECARIES AND THEIR LABORATORIES

It is all too easy, however, in our concern to locate chemical premises associated with distinguished academics, to miss Oxford's huge potential chemical resource: namely, the premises of those druggists, apothecaries, and chemical manufacturers that abounded in mid- and late-seventeenth-century Oxford. Indeed, Anthony Wood further tells us that around 1659, several University gentlemen met 'in Clerk's house an apothecarie in St Marie's parish, [and did] exercise themselves in chimicall extracts.'[26] Was Clerk trying to run a senior members' chemistry course in competition with that of Stahl? Either way, there is a suggestion that it was from the well-equipped and already established private laboratories of the City apothecaries – of which there were over a dozen

From Alchemy to Airpumps: The Foundations of Oxford Chemistry

within a couple of hundred yards of St Mary the Virgin, the University Church, on the High Street alone – that Oxford chemistry really sprang. In fact, we know more about the premises and possessions of these ingenious apothecaries than we do about the 'scientists' proper, for both Oxford city and University archives give their precise locations, lines of business, and often valuations of their estates at death. For this information I am largely indebted to the meticulous archival work of Carole Brookes, sometime of University College, whose excellent 1985 Chemistry Part II thesis it was my pleasure to supervise (see Figures 2.1 and 2.2,).[27]

Carole Brookes was able to identify over 40 apothecaries or commercial chemists operating in Oxford in the latter half of the seventeenth century, most of whose premises (when they could be identified) were on or just off the High Street, Turl Street, and thereabouts. Needless to say, most of them would have possessed furnaces, stills, crucibles, and other standard pieces of apparatus, and

Figure 2.1 Map of the central and eastern part of the city of Oxford from David Loggan, *Oxonia Illustrata* (1675). While apothecaries' premises were to be found in various parts of the city, most were clustered around the High Street, which runs east–west across the map. Photographed by Mr Keith Waters, of the Inorganic Chemistry Laboratory, by kind permission of the Warden and Fellows of Wadham College, Oxford.

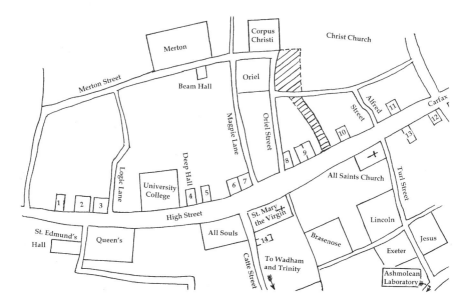

Figure 2.2 Apothecaries' premises on the High Street, Oxford, in the late 17th century. College and church sites remain unchanged. The shaded area (between the buildings marked 9 and 10 on the High Street) was cleared of its original properties *c.* 1873 to make way for King Edward Street and the Oriel Square development. Apothecaries' shops and other laboratories marked on the map are listed below, with the modern number of the premises.

(1) William Wilden, No. 80 High Street
(2) Angel Inn: William Day, Nos. 86–7
(3) Bostar Hall, Thomas Willis's 'hospital'
(4) Deep Hall: John Crosse, Robert Boyle, and friends, No. 88
(5) Arthur Tillyard, No. 90
(6) Thomas Jackson, No. 93
(7) John Fulkes, No. 94
(8) William Potter, No. 102
(9) John Clark (Clerk), William Taylor, William Petty, and friends, 'Bulkeley' or 'Buckley' Hall, Nos. 106–7
(10) Ram Inn: John Bowell and Peter Stahl, Nos. 113–14
(11) Matthew Leech, No. 129
(12) Francis and William Egleton and family, No. 3
(13) Stephen Toone, No. 15
(14) Thomas Adams, 'Haberdashers' Hall', Catte Street (now part of Radcliffe Square)

All locations tracked down and identified by Carole Brookes: see her thesis 'Experimental Chemistry in Oxford' (note 27). Drawing, based on Loggan, by Allan Chapman.

one can tell how simple it would have been for an aspiring University chemist to see chemical operations being routinely performed, purchase basic chemicals and equipment, and even obtain practical instruction. Indeed, it is all too easy to miss this large diaspora of practical working chemists in our pursuit of the origins of academic chemistry in Oxford.

Many of these men, moreover, would have been figures of clout in both Town and University, for apothecaries were 'privileged tradesmen' in Oxford who, because of their valuable skills, enjoyed, along with stationers, printers and the like, the protection of the University, and sometimes even held prestigious non-academic University posts, such as Bedels of Physic.[28] It had been the entrepreneurial apothecary Arthur Tillyard who first sold cups of coffee at his premises on the south side of the High Street, facing All Souls College, in 1655, while the undisputed 'King' of the Oxford apothecaries was Tillyard's near neighbour, John Crosse. Crosse owned Deep Hall, and would therefore have been Boyle's landlord during Boyle's Oxford years, for in addition to the apothecary's trade, Crosse appears to have been something of an Oxford 'hotelier', and was a highly successful businessman. On his death in 1698 he left an estate that seems to have amounted to an incredible £9,200,[29] whereas most Oxford apothecaries left only a fraction of that sum, £320 being around the bottom of the scale. Such a sum as £9,200 would have been deemed substantial for a rich Head of House or even a bishop to leave, at a time when the Fellows of Wadham received annually £20 apiece and the Warden £100.[30] Very few college dons, all of whom were bachelors, or beneficed or even cathedral clergy would have left more than a few hundred pounds (unless they had inherited it or gained it by marriage) in the late seventeenth century, which brings home to us what powers in the land these apothecaries could become. Perhaps Crosse's only rival in the earned wealth stakes was Dr Willis himself, who became one of the most successful physicians in England, making £300 per annum, and leaving £9,000 in cash bequests, plus his 3000-acre estate of Whadden Hall near Bletchley and other properties at his death in 1675.[31] The bachelor Robert Hooke would leave over £10,000 in London in 1703, but mainly on the strength of the lucrative fees that he could claim as Surveyor to the City of London.[32]

Of course, there is no evidence that Crosse, Tillyard, or any of their colleagues were concerned with Aristotle's concept of substance or with van Helmont's ideas on fermentation, but they do remind us that Boyle, Willis, and their academic colleagues were not the only men in Oxford who knew a retort from a bolthead. It would also have meant that a chemically 'literate' community was routinely present in the city, whose members could not only supply necessary materials to the scientists, but also offer practical advice, and even recommend likely boys or men to act as laboratory assistants. One would, for instance, like to know the name and background of the anonymous 'pumper' whom Boyle employed to assist him and Hooke in their groundbreaking airpump experiments at Deep Hall in 1659.[33] One presumes that he would have been the technician who operated the airpump, thereby leaving Boyle and Hooke free to observe what was happening to the experiment set up in the glass receiver.

Indeed, we know of at least one Oxford apothecary of the 1660s who seems not only to have been something of a hotelier like Crosse, but who also, like the above-mentioned John Clerk of St Mary's Parish, gave instruction in chemistry and pharmacy to his lodgers and other visitors. This was Stephen Toone, whose premises were on the north side of the High Street west of the University Church of St Mary the Virgin, on a site now occupied by the covered market's High Street frontage, for the chemically minded Vicar of Stratford-upon-Avon, the Revd John Ward, M.A., recorded in his Diary both staying with Toone and learning laboratory techniques from him when visiting Oxford.[34]

2.5 THE OXFORD AIRPUMP DISCOVERIES

Yet what happened in the mid-seventeenth century to fundamentally redirect chemistry away from its traditional concern with purity, metal chemistry, and pharmacy towards what became a radical inquisition into the composition of nature, and in which Boyle and his friends played an indisputably central role? I would suggest that Evangelista Torricelli's discovery of the vacuum in Florence in 1643 was of primary importance in the creation of the 'new' chemistry. For one thing, Torricelli's discovery flew in the face of Aristotelian science, which argued from logical first principles that vacuums could not exist, because a 'full' universe could not contain areas of 'nothingness' within it. Yet here was something, in the top of Torricelli's newly contrived barometer tube, which displayed all the characteristics of a physical void. This recognition of a physical void above the mercury in Torricelli's sealed tube, moreover, had been made by a procedure that was exactly in keeping with the style of Galileo's telescopic discoveries in astronomy: it was not deduced from *a priori* principles, but come upon by the use of a new specialist scientific instrument that focused or heightened human perception in a very specific physical context, as the telescope did for ocular vision – what Robert Hooke in 1665 would style an 'artificial organ', or an instrument that added a new dimension of perception and investigative power to our normal five senses.[35] The discovery of the vacuum, therefore, was unexpected, made by physical, empirical, and experimental techniques, and necessitated a rethinking of how the universe held together. And very importantly, Torricelli's discovery had disturbing consequences for the accepted notion of Air as a universally ambient element in Aristotelian physics.

Yet Torricelli's discovery only became truly significant for the new chemistry after first Otto von Guericke in Magdeburg and then Robert Boyle in Oxford had been able to evacuate air from large and easily manipulable experimental sites, and not just from the tight confines of a barometer tube. Von Guericke had sucked air out of barrels and hollow iron spheres (immortalised in the 'Magdeburg hemispheres' of school physics), but it had been the 23-year-old Robert Hooke whose mechanical genius had designed and supervised the construction of an airpump with a large 'receiver' out of which the air could be sucked, so that planned experiments could be carefully watched through the glass to reveal how the presence or absence or pressure diminution of air affected a whole range of chemical, physical, and physiological processes (see Figures 2.3 and 2.4,).[36]

In their experiments, conducted probably at Deep Hall, High Street, Oxford, between 1659 and 1661, Boyle and Hooke discovered that air was 'elastic', or capable of a vast range of states, both of compression and of attenuation. Candles went out in vacuums or tenuous air environments, yet they could burn for an uncommonly long duration in sealed volumes of compressed air. Vibrations within a bell suspended in the glass receiver would not pass through the vacuum to the glass walls to convey a ringing sound, yet light traversed the

Figure 2.3 Airpump designed by Hooke and probably constructed by Ralph Greatorex, 1659. The lower brass cylinder had a gear- and crank-handle-operated piston by which air could be sucked out of the large glass 'receiver'. A machined brass stopper, fitting tightly inside a brass ring cemented into the circular aperture in the glass at the top of the receiver, about 4 or 5 inches in diameter, made it possible to set up an experiment before evacuating the air. Even so, vacuums were hard to maintain, and Boyle (p. 8, 2nd edn., 1662) mentions using a coating of 'Sallad Oyl' as a way of trying to preserve air seals. After 1667, Boyle was doing further experiments using a more powerful and efficient pump. Boyle, *New Experiments Physico-Mechanical* . . . (1660). Picture in possession of Allan Chapman.

Figure 2.4 Reconstruction of a scene in the laboratory behind John Crosse's apothecary's shop at Deep Hall, High Street. Boyle and Hooke are conducting an experiment with Boyle's airpump, *c.* 1660. By kind permission of the artist, Rita Greer, 2006.

evacuated space with ease and with no optical distortion. Cold-blooded reptiles could appear dead in a vacuum, yet revive when air was admitted, whereas warm-blooded higher mammals, such as cats, died and would not revive.[37] An open pot of fresh blood from the slaughterhouse began to bubble and froth when placed inside the airpump's receiver.[38] And whilst an extinguished candle could not be relit *in vacuo* by means of a burning-glass, dry gunpowder, strangely, could be made to fizzle and smoke but not explode when exposed to sparks generated by a gun lock in the airpump. Could it not be that fresh blood and the saltpetre – or 'nitre' – crystals in gunpowder had some kind of air bonded inside them, which could be released under low atmospheric pressure or when heat was applied? Did gunpowder explode in the normal way because the air trapped in the saltpetre somehow started a fiery reaction that escalated to explosion in ambient air, whereas *in vacuo*, when no ambient air was present, it only fizzled?[39] Yet if fiery sparks or a source of heat were needed to ignite gunpowder, no matter whether to fizzle or to explode, how did one explain the behaviour of *aurum fulminans*, or 'banging' or 'flashing' gold? This peculiar crystalline substance ($HNAuCl$ or gold(III) have been suggested, amongst others, as a possible formula) did not even need a spark to ignite it. All that it required was a sharp blow. Carefully put a few crystals of *aurum fulminans* in a spoon, for example, then place a heavy coin upon the spoon. Tap the spoon on a table-top and – BANG – the coin is blown up to the ceiling![40]

A good conjuring trick, one might say. But nonetheless, a trick that defied an Aristotelian explanation yet which predicated a *mechanical* one. For in Aristotelian terms, fire, or explosion, needed fire, as like engendered like. Yet in all of the above experiments, both with the airpump and with *aurum fulminans*, something different was happening. Fire, it seemed, could be a property of crystals, and could even be generated *in vacuo* from gunpowder; air could thicken, thin, and evoke different properties in the substances or creatures placed in it; while explosion could be produced by a sharp blow in the absence of any sparks.

The first announcement of some of these remarkable conclusions appeared in Boyle's *New Experiments Physico Mechanical Touching on the Spring of Air* (1661), where he acknowledges Hooke's insights and assistance, and these were further advanced and developed when Boyle had constructed a better and more powerful airpump in 1667. Yet it was Hooke himself, in *Micrographia* (1665), Observation XVI, 'Of Charcoal', who really initiated a radical rethinking of the nature of combustion, in which he built on van Helmont's, Boyle's, and other chemists' emerging concept of 'nitre' or fieriness. Now clearly this 'nitre' was not simply another name for Aristotle's element of Fire. Rather, it was a redefinition of fire as a chemical process, instead of an innate property or element. Hooke, in fact, came to regard fire as a kind of chemical latency present in combustive materials such as wood, but that needed motion-engendered heat to activate it and ambient air to prolong the reaction.[41]

A piece of wood will burst into flames, for instance, if it is made very hot (such as with a red-hot iron, or when focused under solar rays), but it will only do so if abundant air is present to drive on the reaction. Since Hooke saw heat as motion-related – for example, caused by agitated 'corpuscles' – he saw the hot iron or solar rays as teeming with motion. Yet fire will not result if the wood and the red-hot iron are *in vacuo* or even in a limited air space. To take this line of thinking further, Hooke packed fresh wood chips into a tightly sealed iron vessel, and placed the vessel in a hot furnace. If he then pulled the vessel out of the fire while still red-hot, he noticed that at first the hot wood was still unburnt and its cellular structure intact as it slowly toasted to charcoal, but as soon as the ambient air was able to get at it, it burst into spontaneous flame, and rapidly burnt itself to ash. Clearly the wood while still contained in the sealed vessel was intensely hot, but it had insufficient air to break out into flame. But as soon as an abundance of atmospheric air could get at it, up it went in flames.[42]

What therefore was the role of air in facilitating conflagration? Hooke spoke of what he called an 'aerial nitre', a fiery chemical property in the air that made it act like what the alchemist termed a *menstruum*, or powerful solvent, so that when it got access to agitated yet hitherto sealed-up combustive material, it descended upon it in a feeding frenzy of flame. Yet what was this 'aerial nitre'? Was it present in the whole mass of the air, or in a part of it, or was it what we would think of as an allotropic *effect* of air on a hot substance? We must, however, be very careful not to jump the gun, and using our own hindsight knowledge of chemistry, say that Boyle and Hooke had identified oxygen a century before Priestley and Lavoisier, for the seventeenth-century English experimenters had no real concept either of a chemically specific gas or of the atmosphere being composed of a mixture of such gases. Rather, by 1670 they had come to see the mass of atmospheric air as displaying different 'Physico-Chemico-Mechanical' properties depending upon its compression, attenuation, or agitation.

Yet these effects were clearly not just chemical but what we would call *biochemical*, in so far as they were somehow related to what happened inside living bodies, especially when respiration took place. The acceptance of Harvey's experimentally derived theory of the circulation of the blood under the action of the heart was pretty well complete by 1665, though it posed all sorts of questions about the respiratory process and its relation both to mechanical heart action and to observed blood colour changes between veins, lungs, and arteries. In a series of ghastly experiments performed on living dogs at the Royal Society in 1667, Robert Hooke had demonstrated the presence of light-red arterial blood in the lungs. Yet what was the chemical process that took place in the blood as it passed through the lungs? Did the lungs *extract* some general darkening toxin that was exhaled from the mouth as bad breath, did the act of respiration cool the blood by ingesting fresh air (ideas that were closer to Aristotle's and Galen's theories of breathing), or did it force some 'airy' stuff into the blood in the alveoli? Hooke's 1667 dog experiments strongly suggested the latter, as did the escape of air from fresh blood – especially arterial blood – when put into vacuum conditions in the airpump, but nothing was conclusive.[43]

2.6 JOHN MAYOW

It was John Mayow, in the late 1660s and early 1670s, who took things a little further. Mayow had come up to Wadham as an undergraduate in 1658, just before Wilkins left Oxford for the Mastership of Trinity College, Cambridge. He went on to become a Fellow of All Souls before embarking upon a successful medical career in London and Bath, and sadly dying at the age of 38 in 1679. Mayow, who had become something of a protégé of Hooke and had carefully studied Hooke's experimental methods and conclusions when living in London, next devised and executed a brilliant set of simple yet meticulously executed experiments to try to quantify the 'elastic' or chemically active proportion in any given volume of air that was necessary to keep either a candle burning or a mouse alive. This active proportion he termed the 'nitro-aerial spirit'. Mayow proceeded by putting the candle and the mouse into separate glass vessels with equalised levels within a water trough. He noted that when the candle went out, only one-thirtieth part of the air contained within the vessel had actually been consumed and its volume replaced by risen water, yet it was impossible, using a burning-glass, to re-ignite either the extinguished candle or a quantity of camphor and sulphur placed inside the vessel beforehand.[44] And likewise, when the mouse died, it was clear that a measured volume of air had been consumed in respiration – one-fourteenth part, in fact – but that the still large volume remaining after the mouse had expired would support neither life nor combustion.[45] This led Mayow to conclude that this 'nitro-aerial spirit' was not only essential to both respiration and combustion, but was also the very agent that endowed air with its actual elasticity, and that the air became inert without it.[46]

This was brilliant experimental research, and shows that Boyle, Hooke, and Mayow had fundamentally undermined the classical concept of all change being ascribable to an inter-mixing of Earth, Water, Air, and Fire. Within little more than a decade, moreover, they had gone well beyond anything that van Helmont might have deemed possible, and were coming to see chemical action as a process based upon the fundamental rearrangement of basic physical units or active and inert parts within a given volume. And central to the products of these researches had been a radical new piece of technology: the airpump, followed by Mayow's water-sealed and pressure-equalised air jars and water troughs. For while van Helmont, Paracelsus, and others had made an abundance of fresh observations from their experiments, their research technology had ultimately been no different from that available to George Ripley or Chaucer's Canon: furnaces, stills, alembics, balances, and such. And consequently the fruits of these researches remained ambivalent in wider explanatory terms. But the airpump was so important because it created an entirely new physical environment that could take imaginative researchers on to pastures new. Perhaps the most enduring result of these researches, and the one to enter popular chemical understanding, was Boyle's 'Law', which pertained to the relationship between the pressure, volume, and density of a gas (or 'air') under changing experimental conditions.

And working in the same experimental tradition as Boyle and Hooke, Mayow had also used vacuums in his essentially biochemical researches. Not only did fresh blood effervesce and bubble in a vacuum, but fresh *arterial* blood did so with a conspicuous vitality.[47] And if an open vial of dark venous blood was allowed to stand in common air, it became 'florid' at the top, where it was in contact with the air, but remained dark further down. Mayow cites the researches of his Christ Church contemporary and Thomas Willis's pupil Richard Lower to provide substantiation to this discovery. For Lower had found from vivisections that dark blood *entered* the lungs from the veins, but that the blood *exiting* the lungs and passing into the arteries was ruddy in colour. Consequently, Lower had concluded that this colour change took place because of 'the air being mixed' with the venous blood.[48] And from his own subsequent experiments that built upon those of his Oxford predecessors, Mayow concluded that not only did the 'nitro-aerial spirit' give air its mechanical elasticity, sustain flame, and facilitate respiration in animals, but that it also caused the blood colour change in the lungs and played a further crucial role in the generation of body heat.[49] Once again, it is all too easy for us to ask why Mayow did not go on to identify oxygen in 1674, but as indicated above, seventeenth-century chemical culture had no viable concept of chemically specific *gases* as opposed to elastic, variable, spirituous, or wild *airs*. For their model of action was not one of what would later be seen as molecular agency, so much as one of mechanical proportion.

2.7 THOMAS WILLIS

Yet working before these post-1659 airpump researches was the man who had 'studied chymistry in Peckwater Inne chamber' in Christ Church since the late 1640s, and who, more than a decade later, had employed Robert Hooke as a chemical assistant.[50] By the late 1650s, Thomas Willis's chemical investigations had led him to abandon Paracelsus, van Helmont, and what was left of Aristotle, to suggest, in *De Fermentatione* (1659), a five-fold basis for substance and for chemical action. For chemical manipulation seemed to result in five irreducibles, which were: (1) Spirit, or the lightest or most volatile part of a substance that first flew away upon heating; (2) Sulphur, which was connected to flame, inflammability, colour, odour, and the basic form of a body; (3) Salt, which is 'fixed', stable; (4) Water; and (5) Earth.[51] The first three were seen by Willis as 'active' principles, in so far as they gave a sort of defining potency to a substance, whereas the last two were 'passive', as they somehow filled interstices and gave bulk. Yet still, rather like the Aristotelian Elements, it was their respective mixture that made one substance different from another. A piece of granite, for instance, would have a predominance of Salt and Earth, to give it stability, whereas a lump of coal would have received its combustive properties from a predominance of Sulphur and Spirits.

This may not sound very much like modern chemistry, in its concern with what seems like the manipulation of vague properties like volatility or inflammability, but on a closer inspection one can see a chemical aspiration that we can perhaps relate to today. For what Willis is trying to do is break matter apart, identify the several components present in a chemical reaction, and understand what chemical structures are: such as how does one mixture of the five principles produce flesh, another wood, and yet another stones? And while one cannot deny that Willis's concept of substance – if not his laboratory exercises – was still deeply coloured by Aristotelian thinking, rooted as it was in perceived natural properties and causal sequences, we should not be too surprised by the fact. For before the new data of Boyle's airpump researches began to re-order key aspects of chemical thinking, chemistry abounded in intellectual circularity, as the chemist's principal occupation seemed to consist in studying how substances changed, coagulated, flamed, thinned, and solidified when acted upon by intense heat or when they cooled.

Central to Willis's chemical thinking, as it would become in that of Mayow, was fermentation. It was, of course, a very familiar everyday phenomenon on one level, lying at the heart, as it always has done, of brewing, baking, and farming; and let us not forget that seventeenth-century dons would have had a greater familiarity with agricultural pursuits than one might expect to find amongst their modern-day counterparts. For every clergyman, academic, and country gentleman would have received his income from land rentals, and such gentlemen would have watched the profitability of their farms, brew-houses, and dairies with the same assiduity as their modern-day colleagues might monitor the instincts of their stockbrokers. Why, for instance, did snow always melt on a dunghill, no matter how cold the winter; and why did chemists who wanted a long-term gentle heat source for a given reaction – such as the assumed gestation of the Philosopher's Stone – sometimes bury their reactants in sealed vessels within a dunghill for weeks on end? Familiar on an everyday level as fermentation might be, however, on a chemical and medical level it was both inscrutable and endlessly fascinating. For fermentation seemed to bring about spontaneous changes of state – such as from mash to beer, or from health to raging fever – and it produced effervescence, mysterious bubbling, colour changes, and heat without flame.

Of course, Willis had no more idea of the crucial role played by bacteria in fermentation than did anyone else prior to Louis Pasteur 200 years later, as he explained it partly in terms of his above-mentioned five principles, and partly by the mechanical philosophy. For fermentation, to Willis, was an ongoing fight between his 'active' principles of Spirit, Sulphur, and Salt, with their tendency to move and recombine, and the 'passive' ones of Water and Earth. In the ceaseless wrestlings of these 'twin' principles, therefore, lay all chemical change. And as Willis conceived of these principles as being embodied in 'Corpuscles' that collided, inter-penetrated, and mixed, chemical reaction was envisaged by him as a physical, or perhaps mechanical, sometimes violent, and even explosive process, as opposed to the vague mysterious entities of form, potential, actual, and efficient causality that had been assumed to exist in Aristotle's science, and had effectively been paid lip service to by Paracelsus and van Helmont. For woolly and qualitative as Willis's chemistry might strike us today, he was significant in seeing it as a pursuit of the mechanisms of physical action rather than of occult sympathies.[52]

Thomas Willis, in his capacity as a physician, also came to see fermentation as lying at the heart of the life, growth, and disease processes. The heat of the blood was generally ascribed to a sort of fermentation, and when the blood's natural heat became agitated or got out of hand, then it manifested itself as fever. And fever, let us not forget, in that pre-bacteriologically aware age, was one of the commonest forms of, or concomitants to, illness. Smallpox, plague, colic, septic wounds, frights, trauma, insanity, sexual desire, gluttony, drunkenness, immoderate exposure to summer heat or winter cold, and soakings in the rain, all seemed to involve fermentation, as the patient sweated, burned, shivered, thirsted, and suffered as his blood and bodily 'humours' became spontaneously overheated. The throbbing and scabbing of a healing cut, moreover, indicated that good fermentation was going on in the healing process. Indeed, Willis himself was to die in 1675 from a sudden 'fermentation' within his lungs, as a stubborn autumnal cough rapidly turned into a deadly pneumonia.[53]

Like most seventeenth-century doctors, Willis believed that in some ways medicine was analogous to the pursuit of the Philosopher's Stone, and that the discovery of a key principle or drug could have a sudden and universal explanatory or curative power. And to him, the end of the quest lay in acquiring a proper understanding of fermentation, both in the chemical laboratory and in the sickroom.

Chemically and physiologically wrong, in exact scientific terms, as we now know Thomas Willis's ideas to have been, his wider significance in the history of both chemistry and medicine is very great. For in his expunging of mystical properties from reactive chemistry, his mechanical explanations using 'corpuscles' acting upon each other as a way of envisaging chemical change, and in bringing all this to bear on the disease process and the search for rational therapeutics, he was well ahead even of van Helmont. But what matters in trying to make sense of the science of previous ages is *not* to search for what are popularly believed to be 'geniuses ahead of their time' so much as to understand the movement of ideas within specific communities of researchers and how new key pieces of experimental research data – such as Harvey's discovery of the circulation of the blood in 1628 or Boyle's airpump researches 30-odd years later – forced natural philosophers to rethink their definitions of physical reality. And in this respect, the researches of Thomas Willis, published in *De Fermentatione* and *De Febribus* (1659 and 1681) and such writings (not to mention his *Cerebri Anatome* (1664) and other monumental physiological works), were of the highest contemporary and enduring significance.

2.8 THE REVD JOHN WARD: AMATEUR CHEMIST AND PHYSICIAN

The chemical and medical researches of Willis, Boyle, and their circle were also discussed, tested, or elucidated in various ways in the Revd John Ward's diary. For John Ward, sometime of Magdalen Hall and Christ Church, Oxford, and Vicar of Holy Trinity Church, Stratford-upon-Avon, in the years following the Restoration of the Monarchy between 1662 and 1681, belonged to a growing class of scientifically minded clergymen. He made regular visits to Oxford, staying with and receiving instruction from the apothecary Stephen Toone (as we saw above), and in his *Diaries*, now preserved in the Folger Shakespeare Library, Washington DC (he was also a friend of the Shakespeare family, which explains the presence of his MS in a Shakespeare archive) are numerous references to chemical activities and preparations.[54] As was not difficult for a seventeenth-century university graduate and a clergyman, John Ward also obtained permission to practise medicine, and his *Diaries* abound with chemical and medicinal preparations, not to mention discussions of symptoms, prognoses, and post-mortem details, especially arising from the experiences of his medical friends in Oxford. And in addition to his scientific friendships, Ward himself was clearly a very skilled chemist, or 'adept', by the standards of the day, who had access to, and probably owned personally, a well-equipped laboratory. His enclosed furnace, for instance, could be made to work at four grades of heat, though these graduations seem to have been measured not by any thermometric device but by opening successive dampers, increasing air blast, and observing colour changes, as was standard practice at the time.[55] Ward leaves detailed instructions for the preparation of *aurum fulminans*, as well as its medicinal variant *aurum vitae* ('living gold'), which was believed to be a powerful medicine. Several modern scholars have tried to reconstruct these formulae as reactive equations, though there is no firm agreement, partly because of the likely impurity of seventeenth-century ingredients – which were scarcely ever 'pure' by modern-day chemical standards – and partly because of the technical ambiguities in which the written accounts abound. What does seem to have been going on, however, is that Aqua Regia (a mixture of nitric and hydrochloric acids that could dissolve gold) was being used to make some complex chlorides of gold. The reputed extreme touch sensitivity of *aurum fulminans* has discouraged present-day chemists from attempting its actual preparation in a modern laboratory (it was much more sensitive, it seems, than the mercury fulminate used in post-1810 gun cartridges), while volunteers to try out *aurum vitae* are, understandably, hard to find. Michael Osborne, the Oxford research student who has worked with me on the historical chemistry of these substances, however, suggests that *aurum vitae* could, at best, act only as a violent and perhaps stomach-damaging emetic.[56]

So why did seventeenth-century chemists and medical practitioners, even those working at the highest technical and intellectual level, such as Willis, Boyle, Hooke, Ward, Richard Lower (another of Willis's Christ Church protégés) and others, believe that nauseous and deeply toxic metallic brews could possess wondrous curative properties? It all boiled down – as it were – to a fundamentally different concept of the disease process from that which we have today. Yes, Andreas Vesalius, Harvey, the Bartholins of Leyden, and numerous others across Europe's Renaissance medical schools had revolutionised anatomy and laid the foundations of scientific physiology between 1540 and 1680, but the nature of sickness itself remained just as elusive as it had done for Hippocrates 2,000 years before. And in spite of the increasing caution with which seventeenth-century scientists regarded the 'wisdom of the ancients', the classical concepts of disease still seemed as good as any others on offer in 1680. For sickness, in short, was still ascribed to a humoural imbalance, a blockage, or an excess or deficiency of a natural function within the body. And if one wished to shift or unblock the imbalance, then a purge was the most likely way to succeed. And if gold was the noblest metal within God's earthly creation, then it only stood to reason that gold should provide the 'sovereign' of purges that would unblock all manner of obstructions and bring about perfect health. Two or three days and nights of violent retching, or confinement to the privy, should drive out 'consumptions', melancholy, gout, joint pains, the obstructions that caused cancerous lumps, and perhaps even deafness and blindness; and if this failed to work, then it probably indicated that the preparation had not been sufficiently 'pure'. High-profile seventeenth-century hypochondriacs like Robert Hooke spent much of their time collecting, trying out, and recording the effects of such treatments, and Hooke's *Diary* covering the years 1672–1680 abounds in examples of such 'therapies' tried upon himself. Robert Boyle was no better, although both men were M.D.s and Fellows of the Royal Society.[57]

Yet while all of these medical men interpreted the disease process in a very different way from how we think of it now, their strivings towards what one might call a 'rational pathology' are very obvious, and one aspect of that striving was the attempt to understand the nature of the chemical substances that were found in diseased bodies. John Ward, for example, cites the case of an

Oxford woman who had died from a dropsical condition, and whose abdominal cavity was found to contain no less than three bucketfuls of water by Ward's friend Dr Conyers when he performed a post mortem. Conyers then had the idea of distilling the extracted water to see what it contained, but was surprised to find that, from such a large volume, he got only three or four spoonfuls of water out of his retort. The rest seemed to emulsify, and 'turnd to a Kind of slime or mucilage when it was cold'.[58]

2.9 THE ASHMOLEAN LABORATORY, 1683

The last twenty years of the seventeenth century in Oxford were dominated by the presence of Dr Robert Plot and the 'public' chemical laboratory in the basement of the Ashmolean Museum in Broad Street (now the Museum of the History of Science). In 1683, when Ashmole's benefaction to the University was effectively established, Oxford University acquired its second set of scientific premises after the Botanical Gardens of 1619.[59] But unlike the lecture rooms of the Astronomy and Geometry Professors in the Schools Quadrangle of the Bodleian Library, and even the Gardens with their pleasant walks and exotic cultivation beds, the Ashmolean provided experimental and comparative specimen study space that was readily accessible to all members of the University, as opposed to the private laboratories and dissecting and experimental rooms of the earlier chemists. It is also true, and rather sad, that Oxford University did not acquire this wonderful resource until after the most brilliant and creative decades of Oxford chemistry were over. For by the 1680s, Boyle, Willis, Mayow, and Hooke had either died or else gone to live in London, though we should not forget that Oxford's *corps* of City apothecaries and practical chemists continued to increase and to thrive.

In 1683, at the same time as he was appointed to the Museum Keepership, Plot was granted the title 'Professor of Chemistry' by the Vice-Chancellor, becoming thereby the first to profess the science within the formal statutory context of the University. But what role in the life of the University was Ashmole's Museum and laboratory supposed to play, and what exactly were the duties of Dr Plot, who was to be its first Keeper and Professor of Chemistry? No one came to Oxford to read for a degree in 'Chemistry' before the late nineteenth century, as the 1636 Laudian Statutes remained in force with their emphasis upon classical and theological subjects. Yet just as Sir Henry Savile's creation of two Professorships of Astronomy and Geometry in 1619 – confessedly close in spirit to the two subjects of the same title within the medieval *Quadrivium* – had the effect of encouraging wider mathematical studies in the University, so Sedley's Chair in Natural philosophy (post 1618), into which Thomas Willis was ceremoniously inducted at the Restoration in 1660, White's in Moral Philosophy (1621), and Camden's in History (1622) were instrumental in encouraging interest in these essentially non-curricular subjects amongst senior and some junior members. Ashmole's Museum and Plot's Chemistry Chair – like the Botanical Gardens that preceded them – were entirely in keeping with the seventeenth-century University's encouragement of what one might call 'mind-broadening' extra-curricular subjects, though it is true that resident members supplicating for B.M. and D.M. degrees might also find them useful on a more formal level.

Ashmole's Museum, in a beautiful building next to the Sheldonian Theatre and behind its own set of Emperors' Heads, facing on to Broad Street, became one of the intellectual showpieces of Europe, and a bright jewel in the University's crown. On the top floor were the exotic rarities that Ashmole had obtained by somewhat questionable means from the Tradescant family. These included Pocahontas's Cloak, a stuffed Dodo bird, and scores of other natural and manufactured curiosities, Scientific lectures could be delivered and elegant ceremonies performed in the grand entrance or ground floor, while down in the basement, or 'Officina Chimica', was one of the finest custom-built laboratories in Europe. It was stone-vaulted, no doubt as a fire precaution, and equipped with furnaces, stills, and other apparatus, with an adjacent cool, stone-floored and well-lit room that most probably served as an anatomical dissecting chamber. Indeed, it was described by Zacharias Conrad von Uffenbach of Ulm, Germany, when he visited on 28 August 1710, thus: ' . . . the *laboratorium*, I must admit, is very well built. It is as long or deep as the Ashmolean, though not so wide. It is vaulted throughout, and fitted with many really curious furnaces and architectural embellishments of the most costly description.' He was also impressed with the adjacent anatomy room.[60]

The original and early usage of this basement area was clearly demonstrated in 1999, when the Museum was undergoing a major refurbishment and the ditch and ground immediately behind the building were being excavated to provide extra gallery space for the present Museum of the History of Science, in what is now known as the 'Old Ashmolean'. For what came up were numerous bones, human and animal, including two damaged human skulls, the bones of a right leg and pelvis, and numerous teeth: no doubt the detritus of lectures in human and comparative anatomy. And along with them were many pieces of broken or discarded chemical laboratory equipment: charred crucibles, retorts, and such, along with numerous pottery fragments. For quite clearly, the tightly confined area behind the Museum, now abutting on to the Sheldonian yard and part of Exeter College garden, was the Museum's rubbish dump into which the discarded leftovers of scientific research and lecture demonstrations were thrown. The recovered pieces are now conserved as valuable artefacts in their own right, and a selection of them is displayed in a glass case in the old laboratory area of the Museum of the History of Science basement.[61]

In understanding the cultural significance of the Ashmolean after 1683, however, we must remember that the Museum as an institution had by then become one of the necessary technical adjuncts to scholarly and scientific research. For a Museum's collections, especially of natural objects such as crystals, minerals, and anatomical preparations, were now seen as providing yardsticks by which nature could be evaluated and compared, and the 'normal' contrasted with the 'freak' or 'sport'. Crucial, indeed, if science was to be conducted along proper Baconian empirical lines. Similarly, the laboratory, or 'Officina', provided the space where forensic investigations into specific aspects of nature could be conducted, while at the same time the whole building became a teaching resource where the world of learning could not only study the more obviously scientific specimens such as metals, minerals, and bones, but also look at ethnological and man-made products, such as North American Indian artefacts or Roman coins, mainly collected by John Tradescant and his family, and that Ashmole acquired.[62] Of course, the Ashmolean was not the first museum in Europe by a long shot, for since the fifteenth century scholars and connoisseurs had been collecting. But most of them had been private, one outstanding exception being that in Leiden, although since the early 1660s the Royal Society had been assembling its own Museum in the Great Gallery at Gresham College, London. The Ashmolean, however, was not only an ornament of the University of truly international stature, but was also Oxford's first institution specifically designed to look at all of the sciences as they were then understood – chemistry, anatomy, astronomy, experimental physics, optics, mechanics, and so on – as opposed to a single science, such as botany.

Robert Plot, of Magdalen Hall and University College, the original Keeper, selected by Ashmole himself, had wide-ranging interests in Natural Philosophy and was one of its most distinguished practitioners in the late seventeenth century. A man, indeed, whose chemical interests went back to at least 1667, when he, along with the Revd John Ward and John Mayow, attended a course of lectures in that subject given by Oxford apothecary William Wilden, who seems to have been something of a successor to Mr Clerk and to Peter Stahl, in the enduringly popular and, one presumes, commercially profitable business of providing chemical lectures to University gentlemen.[63] A Kentishman by birth, Plot is perhaps best remembered for his two monumental county surveys, *Oxfordshire* (1676 and 1705) and *Staffordshire* (1686), in which he drew up detailed accounts of the natural, human, and economic resources of the two counties. These included chapter studies on airs, waters, earths, fossils, plants, animals, antiquities, and human curiosities. Plot was not a physician, but a Doctor of Civil Laws, though his study of the natural products of the earth inevitably led to full discussions of healthy and unhealthy locations and the medicinal 'virtues' of a whole range of substances, in the tradition of Hippocratic medicine. It is in his discussion of Waters and Earths in particular that one enters the world of a late-seventeenth-century chemical natural philosopher. The medicinal *virtues* of waters especially interested him, and one soon picks up ideas previously discussed by Willis and van Helmont about how one defined good water. For it had been a truism of medicine since Hippocrates' *Airs, Waters, and Places* (*c*. 450 BC) that the presence of good airs, fresh waters, and salubrious locations formed the bedrock of health, and must therefore be studied by the academic physician and, by the 1670s, by the Natural Philosopher.[64]

Although Plot would naturally have known about pure distilled water in the laboratory, his extensive discussion of *natural* waters occurring in the county of Oxford proceeds along fundamentally different lines. For instance, he takes it as axiomatic that all natural waters coming out of, or running on, the surface of the earth contain two inner chemical properties: *saltiness* and *sulphurousness*. The salt was the heavier mineral content, the sulphur the oily and volatile.

Good fresh water, fit for drinking, should be in a vigorous natural motion, to keep both properties well mixed. Stagnant water, such as that found in ships' holds, stank because it was no longer mixed and the volatile sulphurs could 'fly away' into the air and cause a foul smell (which we now know to be the product of bacterial action). Plot frequently cites from 'the Learned Dr *Willis*', his *De Fermentatione*, as a source and as providing an impeccable substantiation for these points.[65]

Salts, in seventeenth-century chemical thinking, could be acidic or alkaline. The river Windrush running through Witney, for example, was strongly impregnated with nitre, which made possible the 'notorious' whiteness of Witney blankets.[66] But the water of Thame was so acidic as to make it worthless for washing (it could never produce a lather) and, even worse, for brewing beer, as it did not allow a fermentation to take place.[67] Waters that allowed mineral deposits (limescale) to form did so because these mineral atoms must be so gross that when not actively agitated they fell to the bottom of whatever contained the water and stuck to it.

River Thames water itself was thought to be so very rich in sulphur, however, that when a ship provisioned with Thames water had been at sea for eight months or so it could become dangerous. For the rocking of the vessel occasioned the volatile sulphur vapours to evaporate out of the water, to occupy much of the air space at the top of the barrel, so that when the bung was eventually drawn, a candle held near the bung hole could result in a loud explosion, a great tongue of flame, 'and sometimes endanger'd firing the ship'.[68]

It is clear that Plot had travelled all over the county (as he had in Staffordshire and elsewhere in England) examining its marls, earths, minerals and waters, and he mentions many local scientifically minded worthies, such as Dr Lane of Banbury and his brother Mr Lane of Deddington, both physicians, who were experts on the properties of their local environments.[69] And inevitably, Plot must have lectured on these subjects in the Museum, and no doubt boiled, sniffed, and tasted a wide variety of specimens brought into his laboratory. One method by which a skilled experimenter or doctor often evaluated the chemical content of a mineral or medicinal spring was to have a quantity poured over a red-hot shovel or piece of iron, and to carefully sniff the resulting steam, for in an age when few reliable chemical tests existed the operator's sense of smell and taste buds were frequently crucial.

It is also plain from his ample citations that Plot saw himself as part of a great Oxford chemical tradition, mentioning as he does Hooke, Mayow, Willis, and 'the Honorable *Robert Boyle Esq.*; the Glory of his Nation and Pride of his Family', who had, in his own study of Oxfordshire waters, used 'Syrup of Violets' as an early acid-alkali indicator. Then in November 1689, upon marrying, Robert Plot retired from the Chemistry Chair and left Oxford. He was succeeded to the post by Sir Edward Hannes, a medical man, who like so many seventeenth-century Oxford scientists (and most notably Robert Hooke) had followed the road from Westminster School to Christ Church.[70]

Unfortunately, however, the Museum did not become a great centre of chemical research, although it was a popular venue for scientific lectures and demonstrations throughout the eighteenth century and down to Dr Charles Daubeny's time as Keeper in the early Victorian period.

Yet one experiment performed in the Museum by the Revd Mr John Whiteside captured the imagination of Dr Edmond Halley who, in 1704, returned to Oxford as Savilian Professor of Geometry. Writing in *Philosophical Transactions* (1716) of the spectacular *Aurora Borealis* visible in London over 6–7 March of that year, Halley speculated as to the nature of the auroral glow. He recalled that some time before he had witnessed an experiment performed by Mr Whiteside in the Museum, when Whiteside had discharged gunpowder *in vacuo*, in an airpump. After the light of the combustion had died away, an eerie glow still radiated out of the glass airpump receiver – the experiment must have been performed in relative darkness – which reminded Halley of the strangely fire-less, insubstantial light of the aurora. As no one knew the cause of auroral light in 1716, one cannot help but admire Halley's ingenuity in suggesting that the two types of feeble glow might have some sort of physical relationship.[71] As Whiteside, who was an assiduously efficient Keeper of the Ashmolean as well as an active experimenter and lecturer in Natural Philosophy, had only taken up office in the Museum in 1714, Halley may have been recalling a recent event, although Whiteside had been active in Oxford before becoming the Museum's Keeper.

But Oxford's great creative flourish of chemical research was clearly in decline by the late summer of 1710, when Zacharias Conrad von Uffenbach spent some weeks in the city.[72] It is a pity that he did not come to Oxford after Whiteside's accession to the Keepership four years later, for then he would no doubt have found a much more energetic ambience about the Ashmolean, its collections, and its laboratory. And as Whiteside was famously convivial and welcoming, he might have formed a better opinion of him than he did of the people working in the Ashmolean laboratory.[73]

Yet by the early eighteenth century that first intense burst of creativity in Oxford chemistry was past its peak. And while one should not for a moment underplay the immense imaginative energy that Willis, Boyle, Hooke, and Mayow brought to the subject between 1650 and 1680, one must not forget that what gave that energy both a driving force and a focus was not simply better organisation, or the proto-Royal Society, but new physical data. The physico-chemical implications of Harvey's circulation of the blood was one such body of data, and leading in some respects from it, but superseding it as far as straight chemical research was concerned, were Boyle's ground-breaking experiments into the nature of air and combustion after 1658. For the new data generated by these experiments demanded a fundamental rethinking of how the natural world was put together, and how 'nitro-aerial spirits' in particular could somehow move between air, crystals of saltpetre, and flame, as well as produce the light colour of arterial blood. The more quantitative physical models that these discoveries seemed to suggest predicated a revival of the ancient idea of atoms and corpuscles that, while still rather vague in themselves and very different from the subsequent atoms of John Dalton, let alone those of today, nonetheless introduced a key concept into chemistry that was to have enormous explanatory potential for the centuries ahead.

I would suggest, however, that this burst of original creativity had run its course by the time of Boyle's death in 1691 not because of a failure of a new generation of brilliant chemists to emerge, but because the experimental technology of the age had been exhausted once the last drop of new data had been squeezed from the airpump and from water-sealed jar experiments. For chemistry, along with physiology, optics, microscopy, telescopic astronomy, and the other new sciences that the early Royal Society's original Fellows had done so much to advance, all hinged upon fresh experimental data extracted from nature either by complex pieces of apparatus, such as the telescope or the airpump, or else by other ingenious procedures. And until significant new data began to emerge from the hands of Joseph Priestley and Antoine Lavoisier in the form of quantitative gas chemistry a century after the Oxford 'golden age', chemistry tended once again to get bogged down in rather circular experiments into the nature of metals, calcination, waters, minerals, and the newly emerging notion of 'phlogiston'. A state of affairs, indeed, which may have caused the neglect of the state-of-the-art Ashmolean Laboratory, and led Professor Frewin, as von Uffenbach had put it, to 'trouble himself very little about it'. For where we do encounter fresh and stimulating scientific information coming out of the Ashmolean Museum in the eighteenth century, it tends to be generated either by experiments in the new sciences of electrostatics, 'Natural Philosophy', and experimental physics, or by the ground-breaking astronomical discoveries and researches of Professor James Bradley, made with what were probably the finest instruments in Europe (by George Graham) at his private observatory at Wanstead, Essex, which gave physical substantiation for certain key aspects of Newtonian celestial mechanics, and that he discussed in professorial lectures in the Museum.[74] For even the energetic John Whiteside, after 1714, had not been a chemist so much as an astronomer and experimental physicist, and something of a protégé of the very long-lived Edmond Halley. And in many ways it was Newtonian astronomy, optics, and physics that seemed to be moving ahead fastest by 1720.

If, however, it was realised that the key to physics and cosmology lay in the action of Newton's Laws, and that those laws were truly universal, could it then be that these might somehow also lie at the heart of chemical action?

ACKNOWLEDGEMENTS

I wish to thank especially my wife Rachel for all the hours of work she has put in typing and proof-reading this chapter, and without whose collaboration things would have moved much more slowly and less accurately. I am also indebted to the artist Rita Greer for permission to reproduce her history painting of Boyle and Hooke performing their airpump experiments in John Crosse's Laboratory. In addition, I would like to acknowledge Tony Simcock of the Museum of the History of Science, for all his archival assistance and for giving me the benefit of his copious erudition over the years. And further thanks are due to all of my History of Chemistry Part II students over three decades, both to those whose names and researches are cited in my text and footnotes, and to the many others whose theses and conversations have taught me so much.

NOTES AND REFERENCES

1. The *Quadrivium*, or 'Four', was part of the medieval and Renaissance Arts syllabus, teaching the sciences of proportion: geometry, astronomy, arithmetic, and music (or scale). See Charles Edw. Mallett, *A History of the University of Oxford*, Barnes and Noble, NY, Methuen & Co., London, 1924, 1968, 182–3; Anthony J. Turner, 'Robert Plot', *Oxford Dictionary of National Biography*, O.U.P., 2004.
2. J. R. Partington, *A Short History of Chemistry*, Macmillan, London, 1957, reprint 1965, 37–9; E. J. Holmyard, *Alchemy*, 1957, Penguin edn. 1968, 117–22; R. T. Gunther, *Early Science in Oxford* vol. I, 'Chemistry', 1920, 1–3.
3. James Hasolle (pseud. for Elias Ashmole), *Fasciculus Chemicus: or Chymical Collections*, London, 1650. Elias Ashmole, *Theatrum Chemicum Britannicum*, London, 1652.
4. Ashmole reprints 'The Canon's Yeoman's Tale' in *Theatrum*. For a modern edition of the Tale see *The Works of Geoffrey Chaucer*, ed. F. N. Robinson, Riverside Press, Cambridge, Mass., 1957, 213–23.
5. William Shuchirch is discussed in 'Yearbooks and plea rolls as sources of historical information', *Transactions of the Royal Historical Society* 1922, **5**, 37–40. I am indebted to my Oxford Chemistry Part II research student Paul Antrobus for this reference: see also Antrobus, 'Practical aspects of medieval English alchemy: Chaucer to Ashmole', unpublished Oxford University Chemistry Part II thesis, 1985, page 38. Copy in History Faculty Library, Oxford.
6. Anthony Gross, 'George Ripley', *Oxford DNB*. Also Antrobus, 'Practical aspects . . . (n. 5), 17, 20. For the names, dates, and manuscript locations of Antrobus's 129 English alchemists, see his 'Index of medieval English alchemists born before 1540', thesis, Appendix 116–20.
7. Leading scholars in this revisionist tradition are: Lawrence M. Principe, *The Aspiring Adept: Robert Boyle and his Alchemical Quest*, Princeton University Press, Princeton, NJ, 1998. Also, William R. Newman and Lawrence M. Principe, *Alchemy tried in the Fire*: [George] *Starkey, Boyle, and the Fate of Helmontian Chemistry*, University of Chicago Press, Chicago and London, 2005. The author would in addition like to express appreciation to his current (2007–8) M.Chem. Part II student Kate Newton (Queen's), who is currently working with him on a thesis on Elias Ashmole, for some interesting insights into the late medieval English 'Chymical' community.
8. On 7 August 2003, the author attempted to make 'butter of gold', in accordance with a medieval recipe, in the Inorganic Chemistry Laboratory, Oxford. A couple of grammes of pure gold filings were melded into a crucible containing 4–5 ccs of mercury, boiling over a Bunsen burner. It 'tinctured' the mercury to a golden colour. When the amalgam was spread with a spatula on to a hot piece of tin, and the mercury driven off by further

heating, it left the tin beautifully gold-plated. The whole was performed in a fume-cupboard, though the evaporating mercury vapour turned a 9-carat-gold ring grey, which needed much rubbing to restore its lustre. The experiment was performed as part of a programme in Terry Jones's *Medieval Tales* series. The programme, 'The Philosopher', dealt with medieval science, and the author was consultant and co-presenter. The BBC paid for the gold (about £85 per gramme), though most of it was subsequently recoverable.

9. R. E. W. Maddison, *The Life of the Honourable Robert Boyle, F.R.S.*, Taylor and Francis, London, 1969, 89–132, for Boyle in Oxford. The influences of the older alchemical tradition on Boyle have been explored by several modern scholars: see Principe, *The Aspiring Adept* (n. 7); Newman and Principe, *Alchemy tried in the Fire* (n. 7).
10. Holmyard, *Alchemy* (n. 2), 174. For Paracelsus's ideas see Charles Webster, *From Paracelsus to Newton: Magic and the Making of Modern Science*, C.U.P., 1982.
11. Partington, *A Short History of Chemistry* (n. 2), 48–9. For tree-growing experiment, see 51–2. Boyle attempted a parallel experiment, successfully growing a 'Squash' or 'Pompion' (marrow or cucumber) in similarly dessicated soil using only water: Robert Boyle, *The Sceptical Chymist: or Chymico-Physical Doubts and Paradoxes*, London, 1661, 107–9.
12. Boyle, *The Sceptical Chymist* (n. 11), 350.
13. Robert Boyle, *Experimental Notes & Co. about the Mechanical Origin of Qualities*, 1675; Michael E. W. Hunter and Edward B. Davis, *The Works of Robert Boyle*, 1999ff., vol. 8, 2001, 367 (for sharp tastes). R. Boyle, 'The History of Fluidity and Firmness', in *Certain Physiological Essays*, 1661. *Works of the Hon. Robert Boyle*, in five volumes, London, 1744, vol. I, 263 (for buttons and loops). Antonio Clericuzio, *Elements, Principles, and Corpuscles: a Study of Atomism and Chemistry in the Seventeenth Century*, Kluwer, Dordrecht, 2000.
14. T. Lucretius Carus, *De Rerum Natura* ('On the Nature of Things'), c. 60 BC., transl. William E. Leonard, 1916, Everyman, Dent, London, 1926, 1949. Book II in particular discusses atoms and their behaviour. For rediscovery of Lucretius see A. C. Crombie, *Augustine to Galileo* 2, Peregrine/Penguin, 1969, 116.
15. John Aubrey, *Aubrey's Brief Lives*, ed. Oliver Lawson Dick, Secker and Warburg, London, 1975: see Life of Boyle, 36–7. Tom Pugh, 'How did Boyle's experiments in early chemistry establish and verify his mechanistic theories of matter?', Oxford University unpublished Chemistry Part II thesis, 2005: pages 27–31 in particular deal with Boyle's religious motivations. Copy on deposit in History Faculty Library, Oxford University. *The Diary of the Revd. John Ward, A.M., Vicar of Stratford-upon-Avon, Extending from 1648 to 1679 from the original Ms. preserved in the Library of the Medical Society of London*, arranged by Charles Severn, M.D., London, 1839: 135 for Boyle's character and lifestyle.

16. For a good account of Willis's early life and researches, see Kenneth Dewhirst, *Willis's Oxford Lectures*, Sandford Publications, Oxford, 1980, 1–37. Also Robert Mortensen, 'Thomas Willis', *Oxford DNB*.
17. Toby Barnard, 'Sir William Petty', *Oxford DNB*.
18. This cultural world is discussed in A. Chapman, *England's Leonardo: Robert Hooke and the Seventeenth-Century Scientific Revolution*, Institute of Physics, Bristol and Philadelphia, 2005, 14–16.
19. Maddison, *Life of . . . Boyle* (n. 9), 68.
20. For an excellent background to the Oxford scientific community 1648–1660 see Robert G. Frank, Jr., *Harvey and the Oxford Physiologists: A Study of Scientific Ideas*, University of California Press, Berkeley, Los Angeles, London, 1980, 45–62. Also W. N. Hargreaves-Mawdesley, *Oxford in the Age of John Locke*, University of Oklahoma Press, 1972, 78–94. Locke, 'a man of turbulent spirit, and clamorous . . . ' was a disruptive influence at Peter Stahl's [Sthael] Chemistry lectures in 1663: *The Life and Times of Anthony Wood, Antiquary of Oxford, 1632–1695, described by himself*, ed. Andrew Clark, Oxford Historical Society, Clarendon Press, 1891, vol. I, 472.
21. Chapman, *England's Leonardo* (n. 18), 20–3. Also *The Correspondence of Robert Boyle II, 1662–1665*, ed. Michael Hunter, Antonio Clericuzio, and Lawrence M. Principe, Pickering and Chatto, London, 2001; 15 Aug. 1665, 512–13; 'Pall Mall', 6 Oct. 1664, 342–4.
22. *Life and Times of Anthony Wood*, vol. I (n. 20), 290.
23. Entry from flyleaf of Thomas Willis's clinical notebook, Wellcome Library MS. 799A, cited in Dewhirst, *Willis's Oxford Lectures* (n. 16), 30: superscript ref. 33.
24. Ibid. 22.
25. Charles Webster, *The Great Instauration. Science, Medicine, and Reform 1626–1660*, Duckworth, 1975, 273–82, 384–402. Webster provides a fascinating analysis of the wider intellectual priorities, interests, and perceived objectives of English scientists in particular in the period before the Royal Society.
26. Ibid. 22.
27. Carole Brookes, 'Experimental Chemistry in Oxford, 1649–c. 1700. Its techniques, theories, and personnel', unpublished Oxford University M.Chem. Part II thesis, 1985. Copy in History Faculty Library, Oxford.
28. The Bedels were non-academic functionaries of the University, who were often 'privileged tradesmen' in the City of Oxford: R. T. Gunther, *Early Science in Oxford*, I, 'Chemistry', Oxford, 1920, 46, mentions Christopher Whyte, the 'University Chemist' or assistant to many researchers, serving as deputy for one Mr Piers, 'superior bedell of arts'. Matthew Cross (d. 1655), father of John Crosse, Apothecary, served as Bedel of Law: Brookes, 'Experimental Chemistry' (n. 27), 182. Modern archival research, indeed, is teaching us more and more about the hitherto anonymous 'technicians' and assistants whose skill and ingenuity facilitated the achievements of the famous scientists. A mathematical instrument-maker by trade, who worked

at Wadham College in Dr Wilkins's Wardenship as 'Manciple' (or kitchen manager), was Christopher Brookes, whose practical scientific skills were found useful by the researchers: see C. S. L. Davies, 'The Mathematical Manciple', *Wadham College Gazette*, January 2003, 73–4.
29. Brookes, 'Experimental Chemistry' (n. 27), 181–2.
30. Thomas Jackson, *Wadham College Oxford*, Clarendon Press, Oxford, 1893, 57–8.
31. Dewhirst, *Willis's Oxford Lectures* (n. 16), 25. Also R. Mortensen, 'Thomas Willis', *Oxford DNB*.
32. A. Chapman, *England's Leonardo* (n. 18), 259–60.
33. Boyle speaks of 'our pumper' as a third party, in addition to himself and to Hooke, in *New Experiments Physico-Mechanical, Touching the Spring of Air* . . . . 2nd edn., Oxford, 1662, 80. He *may* have been John Dwight, who was a young man at Christ Church at this time, and was later to play a major role in the 1680s in the development of salt-glazed pottery at Burslem, Staffs. Dwight and his subsequent pottery work are mentioned by Gunther, *Early Science in Oxford* (n. 28), 29.
34. Brookes, 'Experimental Chemistry' (n. 27), 30, 176, for maps of seventeenth-century Oxford chemical premises. *Diary of . . . . Ward* (n. 15) 8.
35. Robert Hooke, *Micrographia*, London, 1665. 'Preface', unpaginated, sig. a.a. recto.
36. Robert Boyle, *New Experiments* (n. 33), 2–6, describes the airpump as devised by Hooke and the pumping engineer, Ralph Greatorex. Steven Shapin and Simon Schaffer, *Leviathan and the Airpump: Hobbes, Boyle, and the Experimental Life*, Princeton University Press, 1985, discuss the philosophical questions posed by Boyle's airpump experiments.
37. The impossibility of reviving higher animals that had been placed in the airpump even made it into a ballad: 'To the Danish Agent late was showne/ That where noe Ayre is, theres noe breathe/A glass this secret did make knowne/Where[in] a Catt was put to death./Out of the glasse the Ayre being screwed/Pusse died, and ne're so much as mewed', Anon., *The Ballad of Gresham College, c.* 1663, reprinted by Dorothy Simson, *Isis*, 1932, **18**, 103–17.
38. 'The Continuation of the Experiments concerning Respiration . . . ', *Philos. Trans.* 8 Aug. 1670, **62, 5**, 2011–31. Also *Philos. Trans.* 12 Sept. 1670, **63, 5**, 2035–56: 2043. Boyle is not specified as the author, but the researches were his.
39. Boyle, *New Experiments* (n. 33), 40–50, for gunpowder. Boyle further describes firing a ring of gunpowder *in vacuo*, in *New Experiments touching the Relation betwixt Flame and Air*, London, 1673, Expt. V.
40. Boyle, *New Experiments . . . Flame and Air* (n. 39), Expt. V, for *aurum fulminans* fired *in vacuo*. Also J. R. Partington, *A History of Chemistry II*, London, 1961, 308. Michael Osborne, 'The Medical Interests of the Oxford Chemists in the Late Seventeenth Century', unpublished Oxford University M.Chem. Part II thesis, 2002: 41–6 for modern analysis of *aurum fulminans*, and 72–3 for John Ward's recipe for its preparation.

41. Hooke, *Micrographia* (n. 35), 100–6.
42. Hooke, *Micrographia* (n. 35), 103. Thomas Birch, *History of the Royal Society* vol. III, London, 1756, 20 Feb. 1679, 465. *The Posthumous Works of Robert Hooke*, ed. R. Waller, London, 1705, xxi.
43. 'An Account of a Dog dissected, By Mr Hook', in Thomas Sprat, *History of the Royal Society*, London, 1667, 232. Also Hooke, 'An Account of an Experiment . . . of Preserving Animals alive by Blowing through their Lungs with Bellows', *Philos. Trans.* 21 Oct. 1667, **28, 2**, 539–40. Also Frank, *Harvey and the Oxford Physiologists* (n. 20), 186–8.
44. John Mayow, *Medico-Physical Works, being a translation of 'Tractatus Quinque Medico-Physici' 1674*, Alembic Club, Edinburgh and London, 1957, 71.
45. Mayow, *Medico-Physical Works* (n. 44), 73.
46. Mayow, *Medical-Physical Works* (n. 44), 75.
47. 'The Continuation of the Experiments . . . ' (n. 38), and further continued, *Philos. Trans.* 12 Sept. 1670, **63, 5**, 2035–56: 2043.
48. Mayow, *Medico-Physical Works* (n. 44), 102–3.
49. Ibid. 104. John Aubrey, '*Brief Lives' Chiefly of Contemporaries set down by John Aubrey between the years 1669 and 1696*, vol. II, ed. Andrew Clark, Oxford, 1898, 303–6.
50. Mortensen, 'Thomas Willis', *Oxford DNB*. Willis's chemistry is dealt with in some detail in Charles Morris, 'Thomas Willis and the Early Development of Chemistry in Oxford', unpublished Oxford University M.Chem. Part II thesis, 2003. Morris examines Willis's practical and medical chemistry, and his interests in nitre, respiration, and atomism. Copy in History Faculty Library.
51. Thomas Willis, *Diatribae duae medio-philosophicae . . . de Fermentatione . . .* , London, 1659, 1–16.
52. For Willis's chemico-mechanical ideas of 'explosive' muscle action, see Frank, *Oxford Physiologists* (n. 16), 221. Also Hansreudi Isler, *Thomas Willis 1621–1675, Doctor and Scientist*, Hafner, New York, London, 1968, 114–22.
53. Dewhirst, *Willis's Oxford Lectures* (n. 16), 25–6.
54. A microfilm set of the Diaries has been made by the Museum of the History of Science, Oxford, and can be consulted there. See also the published one-volume *Diary of the Rev. John Ward A.M.* (n. 15).
55. Osborne, 'Medical interests of the Oxford Chemists . . . ' (n. 40), 72–3.
56. Ibid., 46–8. Also in discussions with Mr Osborne.
57. Dewhirst, *Willis's Oxford Lectures* (n. 16), 56. Willis makes various references to purging in *Willis's Oxford Casebook 1650–5*, ed. Kenneth Dewhirst, Sandford, Oxford, 1981. See Index. Chapman, *England's Leonardo* (n. 18), 112–13. Lucinda McCray Beier, 'Experience and Experiment: Robert Hooke, Illness, and Medicine', in Michael Hunter and Simon Schaffer, ed., *Robert Hooke: New Studies*, Boydell Press, Woodbridge, Suffolk, 1989, 235–52.
58. *Diary of the Rev. John Ward* (n. 15), 268.
59. Gunther, *Early Science in Oxford* vol. I (n. 28), 43–51. Also Oliver Impey and Arthur MacGregor, *The Origins of Museums: The Cabinet of Curiosities*

in Sixteenth- and Seventeenth-Century Europe, Clarendon Press, Oxford, 1985, 145–57, 158–68, 169–78, for Ashmole and the Tradescant family.
60. Zacharias Conrad von Uffenbach, *Oxford in 1710*, ed. W. H. Quarrell and W. J. C. Quarrell, translated from *Merkwürdige Reisen durch Neidersachsen Holland und Engelland*, Ulm, 1753, Basil Blackwell, Oxford, 1928, 37–8.
61. J. A. Bennett, S. A. Johnston and A. V. Simcock, *Solomon's House in Oxford: New Finds from the First Museum*, Museum of the History of Science, Oxford, 2000, 31–48 for chemical, 48–58 for anatomical remains.
62. Many of the perishable items in the original Tradescant Collection decayed away or else were lost or destroyed due to neglect. Some are now conserved in the Founder's Room of the Ashmolean Museum, Beaumont St, Oxford, while some anatomical pieces – such as the bony remains of the famous Dodo bird – are on display in the University Museum, Parks Road, Oxford.
63. Anthony J. Turner, 'Robert Plot', *Oxford DNB*. William Wilder's apothecary's shop and laboratory stood on the south side of the High Street, around what is now No. 80, just east of Thomas Willis's private hospital at Bostar Hall and the Angel Inn, by Logic Lane. See Brookes, 'Experimental Chemistry in Oxford' (n. 27), 30. Gunther, *Early Science in Oxford* vol. I (n. 28), 50.
64. Hippocrates, 'Airs, Waters, and Places. An Essay on the Influence of Climate, Water Supply, and Situation on Health', transl. J. Chadwick, W. N. Mann, I. M. Lonie, and E. T. Withington, in *Hippocratic Writings* ed. G. E. R. Lloyd, Penguin, 1950, 1983, 148–69.
65. Robert Plot, *The Natural History of Oxford-Shire*, Oxford, 1705, Chapter 2, 26–50; for Plot's theories on the nature of water, 26–9.
66. Plot, *ibid.*, 26.
67. Plot, *ibid.*, 37.
68. Plot, *ibid.*, 27.
69. Plot, *ibid.*, 41, 44.
70. Plot, *ibid.*, 42. J. H. Curthoys, 'Sir Edward Hannes', *Oxford DNB*.
71. Edmond Halley, 'An Account of the late surprizing Appearance of Lights seen in the Air on the sixth of March last . . . ', *Philos. Trans.* 1716, **29**, 347,406–28: 420.
72. Von Uffenbach, *Oxford in 1710* (n. 60), 38. The operator he refers to, 'Mr White', or Whyte, would have been Christopher junior of that name. His father (see n. 28) had been the respected chemical operator of the Oxford Chemists of 40-odd years before. Sadly, the assiduity of the father did not seem to descend to the son.
73. A. V. Simcock, 'John Whiteside', *Oxford DNB*.
74. John R. Fisher, 'Astronomy and Patronage in Hanoverian England: The Work of James Bradley, Third Astronomer Royal of England', unpublished Doctoral Thesis, Imperial College, London, 2004, Chapter 4, 145–95, for Bradley's Oxford lectures and related activities. A. Chapman, 'Pure research and practical teaching: the astronomical career of James Bradley, 1693–1762', *Notes and Records of the Royal Society*, London, 1993, **47, 2**, 205–12.

CHAPTER 3
The Eighteenth Century: Chemistry Allied to Anatomy[1]

PETER J. T. MORRIS

Science Museum, Exhibition Road, London SW7 2DD

3.1 INTRODUCTION

The end of the seventeenth century showed promise for chemistry at Oxford. A series of successful experiments had been completed, a chair had been established and a laboratory built in the Ashmolean Museum. There were or had been a few fellows in Oxford colleges who were interested in chemistry. Unfortunately from the beginning of the eighteenth century much of this promise was lost and chemistry was not created as a university discipline for undergraduates until 1850. Chemistry was seen as a part of medicine often under the heading of anatomy, which at that time included the study of objects such as bones and fossils and their relationship to minerals, which were all housed in the Museum. To a small degree we can also associate the loss of interest in chemistry with the rise of astronomy and medicine, both of which had a higher profile in Oxford. The larger failure, however, was the general low level of academic effort. College activity was concerned with teaching undergraduates, and hence the generalist Bachelor of Arts degree that required no knowledge of chemistry. Most chemistry was taught in the Museum by people not directly employed by the university or colleges and they charged fees. In order to see the lack of development of chemistry in Oxford in full perspective we describe briefly first the progress of chemistry elsewhere during the century. We look next at Oxford as it was in this century and then give the careers of and describe the work of those who did lecture on chemistry in Oxford.

Chemistry at Oxford: A History from 1600 to 2005
Edited by Robert J.P. Williams, John S. Rowlinson and Allan Chapman
© Royal Society of Chemistry 2009
Published by the Royal Society of Chemistry, www.rsc.org

3.2 CHEMISTRY IN THE EIGHTEENTH CENTURY[2]

Chemistry in the eighteenth century was – as always – primarily a service science employed in medicine and pharmacy, in mining and metallurgy, and in agriculture and brewing. The most important 'clients' for chemistry also set the research agenda for chemists in many respects. A key topic was the potability and medical value of spring waters in a period when all drinking water came directly from springs, wells and rivers. Another crucial economic imperative was the study of minerals with the aim of finding new sources of metals, either coinage metals such as silver or industrial metals like iron or copper. There was also the search for new medical treatments and cures, especially for indigestion and the dreaded stones of the kidney, bladder and gall bladder that gave patients so much pain. In a period when every don had 'commons' (beer and cheese) for lunch, brewing – carried on by nearly every self-respecting pub and farm – was an important industry. But other key topics for eighteenth-century chemists were less clearly linked to this service role, above all the central topic of fire and heat, which actually predated the widespread use of steam power. This partly stemmed from the industrial concerns of chemists; the smelting of ores into metals, brewing and the early pump engines all involved heat. It was also a reflection of the medical concerns of chemists in this period; what caused fever and inflammation? Heat was also linked to the intense study of nature by eighteenth-century savants; what was the source of heat in warm-blooded animals and why did volcanoes emit heat and glaciers remain frozen even in warm weather?

In modern-day terms, chemistry was largely about inorganic chemicals, and nineteen elements were discovered (if not necessarily isolated) during the century namely: aluminium, barium, beryllium, boron, carbon, chlorine, chromium, cobalt, fluorine, magnesium, molybdenum, oxygen, platinum, potassium, strontium, tellurium, titanium, tungsten, and zirconium. Most of these elements were discovered as result of the analysis of minerals and chemistry was often divided into the three categories familiar from the quiz game 'Twenty Questions': animal, vegetable and mineral. Organic chemistry only existed as vegetable and animal chemistry, which consisted largely as the decomposition of natural materials, such as wood, spermaceti or ants into simpler but as yet unidentified substances. Antoine Lavoisier (1743–1794) was beginning to make some progress with organic chemistry when he was caught up in the French Revolution (which ironically he supported) and guillotined in 1794 for being a tax collector for the Ancien Regime. Physical chemistry was equally undefined in the eighteenth century, but was a prominent aspect of chemical thought and experimentation in this period. Some chemists concentrated on the phenomenon of combustion, why do certain substances burn readily and why while they are burning do they give off flames and heat? What happens to the burnt substance? Sometimes, as in the case of carbon or sulfur, it seemed to disappear altogether but in the case of metals, it usually produced a powdery ash or 'calx'. Other chemists concentrated on the phenomenon of heat. What was heat, was it a substance with weight? How did heat enter a body and how much heat? Did different materials absorb heat differently? Another topic that gave rise to much speculation was the affinity between different substances, was it possible to predict how two substances would react together and with what degree of avidity? As the eighteenth century progressed other topics came to the fore as new discoveries were made, notably electricity and the chemistry of gases. Both these areas of research were to transform chemistry in the half-century between 1770 and 1820.

In the period up to 1730, there were two key figures in intellectual chemistry: Isaac Newton (1642–1727) and Georg Stahl (1660–1734). Holding to his general view of 'statics', that all physical and chemical phenomena were a result of interactions between particles, Newton believed that powerful short-range attractive forces could explain chemical behaviour, notably the relative affinity of different metals, the elective attractions that were a major topic in early eighteenth century chemistry.[3] Stahl, who became a Professor of Medicine of the University of Halle in 1694, was a typically German chemist in his interest in brewing and the production of metals. He believed in particles and elements (but also the classical four elements), but his major concern was explaining the nature of the conversion of ores to metals (smelting) and why some substances burned readily. He joined the two phenomena together by arguing they both involved the transfer of an invisible element called phlogiston, which he considered to be weightless (or perhaps even having a negative weight). Ores were reduced to metals by the transfer of phlogiston from the charcoal to the calx (the metal oxide) of the ore. Stahl regarded the metal as a compound of the calx (which was hence the element) and phlogiston. Combustible substances such as metals, charcoal and sulfur were phlogiston-rich and when they burnt, their phlogiston was lost to the air. The popularity of phlogiston was increased when it became apparent it could be applied to the colour of substances (strongly coloured substances were usually phlogiston rich) and electrical conductivity. Joseph Priestley, one of the most important and last champions of phlogiston, actually first used the theory in his early studies of electricity.

By the 1730s, Herman Boerhaave (1668–1738), Professor of Medicine and Botany (from 1709) and Chemistry (from 1724) at the University of Leiden in the Netherlands was the leading figure in the field.[4] He published his *Elementa Chemicae* in 1731, the authorised English version of which appeared in 1735, translated by Timothy Dallowe, but his system of chemistry was already well known in Britain thanks to the unauthorised English version of his verbatim lectures published by Peter Shaw and Ephraim Chambers as *A New Method of Chemistry* in 1727.[5] Boerhaave considered heat to be a vigorous weightless subtle substance that was combined with all other substances and influenced chemical reactions. He had a strong influence on the Scottish universities because of the strong links between the two Calvinist nations of the Netherlands and Scotland, especially in the field of medicine and law. The Scottish school of William Cullen (1710–1790) and Joseph Black (1728–1799) – neither of whom had been taught by Boerhaave but were influenced by his doctrines – believed that heat was intimately united to each particle, possibly chemically, which led directly to Lavoisier's postulation of 'caloric' as an element. Priestley, on the other hand, believed that heat was "a subtle vibratory motion of their parts" in keeping with the later kinetic interpretation of heat. Around 1750, Cullen developed a theory of heat and mixture. If a solid and a liquid gave off heat when mixed, a condensation of atoms had been brought about, but if the mixture had a cooling effect, the atoms were more rarefied. It followed, logically, that liquids gave up heat when they became solid and this led to Black's discovery of latent heat when he reflected upon the fact that snow does not immediately melt when the temperature rises, in the late 1750s. Once he had established the latent heat of melting ice, he showed in the early 1760s that boiling also absorbed latent heat. Black's colleague James Watt (1736–1819), the instrument maker for Glasgow University, became very interested in the chemistry of water and its connection with heat, with consequences for both industry and science.

This line of research was eventually overshadowed by the pneumatic chemistry founded by Black when he carried out his research on the 'fixed air' (carbon dioxide) in the mid-1750s but developed largely by the dissenting clergyman Joseph Priestley (1733–1804) in the 1770s. The discovery of chemically distinct 'airs' or gases radically changed the nature of chemistry and moved it away from the study of general phenomena such as heat and fermentation towards the investigation of specific chemical reactions. It was Priestley's discovery of dephlogisticated air in 1774 that was to prove the undoing of the phlogiston theory. At first the new discoveries (including the synthesis of water from oxygen and hydrogen) appeared to offer a way of developing the phlogiston theory and thus proving a comprehensive explanation of combustion. Eventually, however, the oxygen theory of Antoine Lavoisier won the day since the only available alternative by the end of the 1780s, the concept of phlogiston having a negative weight, was unacceptable to most chemists. Priestley continued to support the phlogiston theory while Black refused to take a stand one way or the other. Of course we now know, thanks to Humphry Davy (1778–1829) that Lavoisier's oxygen theory was also incorrect as it stated all acids contained oxygen. Lavoisier also erred, of course, in believing that heat was a weightless element, caloric, similar to Boerhaave's concept of fire, rather than Priestley's subtle vibrations.

In the earlier part of the eighteenth century, chemistry was taught mainly as part of medical training or with the aim of broadening the mind of the student taking a degree in the humanities. Chairs in chemistry were established at several universities in Europe, usually as an auxiliary part of the medical curriculum. The idea one could take a degree in chemistry in its own right was a nineteenth-century development, although chemically inclined students could take a MD on a chemical topic that for all practical purposes was an advanced degree in chemistry albeit with a medical bias. Black's 1754 MD at Edinburgh was a typical example. Nominally it was about the treatment of indigestion but in fact was largely a chemical study of magnesia and its associated 'fixed air' (carbon dioxide). Of course many modern doctorates in organic chemistry claim a medical rationale for their research. The Cambridge chair was founded in 1703 and the chair at Edinburgh followed ten years later.[6] In Edinburgh, unlike Oxford or Cambridge, chemistry was in the medical school. The teaching of chemistry at Glasgow, again within the medical school, began in 1747 but the chair was not established until 1819. Aberdeen set up its chair of chemistry in 1793.

It is clear that academic chemistry teaching was only a minor activity in the eighteenth century. By contrast public lecturers were able to reach many more people, often having an audience of over a hundred, sometimes even three hundred people. The physician Dr Peter Shaw gave public lectures in chemistry in London and at Scarborough, where he had a practice at the spa, in the 1730s.[7] Shaw was succeeded in London by Dr William Lewis in the 1740s.[8] It was towards the end of the century, however, that chemistry became a fashionable subject for public lectures, largely as a result of Joseph Priestley's striking discoveries in gas chemistry.[9] John Waltire, Dr James Arden and Adam Walker gave lectures in the leading manufacturing centres, for instance Waltire lectured in Birmingham in 1780–2, while the London scene was dominated by Dr George Fordyce and the Irish chemist Dr Bryan Higgins who opened his 'School of Practical Chemistry' in 1774.[10] Thomas Henry gave courses in Manchester between 1784 and 1794.[11] Perhaps the most successful lecturer of all, however, was the blind lecturer Dr Henry Moyes of Edinburgh.[12] He travelled through England in the early 1780s, before touring the United States of America in 1784–6, then visited Ireland in 1789.

3.3 OXFORD IN THE EIGHTEENTH CENTURY

It is easy to dismiss Oxford University in the eighteenth century as the home of port-swilling dons and lazy students. While is true that this century marks a low point in the intellectual history of the university, it was nowhere near as bad as it is often painted.[13] The university itself was certainly in a weak state, many professors did not give regular lectures and the system of verbal disputations inherited from the Middle Ages was falling into disuse. This meant that the colleges were at the basis of an Oxford education in this period and it is clear that at least some colleges, notably Christ Church, did offer a good education checked by a system of termly examinations called collections. Oxford offered a very broad classical education, with an emphasis on gentlemanly behaviour and civility, with the aim of training the many students who were preparing to become clergymen in the Church of England. This system provided an excellent liberal education, but it was up to the student whether he took full advantage of it or not. The keen student could go beyond his basic college-based education and attend extra-mural lectures, which is where chemistry fitted into the Oxford system.

The Bachelor of Arts degree, which was essentially a classics degree, was a four-year course, tested by an oral examination at the end. There were also degrees in divinity (theology), law, medicine and music, but the system was set up in a way that discouraged undergraduates from taking these degrees instead of the BA. Oxford University was cut off from the mainstream of national (especially political) life in this period as it was High Tory and High Church, and suspected (with some justification) of Jacobite sympathies. This political

The Eighteenth Century: Chemistry Allied to Anatomy

Table 3.1 Teachers of Chemistry in Oxford, 1683–1803.

Dates	Position
Robert Plot 1683–1690	University Professor, Keeper of the Museum
Edward Hannes 1690– (1704–10)?	University Professor, Student of Christ Church
John Freind 1704	*Professor*[a], possibly as locum for Hannes
Richard Frewin 1708?–1740?	*Professor*[b]
Johann Lavater *ca.* 1710	
John Whiteside *ca.* 1715–20	Keeper of the Museum[c]
Thomas Hughes 1740	Official appointment but overshadowed by Alcock
Nathan Alcock 1740–1757	*Lecturer*, later officially recognised
John Smith 1758–1766	*Lecturer*, primarily in anatomy
John Parsons 1770–?	Dr Lee's Reader in Anatomy, Christ Church
Martin Wall 1781–1785	Public Reader in Chemistry
William Austin 1785–1786	Public Reader in Chemistry
[William Higgins 1786–92]	[Assistant to Austin and Beddoes]
Thomas Beddoes 1787–1793	*Reader in Chemistry*
[James Sadler 1790–92?]	[Assistant to Beddoes]
Robert Bourne 1793–1801	*Reader in Chemistry*, Fellow of Worcester
John Kidd 1801–1803	*Reader in Chemistry*, Student of Christ Church

Titles in italics are unofficial ones used informally by the holders of the posts or ascribed to them by their contemporaries.
[a]Title used by Freind on the title-page of the English edition of his lectures.
[b]Frewin's use of this title is mentioned by A.V. Simcock, *The Ashmolean Museum and Oxford Science, 1683–1983*, Museum of the History of Science, 1984, pp. 9 and 34.
[c]Simcock, pp. 11–12, 19, and 38.

tension was relevant to the teaching of chemistry as both Nathan Alcock and Thomas Beddoes – two of the better teachers of chemistry in Oxford in this period – were Whigs (Table 3.1).

With the emphasis on a general education, students could attend lectures on a variety of subjects including what we would now call physical sciences. However, the emphasis was on astronomy and natural philosophy (physics) rather than chemistry.[14] There were several reasons for this. Natural philosophy (including mathematics and astronomy) was part of the traditional curriculum for the Master of Arts degree; chemistry was not. In the early eighteenth century, astronomy and natural philosophy were the leading sciences as a result of the work of Newton. Astronomy and the concept of an ordered universe appealed to a clerical (and conservative) audience in a way that was not possible for chemistry with its materialist implications. And in Oxford, astronomy and natural philosophy benefited from two excellent teachers, who were both Savilian Professors of Astronomy: James Bradley (1693–1762) between 1721 and 1762 and Thomas Hornsby between 1763 and 1810 (and also Sedleian Professor of Natural Philosophy between 1782 and 1810)[15]. Both men also enjoyed the support of the second Earl of Macclesfield (1697–1764), President of the Royal Society between 1752 and 1764, and the owner of an excellent observatory at Shirburn Castle near Oxford. Bradley – who became Astronomer

Royal in 1742 – gave lectures on mechanics, optics, hydrostatics and 'pneumaticks' and attracted large audiences. It is recorded that 1,221 men attended his courses between 1746 and 1760, roughly 40% of the undergraduate population during this period. Thomas Hornsby continued these lectures after Bradley's death. In May 1769 he gave a series of lectures leading up to the transit of Venus on 3 June and an additional course on *The different Kinds of Air, Natural and Factitious, in which the principal Discoveries of Dr Priestley and others will be introduced and proved by actual Experiment* in May 1785.[16] Hornsby founded the Radcliffe Observatory in 1772 (although the building was not completed until 1794) which was at least according to a Danish visitor, Thomas Bugge (1740–1815), one of the best observatories in Europe.[17]

Another important scientific group in the university in the eighteenth century – certainly as far as chemistry was concerned – was that of the physicians. The medical faculty was not very distinguished in this period but nonetheless produced 326 Bachelors of Medicine and 206 Doctors of Medicine.[18] Although the resident MDs formed an intellectual core at a time when the university was generally weak, the leading physicians in Oxford, led by Richard Frewin, opposed any intrusion by physicians from outside the university, including the Leiden-educated Nathan Alcock, an opposition that was at least partly political in nature. Matters improved somewhat after the death of Frewin in 1761, the establishment of Dr Lee's Readership in 1767, and the founding of the Radcliffe Infirmary in 1770. Superficially the Radcliffe Infirmary was simply a provincial hospital like many others founded in this period; it had no institutional connection with the medical faculty of the university.[19] Yet it rapidly became one of the best hospitals in Europe, at least according to the German visitor Johann Grimm (1737–1821),[20] and its honorary physicians were among the leading medical figures in the university. In particular it had a close association with chemistry (Smith, Parsons, Wall, Bourne) and botany (John Sibthorp).

3.4 THE TEACHING OF CHEMISTRY IN EIGHTEENTH-CENTURY OXFORD

Given the general situation in Oxford in the eighteenth century, why would anyone want to study chemistry? It was not needed to take any degree, there was no established chair after the first decade of the century, it would not advance anyone's academic career (in contrast to Cambridge), and it was not fashionable until the 1780s when the situation changed markedly as we will see. However, it was considered to be part of a liberal education, an accomplishment that an educated gentleman should cultivate to a modest degree. For the many Oxford undergraduates destined for the Church, it was a minor element of natural theology, revealing the divine hand in the material world and explaining some passages of the Bible, such as the miracle of the Golden Calf.[21] For the aspiring physician, it was part of the medical curriculum, but to a much lesser extent than at the Scottish Universities and at Leiden. The future landowner, perhaps to his surprise, could gain much from the chemistry of this period, helping him to test his soils, test the springs on his estates for their value as mineral waters (or their potability as drinking water), and to analyse potentially useful minerals that might be found on his land.

How could a student learn chemistry in eighteenth-century Oxford? As we shall see, the official teaching of chemistry in the university was intermittent during this century. Oxford was off-limits to the itinerant lecturer since only MAs of the University could give lectures, as the young Nathan Alcock found to his cost in 1740. This probably more than anything else limited access to any kind of chemical education in Oxford in the first eight decades of the eighteenth century. He could go to another university – most likely Cambridge or Edinburgh – which one could argue rather defeated the purpose of going to Oxford. But chemistry was a highly specialised form of training in the eighteenth century and Oxford provided a more general education. That at least some students took this route is confirmed by a manuscript book of lecture notes preserved in the Museum of the History of Science in Oxford.[22] It appears they are notes of Cullen's lectures in Edinburgh taken by Francis Bayley of Shrewsbury and Balliol College, just before and after obtaining his BA in 1758 as there is a footnote at the end of the volume in the same hand as the rest of the notes signed "F. Bayley".[23] That they are Cullen's lectures is indicated by the title of the second part of these notes, "Cullen's Lectures on Fire" and a final footnote that states that "Dr Cullen afterward added another class of bodies, viz: aerial". This notebook also sheds some light on the activities of the chemically inclined student in this period. Halfway through the book, there is a set of analyses of the water of various springs at Hulme Hall, near Manchester, dated August 8, 1758. For a young amateur they are fairly sophisticated, involving the use of syrup of violets, mercury(II) chloride and lead acetate. He concluded that they were almost pure soft waters containing a small amount of Glauber's salt (sodium sulfate) and perhaps a little marine salt (sodium chloride). It is likely that Bayley's interest in chemistry was medical in origin as both his father, Charles Bayley, and his grandfather were physicians.[24] Sadly Francis never became a physician himself as he died at the age of 25 in 1761 and is buried at Clungunford, Shropshire.[25]

If our putative student was unable to attend lectures in Oxford or unable to afford to spend a few terms elsewhere, he could borrow or buy the notes of students (or a professional amanuansis) who had attended the lectures, which were even published with or without the permission of the lecturer. As Cullen and Black never published a text of their lectures in their lifetime, this was the only way their chemical doctrines reached a wide circle, as shown by the unfortunate Francis Bayley's notes of Cullen's lectures that may have come into the hands of Thomas Hornsby after Bayley's death. If all else failed, the student could always teach himself chemistry from books. Some modern chemists might consider this to be a poor form of teaching but it was the predominant mode of learning chemistry in this period and indeed well into the nineteenth century (I may add a personal note here and remark I learnt 85% of my chemistry from textbooks and only 15% from lectures and practical work even in the 1970s). Until the 1730s, the leading textbooks were George Wilson's

Complete Course of Chemistry[26] and Nicholas Lemery's *Course of Chemistry*[27] both of which had a medical bias. The field was then dominated by Boerhaave, usually transmitted through the works of Peter Shaw, namely his *Philosophical Principles of Universal* Chemistry,[28] his *New Method of Chemistry*,[29] and his translation of Boerhaave's own textbook, the *Elementa Chemicae*.[30] By the 1760s, the works of two other major Continental chemists, Caspar Neumann and Pierre Macquer were available in English translations;[31] Macquer's *Elements of Chemistry* was particularly popular. James Keir's translation of Macquer's *Dictionary of Chemistry*[32] followed in 1771 and was used by Priestley's friend Josiah Wedgwood.[33] One of the first specialist works in chemistry was Joseph Priestley's *Experiments and Observations upon different kinds of Air*.[34] Richard Watson's *Chemical Essays*[35] was one of the most widely praised chemistry books in the late eighteenth century – Edward Gibbon described it as "a classic book, the best adapted to infuse the taste and knowledge of chemistry"[36] – and it was one of the first books to popularise chemistry. Once the Chemical Revolution was underway in the late 1780s, the first recognisably modern chemical textbooks proliferated, including a translation by William Nicholson of Antoine Comte de Fourcroy's *Elements of Natural History and Chemistry*,[37] Jean Chaptal's *Elements of Chemistry*,[38] Lavoisier's *Elements of Chemistry* (translated by Robert Kerr)[39] and Nicolson's own *First Principles of Chemistry*.[40] Instructions on practical chemistry were available from Peter Shaw and Francis Hawksbee's *Essay for Introducing a Portable Laboratory*[41] and Antoine Baumé's *Manuel de Chymie* (1763) which was used by Richard Watson when he became professor of chemistry in Cambridge.[42]

How did chemistry students carry out practical work in this period? It has to be borne in mind that as late as the 1820s, even experienced chemists such as Joseph Priestley or William Hyde Wollaston (1766–1828) carried out experiments in their kitchen or sitting room. Wollaston developed apparatus that could be used on a tea tray[43] and Black carried out his research on latent heat using wine glasses and a fish kettle.[44] There was a chemistry laboratory, the Officina Chemica, in the basement of the Ashmolean Museum, and it appears that it may have been used by dons and students for experimental work even in the absence of an official chemistry lecturer. An eighteenth-century laboratory was simply a large room with benches and a furnace or stove, not very different from the college kitchens. There was no running water, gas or fume-cupboards! The German traveller, Zacharias Conrad von Uffenbach (1683–1734), whose description of the building is in the previous chapter, also went on to give his impression of the state of the laboratory in 1710:

> But to return to the Laboratory. I must say it is right well built. It is as long as deep as the [old] Ashmolean but not as broad. It is completely vaulted over and is provided with many kinds of furnaces, some quite remarkable, all of which are decorated in the most costly manner with architectural decorations and the like . . . The present professor of Chemistry does not trouble about it, and the operator (said to be a good-for-nothing man) still

less. And so although the furnaces are in fairly good order, they look very much worse for wear. Not only are the finest instruments, tiles and suchlike almost all broken to pieces, but the whole place is filthy. Is it credible that so little attention should be given to so costly and beautiful a work? Who would have believed it in England which we foreigners hold in so much in awe that we believe all subjects and chemistry in particular to be passing through a Golden Age?[45]

The "present professor" was Richard Frewin (see below) and the "good-for-nothing" operator was one of the sons of Christopher White, the original operator who had died in 1696.[46] He may well have got the position simply to keep the equipment in the museum as the elder White has considered it to be his own property (despite the fact it had all been bought by the university for the laboratory) and was presumably inherited by his sons.[47] Uffenbach was perhaps excessively critical (as he was with the laboratory of the Royal Society) and a later Dutch visitor in 1749 described the laboratory with its furnaces as magnificent.[48] Most of the experimental work would have been simple analysis or the preparation of inorganic compounds, although as we will see, some sophisticated work was carried out in Oxford later in the century on high-temperature chemistry and freezing mixtures.

There are two important points to bear in mind about the history of chemistry in eighteenth-century Oxford. Chemistry was almost invariably treated as part of the medical curriculum, rather than the liberal arts, and in Oxford it was linked to anatomy, either being taught by the anatomy lecturer, in the same place as anatomy or at least alongside anatomy. Anatomy included much of the study of outside soft and hard tissues. Bones, fossils and stones were not easily distinguished and lay at the boundary between chemistry – then still seen as the taxonomic study of animal, vegetable and mineral matter – and anatomy. Since the introduction of iatrochemistry by Paracelsus in the mid-sixteenth century, chemical medicines had become increasingly important and many early eighteenth-century chemistry textbooks were concerned with the manufacture of these new medicines. Anatomy was concerned with the use of external medications and hence chemistry. Christ Church was the leading college, financially and intellectually, and its dons were dominant in many fields including chemistry. This was partly because Westminster School was the major source of undergraduates for Christ Church in this period. Richard Busby, its long-serving headmaster between 1638 and 1695 developed self-confidence in his pupils and encouraged the sciences (as evidenced by his famous pupils Robert Hooke and John Locke) as well as classics.

One of Busby's pupils who went to Christ Church was (Sir) Edward Hannes (1663/4–1710), who was the Professor of Chemistry at the beginning of the eighteenth century.[15] But the chair lapsed after he left Oxford to practise medicine in London in 1699, and become a physician to Queen Anne in 1702. He resigned his Studentship (fellowship) at Christ Church, but that was because he married not because he was leaving Oxford. It is even debatable if he ever vacated his professorship before his death. Thereafter for the rest of the eighteenth century, there was no official university position in chemistry as far as we can now tell. There were various lecturers and readers, as we shall see, but they were officially sanctioned extra-mural lecturers. Martin Wall, in the early 1780s, held the only official title in the whole century as the 'Public Reader', in other words a lecturer to the public at large not a university teacher.

Certainly Hannes's fellow "medical poet" John Freind (1675–1728) appears to have been acting as a locum when he gave his excellent course of lectures on chemistry in the Ashmolean in 1704. Freind was the son of the rector of Croughton, Northamptonshire, and after being educated by Busby at Westminster was initially a classicist at Christ Church with the encouragement of the Dean Henry Aldrich, although he also studied mathematics under David Gregory.[49] He then turned to medicine and published his first medical work, *Emmenologia*, which described menstruation in terms of Newtonian hydraulics, in 1703. Through Gregory, he was influenced by the thinking of James and John Keill and their Newtonian views of matter. It is therefore not surprising that Freind's lectures – published as *Praelectiones chymicae* in 1709 – expounded the principles of nine chemical operations in terms of Newtonian attraction and statics.[50] Freind left Oxford for Spain as the Earl of Peterborough's physician in 1705 and he then became a leading physician in London and was elected a Fellow of the Royal Society in 1712, the same year that he published an English version of *Praelectiones chymicae*. Ten years later, he was elected the MP for Launceston in Cornwall, as a Tory. Soon afterwards, thanks to his closeness to his Christ Church friend, Lewis Atterbury, he was exposed as a Jacobite conspirator in 1722 and was imprisoned briefly in the Tower of London. In spite of his Jacobite activities, he was made official physician to Queen Caroline, wife of George II, in 1727.

Freind's successor as the chemistry lecturer was Richard Frewin (1681–1761) who was born in the City of London.[15] He studied medicine at Christ Church, although he did not take his BM until 1707 (he took the MD in 1711). He assisted his colleague Freind during his 1704 lectures. He also studied in Cambridge and may have attended the lectures of John Vigani. Dr Johann Rudolph Lavater, the lecturer in anatomy, gave chemistry lectures, presumably in the Ashmolean around 1710. His lectures were largely based on Nicolas Lemery's *Course of Chemistry*, the third English edition of which – a translation of the eighth French edition by James Keill – had been published in 1698.[51] Uffenbach described Frewin as "Professor of Chemistry" in 1710, but it is unclear if he ever held a formal position in chemistry at Oxford or if he was the successor of Hannes or Freind or even when he was appointed (given that Hannes could have been the nominal professor until his death in 1710). It is not even known if Frewin ever gave any lectures on chemistry, he certainly never published any. He did, however, have a fine herbarium and was a considerable benefactor of the University. But likewise when he was elected Camden Professor of History in 1727, he did not teach or carry out any research on history. Like most eighteenth-century dons, he appears to have regarded university posts as sinecures. He may have taken his readership in rhetoric at Christ Church more seriously but even that is doubtful, but it appears that he was a

Student (fellow) of Christ Church. Frewin earned his living as a highly regarded (and presumably well-paid) physician in Oxford, although he never published any medical work. The sole reason for assuming that he was the official reader (or professor) in chemistry is his strong opposition to the lectures of Nathan Alcock in 1740 which appears to have stemmed from the infringement of his monopoly on chemistry teaching in the university.

Nathan Alcock (1709–1779) was born in Runcorn, Cheshire; he was a descendant of Bishop John Alcock who had founded Jesus College, Cambridge, in the fifteenth century.[15] As Nathan was not keen on the usual classical education, his father bribed him with an estate in the Wirral to take up medicine. After studying in Edinburgh, Alcock read medicine at Leiden under Boerhaave, and chemistry under Gaubius and Boerhaave. He eventually took his MD at Leiden with a thesis on pneumonia in 1742. In 1739, he spotted a potential opening in Oxford. Francis Nicholls, the lecturer in anatomy, had already left Oxford for London and his successor, Thomas Lawrence, had not been confirmed by the elderly and absent Regius Professor of Medicine, William Woodforde.[52] Alcock also decided to teach chemistry, presumably because of Frewin's inactivity and the popularity of Boerhaave's doctrines, he had learnt in Leiden. These lectures were not authorised (he did not have the Oxford MA) and they were strenuously opposed by the University authorities in general and by Richard Frewin in particular (Table 3.2). They set up Thomas Hughes of Trinity College as the lecturer in chemistry in opposition to Alcock.[53] However, Alcock's lectures were much more popular and his powerful friends among the MAs of the university – including the jurist Sir William Blackstone and a future bishop of London Robert Lowth – pressed the University to grant him a MA in 1741. He then became the officially sanctioned (if not officially appointed) lecturer in chemistry instead of the unfortunate Hughes. Alcock was then awarded the MB in 1744 and the MD in 1749, the same year that he was elected a Fellow of the Royal Society. It appears that he was never a fellow of an Oxford college. He was still teaching chemistry and anatomy in the Ashmolean when the Dutchman Pieter Camper (1722–1789) visited Oxford in 1749.[48] But he was not appointed a reader or professor in chemistry and, like Frewin, he relied on his medical practice for his income. This medical competition may also explain Frewin's hostility and politics was another cause of friction as

Table 3.2 Nathan Alcock's lectures given in Jesus College in 1756.[a]

Lecture 1, Introduction; **2–4**, Fire; **5**, Fire and Air; **6–7**, Air; **8**, Water; **9–10**, Mineral Waters; **11**, Earth; **12**, Salts, acid salts in particular; **13–15**, Salts: neutral salts; **16**, sulphur; **17**, Oils and Vegetable Matters; **18**, On the analysis of Animals and Vegetables; **19**, Temperature; **20**, Animal Juices; **21**, Metals: gold (?lead, platinum), silver; **22**, metals: copper, tin; **23**, Metals: antimony; **24**, Metals: bismuth, zinc, cobalt, arsenic; **25**, Colours, prisms; **26**, On fermentation; **27**, Artificial gums; **28**, On Docinay[b] and some mercurials; **29**, Poisons; **30**, Phosphors; **31**, [Incomplete]

[a]From the notebook of John Thomas, 12 May 1756, cited by P.W. Kent, *Some scientists in the life of Christ Church, Oxford*, printed privately, 2001, Appendix 2.
[b]docimasy=the extraction of metals from their ores, *Oxford English Dictionary*.

Alcock was a Whig, whereas Frewin, like his colleague Freind, was a High Tory.

Alcock left Oxford for Bath in 1757 and he was replaced as the anatomy lecturer by John Smith (1721–1796). It is not clear if Smith also replaced Alcock as the chemistry lecturer – it is significant that Francis Bayley went to Edinburgh in exactly this period – but he gave chemistry lectures as part of an anatomy course in 1759.[54] Chemistry came after agriculture, and was only a small part of the course. Smith drew heavily on other authors and he kept the content of his lectures within the medical remit of the course. For example, 'alkaline salts' was largely a discussion of bile and limewater was mentioned for its ability to dissolve kidney stones. This is not to say that he was completely out of touch with contemporary developments. He mentioned Black's MD thesis (which had been published only five years earlier) and Joshua Ward's process for the manufacture of sulfuric acid. It appears that Smith continued to teach chemistry alongside anatomy at least until he was appointed Savilian Professor of Geometry in 1766.[55]

The link between chemistry and anatomy was maintained when Dr Lee's Readership of Anatomy was established at Christ Church in 1767.[56] Dr Matthew Lee (1694–1755) had practised medicine in London, becoming the physician to Frederick, Prince of Wales. In his will, he left money to found an anatomy school in Christ Church, see Figure 3.1, and a university readership in anatomy, thereby fulfilling a plan of his former medical colleague and fellow member of Christ Church, John Freind. It was a condition of Dr Lee's will that the reader had to

Figure 3.1 The Anatomy School at Christ Church, built in 1766, and used also for chemical lectures from about 1770. It became solely a chemical laboratory from 1863, when the anatomy teaching had moved to the new Museum, until its closure in 1941. It is now part of the Senior Common Room. From J. and H.S. Storer, *The University and City of Oxford, displayed in a Series of Seventy-Two Plates*, London, 1821.

come from Westminster and Christ Church and not be a clergyman. However, the first condition appeared to have lapsed as early as 1785. The first reader John Parsons (1742–1785), who was born in York, and was educated at Westminster School and Christ Church as required by Dr Lee's will.[15] He was appointed in 1767, although he did not take his BM until 1769 and his DM in 1772. In addition to taking the essential MA at Christ Church in 1766, Parsons also studied medicine at Edinburgh, where he probably attended Cullen's last chemistry lectures before he handed the teaching of chemistry over to Black. Parsons delivered a lecture course in chemistry and his first syllabus, entitled *Plan of a Course of Lectures in Philosophical and Practical Chemistry*, was published anonymously in the early 1770s.[57] Notes of his lectures[58] were made by the lawyer William Scott, later Baron Stowell. William Scott (1745–1836) took his BA in 1764 at Corpus Christi College and then became a Durham fellow at University College (he was born in County Durham) until 1776 when he took chambers in London.[15] Parsons became physician to the Radcliffe Infirmary in 1772 and the first Lichfield Professor of Clinical Medicine in 1780.

3.5 THE REVIVAL OF CHEMISTRY AFTER 1775

In the latter part of the eighteenth century chemistry became increasingly popular with the public at large. Whether because of this rising interest or not, chemistry at Oxford entered a new more positive phase in the late 1770s. William Adams, Master of Pembroke between 1775 and 1789, was interested in chemistry, probably as a result of his close friendship with Samuel Johnson who was an enthusiastic amateur chemist. Four notable chemists – William Higgins, Thomas Beddoes, James Smithson and Davies Gilbert – matriculated at Pembroke College during this period.[59] On the other side of St Aldates, Dean Cyril Jackson of Christ Church (and an Old Westminster) was a generous host to Joseph Black in 1788[60] and had Thomas Beddoes admitted to Christ Church's Common Room.[61]

Parsons's successor as the lecturer of anatomy, William Thomson, did not teach chemistry[62] although he attended Black's chemistry lectures in Edinburgh in 1781–2 and became a friend of James Hutton. As an avid geologist, he gave a course in mineralogy and did research on the action of heat in geology in support of Hutton's theories of slow gradual geological change.[15] Probably because Thomson was not appointed as Dr Lee's Reader until 1785, Martin Wall (1747–1824) was appointed Public Reader in chemistry in 1781. He was the son of John Wall, a Worcester physician (and a graduate of Worcester College) who had been one of the founders of the Worcester Porcelain Company in 1751.[15] He developed a new process for the manufacture of porcelain and his name is closely associated with the company. Wall took his MA at New College in 1771, studied medicine at St Bartholomew's hospital and under Black at Edinburgh. He took his BM at Oxford in 1773 and had been made physician to the Radcliffe Infirmary in 1775 and a Fellow of New College in 1778.[15] He may have been suggested as the Public Reader of Chemistry by Parsons. His father's work on porcelain may have also given him an interest in chemistry. He charged a course fee of two guineas (compared with the standard fee of three guineas at Edinburgh) and it appears that he aimed his lectures at a clerical rather than a medical audience. According to the *Minutes of Dr Wall's Lectures* (1781) he stated that chemistry was "an immediate revelation from Heaven to Adam, and had its name from Cham the progenitor of the Egyptians".[63] Later in the course, he said that "Moses was probably acquainted with the method of fusing metals witness the G.[olden] calf".[64] There was little on theory in Wall's course, but he mentioned phlogiston. Intriguingly, "oxygene" is written above it and in the same hand, so Wall may have at least mentioned Lavoisier's theory in his second or later course of lectures. Wall considered that "Chymistry in England was neglected for Newtonism. In Germany [chemistry] was cultivated by Stahl, Hoffman, Boerhave [sic], von Helmont, Rouelle in Paris was the founder of Philosophical Chemistry".[65]

Wall succeeded Parsons as Lichfield Professor of Clinical Medicine in 1785, following the latter's early death from fever, and his successor as the Public Reader of Chemistry and physician at the Radcliffe Infirmary was William Austin (1754–1793). Austin was born in Wotton under Edge, Gloucestershire, and came from a family of clothiers.[15] He was very energetic – walking from Oxford to London in a day – and studied botany and Arabic at Wadham College before studying medicine under Percivall Pott at St Bartholomew's Hospital, taking his MD at Oxford in 1783. Meanwhile, he gave lectures in mathematics for John Smith, now Savilian Professor of Geometry, during the latter's absence. Smith may have suggested his appointment, although it also seems to have been closely linked with the position of physician at the Radcliffe Infirmary in this period. Austin was only Reader for a year before being appointed physician at St Bartholomew's Hospital, where he continued to give chemistry lectures. Like Parsons, he died at a fairly early age of fever, which must have been an occupational hazard for physicians in the eighteenth century.

William Higgins (1763? –1825), the nephew of the London chemistry lecturer Bryan Higgins, and the son of Thomas Higgins, a physician in County Sligo, was briefly at Magdalen Hall in 1786, before moving (like Thomas Beddoes) to William Adams's Pembroke College but he never took a degree. He was made 'operator' to William Austin just before the latter's resignation in 1786.[15] Higgins was one of the first converts to Lavoisier's system of chemistry in England, probably under the influence of his uncle and he in turn converted Austin. After a financially difficult period following his return to Dublin in 1792, Higgins became Professor of Chemistry at the Royal Dublin Society and chemist to the Irish Linen Board.

Much of Higgins's subsequent research under Beddoes in 1787-8 was published in his *Comparative View of the Phlogistic and Antiphlogistic Theories*[66] that took issue with his fellow Irish chemist Richard Kirwan's attempt to save the phlogiston theory in his *Essay on Phlogiston* (1787).[67] Higgins's book later became famous as an unwitting anticipation of John Dalton's Law of Proportions and his atomic theory, although it actually owed more to the same Newtonian ideas used by Freind. Higgins also carried out an analysis of a human calculus (kidney stone).[68] He remarked that while he was "busy in assisting at a public course of Chemistry in Oxford" he found that when animal blood was heated with nitric acid, it became the colour of bile and very bitter.[69] In the *Comparative View*, Higgins gives a description of the Officina Chemica that can be compared with Uffenbach's description eight decades earlier:

> The Chemical elaboratory at Oxford is near six feet lower than the surface of the earth. The walls are constructed with common limestone, and arched over with the same; the floor is also paved with stone. It is a large room and very lofty. There are separate rooms for the chemical preparation, so that nothing is kept in the elaboratory, but the necessary implements for conducting experiments. There is an area adjoining it on a level with the floor, which though not very large is sufficient to admit a free circulation of air. The ashes and sweepings of the elaboratory are deposited in it. There is a good sink in the centre of this area so that no stagnated water can lodge. The pr[iv]y which is seldom frequented is overground and unconnected with the elaboratory. Notwithstanding all this, the walls of the room afford fresh crops of nitre every three or four months.[70]

Dr Wall had taken the deposit to be fixed vegetable alkali (potassium carbonate) but Higgins showed it to be mainly nitre, an analysis confirmed by John Kidd, the Aldrichian Professor of Chemistry, in 1814 (see the next chapter for further details).[71]

During the archaeological excavations in 1999, to the rear of the basement of the Museum of the History of Science, the site of the Officina Chemica, described briefly in the previous chapter, several ceramic crucibles (and a couple of retorts) were found. Although they were of a Hessian style, chemical analysis of the ceramic hints that they were British made. According to a chemical analysis of the residues, these crucibles appear to have been used for three different kinds of experiments. Some of them contained high-temperature glazes, which suggests that Martin Wall may have carried out experiments to produce ceramic glazes, possibly in connection with his father's celebrated porcelain factory at Worcester (although his father had died in 1776). Others had been used for experiments with alkaline metal sulfates, possibly an early attempt to make synthetic alkalis along the lines of the work of James Keir in the 1770s, who used the double-decomposition of calcium hydroxide and sodium sulfate. There was much interest in making synthetic alkalis following the loss of the American wood ash as a result of the American war. Bryan Higgins also tried to produce synthetic alkalis[72] and this work may have been carried out by his nephew. Other crucibles seem to have been used to heat zinc and lead. It has been suggested that they were perhaps linked to alchemical experiments in the early days of the Museum, but as they are very much alike it seems more likely that they all date from the mid-1780s, from the work of Wall, Higgins and possibly Beddoes. The zinc and lead may have been linked to research on alloys – Keir patented an alloy of copper and zinc in 1779 – or the work on synthetic alkalis as one method involved heating sodium sulfide with lead. Thus, the discovery of these crucibles suggests that the laboratory was active again in Wall's time after almost a century of disuse.[73]

The best-known chemist in Oxford in the eighteenth century, Thomas Beddoes (1760–1808), was Austin's successor, although it is very unclear if he was ever appointed officially; certainly no record seems to exist.[74] However, nearly all the appointments as 'chemical reader' appear to have been informal, and in at least some cases, for instance Alcock and probably Beddoes, to be no more than unwritten leave to give private lectures. Born at Shifnal in Shropshire, Beddoes was educated under Adams at Pembroke and it is claimed that he attended chemical demonstrations at the Old Ashmolean, but it is unlikely that there was any chemistry teaching at the Ashmolean in the late 1770s.[15,75] He did, however, learn chemistry at Bryan Higgins's School of Practical Chemistry in Soho while working at the nearby school of anatomy in Great Windmill Street. He then studied chemistry under Joseph Black in Edinburgh. He appears to have been one of the first Oxford students to study chemistry for its own sake. In 1787, he visited Guyton de Morveau and Lavoisier in France and also became imbued with the democratic ideas that were popular in France just before the revolution. On his return, he took up the Readership of Chemistry. Beddoes gave very popular lectures for which, like Wall, he charged two guineas per course.[76] He taught chemistry five days a week. The quality of his lectures is uncertain. Beddoes was an excellent self-promoter and his later assistant at the Clifton institution, John Stock reported in 1811 that:

> Indeed one of his pupils and friends, who was long connected with Oxford, speaks of the effect of his lectures in the following terms – "The time of Dr Beddoes' residence at Oxford was a brilliant one in the annals of the University. Science was cultivated more than it has been since and I believe I may say of the period preceded. Dr Thomson's lectures on anatomy, and Dr Sibthorp's on botany, were delivering at the same period; and produced a taste for scientific researches which bordered on enthusiasm.[77]

On the other hand the well-known radical politician James Fox, a friend of Joseph Priestley, criticised Beddoes's lectures, admittedly at second hand.[78] Beddoes also had problems with his demonstrations[79], which may have prompted him to hire James Sadler (1751–1828) who had given demonstrations of 'philosophical fireworks' at the town hall in 1789 and 1790.[15] Sadler, who had originally trained as a pastry cook under his father who was a confectioner in the High Street. Sadler is now best known as the first British person to ascend in a balloon. He made his first ascent on 4 October 1784 in a hot-air balloon of his own manufacture, similar to the one used by the Montgolfier brothers, almost a year after the first French ascents and three weeks after Vincenzo Lunardi's flight in London. He travelled six miles. Five weeks later, he made a second ascent in a hydrogen-filled balloon that travelled fourteen miles in a flight that was said to have lasted only twenty minutes. Beddoes described

Sadler as a "clever, practical and experimental manipulator in chemistry" who had helped him to refurbish the laboratory. He was mechanically inclined and very interested in steam engines, including steam-powered vehicles, despite efforts by Boulton and Watt to hinder his work. In the late 1790s he set up a mineral-water factory near Golden Square, London, using the method published by Joseph Priestley in 1772. Sadler used an image of a balloon as an early version of a corporate logo to sell his soda water, but in contrast to the Drury Lane mineral water factory of the Swiss Johann Jacob Schweppe founded in 1790,[80] it was not very successful. He also acted as a chemical advisor to the Royal Navy, but he was not often consulted and he remained in financial difficulties.

Unfortunately, Beddoes's zeal to reform both the University and the state brought him into conflict with leading members of the University, notably Bodley's Librarian, the Revd John Price of Jesus whom he had annoyed by complaining about the lack of chemistry books in the library and the closure of the Bodleian on his only free days, Saturday and Monday (when Price travelled to and from his parish at Wilcote), in his Memorial concerning the state of the Bodleian Library.[81] Unlike his predecessors and successors, Beddoes also lacked a power base in the medical faculty – although he had been awarded an Oxford MD in 1786 – or at the Radcliffe Infirmary. The latter is a striking distinction between Beddoes and the other late eighteenth-century lecturers in chemistry at Oxford.

In the summer of 1792, the highly regarded Sherardian Professor of Botany, John Sibthorp (1758–1796), approached the Crown, through the Chancellor of the University Lord North and the Home Secretary Lord Dundas, to have the stipend that was funded by the Crown restored from £100 a year to the £200 a year that been given a century earlier.[82] This request was granted but almost immediately senior figures in the university, alarmed at the growing strength of chemistry at Cambridge, wondered if it would be better if the extra £100 a year was used for a new professorship in chemistry. Beddoes was an obviously well-qualified candidate for the post and hopes were high. At this point, however, concern was growing in the Home Office about increasing democratic agitation, in which Beddoes paid a leading part. Indeed only eleven days after Lord North had forwarded Beddoes's memorandum on the chemistry chair, Beddoes was placed on the Home Office list of "disaffected and seditious persons". A new idea was put forward by the university: it was understood that Beddoes would soon give his last course of lectures and then leave, allowing the Crown to appoint someone more reliable as the new professor. After Revolutionary France had guillotined Louis XVI and declared war on Britain in early 1793. Beddoes's position became untenable. He resigned and moved to Bristol to develop his ideas of social medicine. He was also interested in the use of gases in medicine and set up the famous Pneumatic Institute at Clifton in 1799. This was short-lived; his assistant Humphry Davy left for the Royal Institution in 1801 and soon afterwards, the Pneumatic Institute became the Preventive Medical Institution for the Sick and Drooping Poor. Beddoes died in 1808 of a 'dropsy of the chest', possibly the result of inhaling chlorine.

Whilst he was still in Oxford, Beddoes publicised the research on cooling mixtures by the apothecary of the Radcliffe Infirmary, Richard Walker, by reporting his early research in the leading scientific journal of the day, the *Philosophical Transactions*, in 1787.[83] Cooling mixtures were of great interest in this period as they were effectively the only way – extreme weather excepted – of obtaining very low temperatures and thus determining what was the lowest temperature that could be achieved artificially. Little is known about Walker. He was appointed apothecary to the Radcliffe Infirmary on 7 February 1781, at which time he must have been more than twenty-five years of age.[84] In 1795, it was alleged that he was capricious towards the patients, casual towards to the physicians and treated religion with ridicule. It was further alleged that the medicine bills were too high. He denied the charges but was reprimanded and lost his gratuity.[85] After similar complaints were made ten years later he was dismissed on 9 October 1805.[86]

If his career at the Radcliffe was undistinguished, Walker's research on cooling mixtures was impressive – it culminated in his successful attempt to freeze mercury at an ambient temperature of 30 °F (−1 °C) – and that were published in his book, *An Account of Some Remarkable Discoveries in the Production of Artificial Cold*.[87] He added nitric acid to finely powdered and dried Glauber's salt (sodium sulfate), then stirred in a similar mixture of nitric acid and sal ammoniac (ammonium chloride). This cooling mixture was notable for the lack of snow that was commonly used. This mixture was used to freeze mercury for the first time in England, the end goal of low-temperature research at the time comparable to the later quest for absolute zero. The experiment was performed at the Anatomy School (Dr Lee's building) in Christ Church in January 1789, when several notables including Dean Jackson of Christ Church were present.[88] Although it was erected mainly for the study of anatomy and the practice of dissections, given Parsons's chemical interests, it is very likely that Dr Lee's building was also regularly used for practical chemistry. As it became the Christ Church chemical laboratory in 1863, it can claim to be the oldest college chemical laboratory in Oxford, and indeed the longest lasting, as it was one of the last to close in 1941. It is just possible that Richard Walker may have had access to Francis Bayley's notebook through Thomas Hornsby and hence the section on cooling mixtures in his notes may have prompted this research. Dr Robert Bourne, Beddoes's successor, alluded in his lectures to Walker's researches saying that "they were not the effect of chance ... but of a regular chain of reasoning".[89]

The Vice-Chancellor's hopes that the Crown might fund a chair in chemistry once Beddoes was out of the way turned out to be ill-founded, probably a consequence of Dundas's irritation at the whole bungled affair and the death of the supportive Chancellor Lord North on 5 August 1792. Beddoes supported the appointment of the Revd Henry Peter Stacy (1760–1818), because he had a large family to support. Having failed to gain a position at Oxford, Stacy became curate-in-charge at St Paul's, Hammersmith, in 1794 where he was friendly with the chemist and coachbuilder Charles Hatchett, who supported his election to the Linnean Society in 1797.[90] It appears that Stacy then lost his money supporting gun-barrel boring experiments by Sadler, which may explain why he left Hammersmith in 1801 and became a chaplain in India. Although he had a reputation as a botanist, it is not clear how Stacy learnt chemistry or even if he was proficient in chemistry; his only publication was on the failure of turnip crops.[91] The new Reader of Chemistry was Robert Bourne (1761–1829). Born at Shrawley in Worcestershire, Bourne went to Worcester College, where he took his MA in 1784 and became a fellow of the college.[15] He then studied medicine at St Bartholomew's Hospital, possibly partly under Austin who had also been at Worcester College, and it is said he also studied in the Netherlands for a year. He took his MD in 1787 when he also became an honorary physician at the Radcliffe Infirmary. Bourne has been completely overshadowed by his colourful predecessor. Yet the notes of his lectures taken by Stephen Peter Rigaud (1774–1839)[92] of Exeter College, preserved in Magdalen College Library, show that he was a perfectly competent chemistry lecturer.[93] Clearly the University had confidence in Bourne's abilities since he became the first Aldrichian Professor of Medicine in 1803. He succeeded Wall as the Lichfield Professor of Clinical Medicine in 1824.

Based largely on Antoine-François Fourcroy's *Elementary Lectures on Natural History and Chemistry* (the translation by Thomas Eliot was published in 1785), Bourne's lectures were descriptive and with little theory. He shared Black's distaste for theorising and also his scepticism about the state of chemistry: "chemistry has not long vindicated its claim to rank among the circle of sciences".[94] On the other hand, Bourne scorned the material doctrine of heat, accepted by the Scottish and French chemists, as the imagination of some philosophers.[95] Although he was happy to recommend Lavoisier's *Elements of Chemistry* (presumably in the translation by Robert Kerr published in 1790), Bourne, like Black, was reluctant to abandon phlogiston altogether, but "the former supposition [Lavoisier's theory] though not perfect is certainly the more simple and appears to be gaining more ground".[96] It is interesting to note that his lectures were not medically oriented, there is little mention of materia medica, despite R. T. Gunther's remark that they took place "in a period when the new Radcliffe Infirmary doubtless gave a great impetus to the study of medicine and thereby to chemistry".[97] Thus by focusing on chemistry on its own, Bourne may be regarded as one of the first modern lecturers of chemistry in Oxford, alongside Thomas Beddoes.

3.6 CONCLUSION

Although Oxford had established one of the earliest chairs in chemistry in the world in 1683, the teaching of chemistry in Oxford up to 1781 was fragmentary and largely derivative. Why the teaching of chemistry at Oxford faltered while the later chair of chemistry at Cambridge has continued without a break up to the present is not entirely clear. The religious bar of the Test Act of 1662, which is often given as the reason for the poor teaching of chemistry at Oxford, applied equally to Cambridge and assumes for no good reason that only dissenters were proficient at chemistry in this period, a generalisation that is only rendered plausible by conflating members of the established Church of Scotland with English dissenters. Certainly the functional insignificance of academic appointments in the earlier part of the century was one factor. If Richard Frewin could not be troubled to teach history, far less do historical research, as the Camden Professor of History, why would he have bothered to take his presumed position as Reader of Chemistry any more seriously? But was Frewin ever formally the Reader of Chemistry and did he perhaps lecture more frequently and over a longer period that has been claimed by historians? Part of our problem in assessing the state of chemistry in early eighteenth-century Oxford is the lack of documentary evidence for what actually happened. There was also uncertainty about the desired audience for chemistry at Oxford. Was the aim of teaching chemistry the improvement of medicine and pharmacy, to broaden the mind of undergraduates studying classics or simply to provide useful instruction in an upcoming subject for the university at large? And if the university's ambitions for chemistry were so vague, it is hardly surprising that the Readership did not attract the attention of ambitious young academics eager to climb the academic and clerical ladder. Indeed the continuity of the Chair at Oxford's Fenland rival is largely the result of the ambition of Richard Watson, who, lacking any knowledge of chemistry, got himself elected as Professor of Chemistry at Cambridge in 1764. He may have been a careerist and an opportunist of the first water – he moved from the chair of chemistry to the Regius Professorship of Divinity in 1771 and onwards to the See of Llandaff in 1782, his ambition to ascend to York or Canterbury thwarted by the political upheavals of the 1780 and 1790s – but he was clearly bright, being second Wrangler in his year, and he became a very good chemist.[98] He had no parallel in Oxford with its grounding in classics and humanities rather than the mathematics of Cambridge.

For all practical purposes, chemistry at Oxford University began with the appointment of Martin Wall in 1781. Since that time, there has always been some form of chemical teaching and some chemical research, however, rudimentary, carried out within the University. But chemistry never completely disappeared: students read textbooks and carried out experiments in the long vacation, medical dons gave the occasional course of lectures and the practical side of chemistry was doubtlessly carried on by apothecaries such as Richard Walker and the members of the new profession of chemists and druggist who were taking over from apothecaries during this century. Yet this lack of activity had no long-term effects. Modern chemistry was largely a nineteenth-century creation, and was developed largely in France and Germany. So when Oxford began to take chemistry seriously in the mid-nineteenth century, the lack of any historical lineage, such as Cambridge or Edinburgh enjoyed, was of no real consequence and this lacuna did not prevent Oxford for becoming one of the leading chemistry schools in the world in the early twentieth century, just as the lack of any historical heritage in chemistry was no barrier to Stanford's rise to eminence in the mid-twentieth century.

A very notable feature of the studies in chemistry in the eighteenth century was the failure of the university to follow up the initiatives of the previous century although much was done in other sciences. To a large degree it was left to individual colleges to follow the earlier lead and this continued even toward the end of the next century. In this chapter we have shown that even amongst the colleges only Christ Church made a serious contribution even building a laboratory albeit for anatomy, although Pembroke trained a few notable chemists towards the end of the century. Christ Church was also the home, often more or less loosely, of those who lectured, who later took the post of Public Reader. Throughout this time there was no chair and the professors of chemistry only become a serious power in the development of chemistry in the twentieth century. By then the division of responsibility between the University and the Colleges had become entrenched. Note in the next chapter that while in the nineteenth century Oxford University contributed little to inorganic and organic chemistry and continued in that vein for another hundred years, it began to play a leading part in physical chemistry through its colleges.

REFERENCES AND NOTES

1. This chapter is based largely on Peter J. T. Morris, *Education of Chemists in the Eighteenth Century*, Chemistry Part II thesis, Oxford, 1978, supplemented by G. L. E. Turner, 'The Physical Sciences', p. 659, in *History of the University of Oxford*, v. 5, 'The Eighteenth Century', ed L. S. Sutherland and L. G. Mitchell, Clarendon Press, Oxford, 1986, p. 659, and C. Webster, 'The Medical Faculty and the Physic Garden', *ibid.*, p. 683.
2. W. H. Brock, *The Fontana History of Chemistry*, Fontana Press, London, 1992, pp. 74–127; A. L. Donovan, *Philosophical Chemistry in the Scottish Enlightenment*, Edinburgh University Press, 1976; J. Golinski, 'Chemistry' in the *Cambridge History of Science*, v. 4, 'Science in the Eighteenth Century', ed R. Porter, Cambridge University Press, 2003.
3. For the chemical views of Newton and his followers, see A. Thackray, *Atoms and Powers: an Essay on Newtonian matter-theory and the Development of Chemistry*, Harvard University Press, Cambridge MA, 1970.
4. G. A. Lindeboom, *Herman Boerhaave: the Man and his Work*, Methuen, London, 1968; R. Knoeff, *Herman Boerhaave (1688–1738): Calvinist Chemist and Physician*, Koninklijke Nederlandse Akadamie van Wetenschappen, Amsterdam, 2002.
5. P. Shaw and E. Chambers, *A New Method of Chemistry: including the theory and practice of that art— Written by the very learned H. Boerhaave—*, London, 1727.
6. For Cambridge, see L. J. M. Coleby, *Ann. Sci.*, 1952, **8**, 46, 165, 293; 1953, **9**, 101; 1954, **10**, 234, and the first four chapters of *The 1702 Chair of Chemistry at Cambridge: Transformation and Change*, ed. M. D. Archer and C. D. Haley, Cambridge University Press, 2005. For Edinburgh and Glasgow, see Donovan, ref. 2, and *An Eighteenth Century Lectureship in Chemistry: essays and bicentenary addresses relating to the Chemistry Department (1747) of Glasgow University*, ed. A. Kent, Jackson, Glasgow, 1950.

7. F. W. Gibbs, *Ann. Sci.*, 1951, **7**, 220.
8. F. W. Gibbs, *Ann. Sci.*, 1952, **8**, 122.
9. The relationship between the new pneumatic chemistry and public lecturing is discussed by J. Golinski, *Science and Public Culture: Chemistry and Enlightenment in Britain, 1760–1820*, chap. 4, Cambridge University Press, 1992.
10. F. W. Gibbs, *Ambix*, 1960, **8**, 111; N. Hans, *New Trends in Education in the Eighteenth Century*, chap. 7, Routledge & Kegan Paul, London, 1951.
11. W. V. Farrar, K. R. Farrar and E. L. Scott, *Ambix*, 1973, **20**, 183, on 190.
12. J. A. Harrison, *Ann. Sci.*, 1957, **13**, 109.
13. L. S. Sutherland, 'The Curriculum', p. 469, and P. Quarrie, 'The Christ Church Collection Books', p. 493, in Sutherland and Mitchell, ref. 1.
14. Turner, ref. 1, see p. 673. Turner himself does not make any claim for the predominance of astronomy and natural philosophy in this period, but the situation is clear enough if the teaching of these subjects is compared with that of chemistry.
15. *ODNB*.
16. In the archives of the Museum of History of Science, Oxford, but with no reference number.
17. Turner, ref. 1, p. 681, citing MS Ny Kgl. 377e, Kongelige Bibliotek, Copenhagen, 67 right and 71 left.
18. Webster, ref. 1, p. 685.
19. Webster, ref. 1, p. 708.
20. J. F. C. Grimm, *Bemerkungen eines Reisenden durch Deutschland, Frankreich, England und Holland in Briefe an seines Freunde*, Altenberg, 1775, quoted by W. D. Robson-Scott, *German Travellers in England, 1400–1800*, p. 153, Blackwell, Oxford, 1953, and thus cited by Webster, ref. 1, p. 708.
21. As Martin Wall did 1781, see 'Minutes of Dr. Wall's Lectures in Chemistry', 1781, MS Radcliffe Trust e.9, p. 5, Bodleian Library, Oxford.
22. MS Radcliffe 17, Museum of History of Science, Oxford. When I first saw this in 1978 it was catalogued with Thomas Hornsby's papers, although "not in Hornsby's holograph". It is now catalogued as 'Volume entitled Earth | Water | Chemical History of | Animals | Vegetables | Elective Attractions', containing MS notes from chemistry lectures taken F. Bayley whose signature appears at the end. The date, August 8, 1758, occurs near the middle of the volume. Some or all of the lectures are probably by William Cullen, and therefore given at Edinburgh, including lectures on 'The Nature of Fire' towards the end of the volume.
23. J. Foster, *Alumni Oxoniensis,—*, Parker, Oxford, 1887–92 and the Buttery Books of Balliol College.
24. As ascertained from the records of Balliol College.
25. http://www.gravestonephotos.com/public/gravedetails.php?available=no &grave=15399, accessed 11 February 2008.
26. George Wilson, *A Complete Course of Chymistry— The third edition*, London, 1709.

27. Nicholas Lemery, *A Course of Chymistry,— The Third Edition, taken from the Eighth in French*, trans. James Keill, London, 1698.
28. Peter Shaw, *Philosophical principles of Universal Chemistry—. Drawn from the collegium jenense of Dr. George Ernest Stahl*, London, 1730.
29. Peter Shaw, *A New Method of Chemistry— from the original Latin of Dr. Boerhaave's Elementa Chemicae*, 2 v., London, 1741.
30. Herman Boerhaave, *Elementa Chemicae—*, Lugduni Batavorum [Leiden], 1732. The first official translation was by Timothy Dallowe in 1735, but Shaw's of 1741 was more popular.
31. William Lewis, *The Chemical Works of Caspar Neumann—*, 2 v., London, 1759; P. J. Macquer, *Elements of the Theory and Practice of Chemistry*, 2. v., London. 1764.
32. James Keir, *A Dictionary of Chemistry—Translated from the French*, London, 1771.
33. R. E. Schofield, *Chymia*, 1959, **5**, 180, on 187.
34. They appeared as six sequential volumes between 1775 and 1786. The three-volume summary of 1790 was widely read: J. Priestley, *Experiments and Observations on different kinds of Air—*, 3 v., London, 1790.
35. Richard Watson, *Chemical Essays*, 5 v., London 1781–1787.
36. Edward Gibbon, *Decline and Fall of the Roman Empire*, chap. 52, London, 1776–1788, in a footnote on bitumen in the context of Greek fire.
37. Antoine, Comté de Fourcroy, *Elements of Natural History and Chemistry —translated from the last Paris edition, 1789, being the third—*, 4 v., Edinburgh, 1790.
38. M. I.A. [sic] Chaptal, *Elements of Chemistry*, 3 v., London, 1791. The 'M' is presumably for 'Monsieur', and the 'I' is a 'J', since his name was Jean-Antoine-Claude Chaptal, Comte de Chanteloup.
39. Antoine Lavoisier, *Elements of Chemistry—translated from the French by Robert Kerr*, Edinburgh, 1790; reprinted by Dover Publications, New York, 1965.
40. William Nicholson, *First Principles of Chemistry*, London, 1790.
41. Peter Shaw and Francis Hauksbee, *Essay for introducing a Portable Laboratory,—*, London, 1731.
42. This was later available in an English translation by John Aikin: Antoine Baumé, *A Manual of Chemistry—*, London and Warrington, 1778. For Richard Watson's use of this manual, see C. A. Russell, 'Richard Watson: gaiters and gunpowder' in Archer and Haley, ref. 6, p. 57, on p. 61.
43. John Scoffern, *Chemistry No Mystery, or a Lecturer's Bequest*, London, 1839, p. 2. Some of the small pieces of apparatus that Wollaston used are in the Science Museum, thus showing that he genuinely worked in this way, and that it was not a ruse to prevent visitors seeing his laboratory, as some authors have claimed.
44. H. Guerlac, 'Joseph Black's Work on Heat', in *Joseph Black, 1728–1799. A Commemorative Symposium*, ed. A. D. C. Simpson, p. 17, Royal Scottish Museum, Edinburgh, 1982, citing Joseph Black, *Lectures on the Elements of Chemistry*, ed. John Robison, Edinburgh, 1803, v. 1, pp. 157–8. Since it

was a Scottish kitchen, I think that it is reasonable to describe the flat-bottomed tin vessel as a Tweed fish kettle.

45. [Z. C. von Uffenbach], *Oxford in 1710, from the Travels of Zacharias Conrad von Uffenbach*, ed. W. H. and W. J. C. Quarrell, Blackwell, Oxford, 1928, p. 37.
46. A. G. MacGregor and A. J. Turner, 'The Ashmolean Museum' in Sutherland and Mitchell, ref. 1, p. 639, on p. 649.
47. R. T. Gunther, *Early Science in Oxford, Part 1, Chemistry*, printed privately, Oxford, 1920, pp. 50–1.
48. Pieter Camper, *Petri Camperi itinera in Angliam, 1748–1785*, Nederlandse Maatschappij tot Bevordering der Geneeskunst, p. 83, cited by Turner, ref. 1, p. 662.
49. *ODNB*; *DSB*; J. S. Rowlinson, *Notes Rec. Roy. Soc.*, 2007, **61**, 109.
50. John Freind, *Praelectiones Chemicae—ann. 1704, Oxonii, in Museo Ashmoleano habitae*, Londini, 1709; translated as *Chymical Lectures—Read at the Museum at Oxford, 1704*, London, 1712. For discussions of this book see Thackray, ref. 4, and J. S. Rowlinson, *Cohesion: a Scientific History of Intermolecular Forces*, Cambridge University Press, 2002, pp. 19–20, 26–27.
51. Gunther, ref. 47, pp. 55–56.
52. [Thomas Alcock], *Some Memoirs of the Life of Dr Nathan Alcock, lately deceased*, London, 1780, pp. 8–9, cited by Turner, ref. 1, p. 664.
53. Turner, ref. 1, p. 665.
54. Gunther, ref. 47, p. 58; George Wingfield, *Anatomical Lectures just as they were taken at a Course read by Doctor Smith of St Mary's Hall, Oxford in the laboratory there, 1759*, Bodleian Library, MS Add A 302. The word '*there*' refers to Oxford, not to St Mary's Hall.
55. Turner, ref. 1, p. 665–6. I disagree with Turner's assertion that Smith would have continued after 1766 as it seems clear to me that Parsons took over.
56. P. W. Kent, *Some scientists in the life of Christ Church, Oxford*, printed privately, 2001, pp. 20–1.
57. In the Library at Christ Church. A copy of these lectures from 1776 is printed as Appendix III of Kent, ref. 56.
58. A notebook containing Baron Stowell's notes of a series of chemical-mineralogical lectures, Museum of the History of Science, MS University 1. The back pages contain a medical poem and an undecipherable reference to 'Dr Beddow' in a different hand-writing. The lectures were attributed to Parsons by William Buckland who was given the notebook by Viscountess Sidmouth, Stowell's daughter.
59. J. R. Partington and T. S. Wheeler, *The Life and Works of William Higgins, Chemist, 1763–1825—*, Pergamon Press, Oxford, 1960, p. 35; *ODNB*; *DSB*.
60. William Ramsay, *The Life and Letters of Joseph Black, M.D.*, Constable, London, 1918, p. 91.
61. E. Robinson, *Ann. Sci.*, 1955, **11**, 137.
62. Implied but not stated by Webster, ref. 1, p. 707.
63. 'Minutes of Dr Wall's Lectures', p. 1.

64. 'Minutes of Dr Wall's Lectures', p. 5.
65. 'Minutes of Dr Wall's Lectures', p. 7.
66. William Higgins, *Comparative View of the Phlogistic and Antiphlogistic Theories—*, London, 1789, reprinted by Partington and Weaver, ref. 59, who argued that most of the experiments in this book must have been carried out in the Officina Chemica, see pp. 5 and 66.
67. Richard Kirwan, *Essay on Phlogiston and the Constitution of Acids*, London, 1787.
68. Partington and Wheeler, ref. 59, p. 5.
69. Ibid., pp. 109, 162, and 164.
70. Ibid., pp. 66 and 76.
71. Ibid., p. 67.
72. L. Gittins, *Ann. Sci.*, 1966, **22**, 175.
73. J. A. Bennett, S. A. Johnson and A. V. Simcock, *Soloman's House in Oxford: New Finds from the first Museum*, Museum of the History of Science, Oxford, 2000.
74. F. W. Gibbs and W. A. Smeaton, *Ambix*, 1961, **9**, 47.
75. Partington and Wheeler, ref. 59, pp. 3 and 33; Robinson, ref. 61; Gibbs and Smeaton, ref. 74; D. A. Stansfield, *Thomas Beddoes, M. D., 1760–1808: Chemist, Physician, Democrat*, Reidel, Dordrecht and Lancaster, 1984; *ODND*; *DSB*.
76. F. S. Taylor, *Philos. Mag.*, 1948, **133**, 154, on 160.
77. John Edmonds Stock, *Memoirs of the Life of Thomas Beddoes, M. D., with an analytical account of his writings*, London, 1811, p. 24.
78. James Fox, *Further Memoirs of the Whig Party, 1807–1821,—*, ed. Lord Stavordale, John Murray, London, 1905, p. 324. I am grateful to Professor Trevor Levere for this reassessment of Thomas Beddoes and the relevant references here and in the following note.
79. In his letter to Joseph Black, 23 February 1788, Edinburgh University Library, Beddoes writes about his difficulties in performing demonstration experiments.
80. http:/www.cadburyschweppes.com/EN/AboutUs/Heritage/schweppes.htm accessed 7 March 2008.
81. [Thomas Beddoes], *A Memorial concerning the state of the Bodleian Library— by the Chemical Reader*, printed privately, Oxford?, 1787. Also see
G. Philip, 'Libraries and the University Press' in Sutherland and Mitchell, ref. 1, p. 725 on p. 742. Philip shows that the Curators of the Library took Beddoes's complaints seriously; they resolved to take on extra library staff in order to expand the opening hours and to obtain all useful book catalogues, for example, those of the Leipzig Book Fair, in order to consider a wider range of books for purchase. These reforms, however, were soon mired in controversy.
82. T. H. Levere, *Ambix*, 1981, **28**, 61. I would not consider the proposed chair to be a Regius Chair but rather a chair funded, but not appointed, by the Crown.

83. T. Beddoes, *Philos. Trans. Roy. Soc.*, 1787, **77**, 282.
84. Radcliffe Infirmary Registers, Oxford Health Archives, Warneford Hospital, Headington, Oxford: Gen. Ct. 11/1/1781 and Spec. Gen. Ct. 7/2/1781.
85. Ibid., Weekly Board, 19/3/1795.
86. Ibid., Gen. Ct. 9/10/1805.
87. Richard Walker, *An Account of Some Remarkable Discoveries in the Production of Artificial Cold, with Experiments on the Congelation of Quicksilver—*, Cambridge and Oxford, 1796. Walker had published the first paper on the topic under his own name in 1788: R. Walker, *Philos. Trans. Roy. Soc.*, 1788, **78**, 395.
88. R. Walker, *Philos. Trans. Roy. Soc.*, 1789, **79**, 199.
89. MS notes taken of Dr Robert Bourne's lectures by S. P. Rigaud, Magdalen College Library, MSS 522-5, volume 2.2.
90. Information taken from a file on H. P. Stacy assembled by Hugh Torrens and now at the Museum of Natural History, Oxford, in particular a note by A. F. Benson, *Newcomen Soc. Bull.*, August 1976, **105**, 4, and a personal communication from Gina Douglas, Librarian and Archivist of the Linnean Society, London, of 2 July 1997. I thank Professor Hugh Torrens and Stella Brecknell, of the Museum of Natural History, Oxford, for their assistance.
91. H. P. Stacy, *Observations on the Failure of Turnip Crops—*, London, 1800.
92. Rigaud took his BA in 1797, his MA in 1799, became Savilian Professor of Geometry and Reader in Experimental Philosophy in 1810 and Savilian Professor of Astronomy in 1827.
93. See also [Robert Bourne], *A Syllabus of a Course of Chemistry Lectures, read at the Museum, Oxford, in seventeen hundred and ninety four*, [Oxford, 1794]. The copy in the Museum of the History of Science was owned by Rigaud and is bound with Robert Bourne, *An Introductory Lecture to a Course of Chemistry, read at the Laboratory at Oxford, on February 7, 1797*, London and Oxford, 1797. This conjunction is strong proof that the anonymous *Syllabus* was by Bourne. I thank Tony Simcock, Archivist of the Museum of the History of Science, for his assistance with this and other material in the archive.
94. The opening statement of his lectures, Magdalen College MS 522-5.
95. Ibid., Footnote on the first page of the second volume.
96. Ibid., volume III, p. 82.
97. Gunther, ref. 47, p. 68.
98. Coleby, ref. 6, 1953; Russell, ref. 42.

CHAPTER 4
Chemistry Comes of Age: The 19th Century

JOHN S. ROWLINSON

Department of Chemistry, Physical and Theoretical Chemistry Laboratory, Oxford University, South Parks Road, Oxford OX1 3QZ

4.1 THE ALDRICHIAN CHAIR

George Aldrich, a son of Thomas and Grace Aldrich of Holborn, was a King's Scholar at Eton who came up to Merton College, Oxford in 1739. He took his BA in 1742 and, after completing his medical training, he took his BM and DM in 1755. It is not known why he chose to work in Nottinghamshire but in 1755 he leased some land from his neighbour and acquaintance, the Duke of Portland, to build himself an elegant house, Cocklode Hall,[i] near Mansfield Woodhouse, not far from the Duke's seat at Welbeck Abbey.[1] He died in 1798, apparently without children, but with provision in his will "for my dear wife Sibylla". The residue of his estate was left in trust to establish three chairs at Oxford, two in medicine and one in chemistry. The Duke had become Chancellor of the University in 1793 and later, as Home Secretary, he would have known of the abortive negotiations to establish a chair in chemistry at Oxford, negotiations that foundered when the University took fright at the radical politics of Thomas Beddoes, the intended incumbent.[2] It may have been the influence of the Duke that induced Aldrich to add chemistry to his two medical chairs, whose duties he specified in some detail. He felt less competent to set out what the chemist should teach, saying only that he should cover

---

[i] The hall was demolished in the 1950s to make way for a waste tip from the nearby Thoresby Colliery.

'Medicinal and Philosophic Chemistry'. The medical posts were to be a 'Prelector in Anatomy' and a professor in the practice of physic. It was common practice then for one man to hold several posts at Oxford since few chairs were adequately endowed, and that in anatomy came to be attached to the Regius Chair of Medicine, presumably to ensure that the professor taught anatomy adequately.[3] The trustees were the Vice-Chancellor, the Warden of Merton, and the Dean of Christ Church, then the dominant college in the university, particularly in scientific and medical matters. Today, the trustees' allocation of the money still conforms roughly to Aldrich's intention; one third of the money is paid to the Regius Professor, one third to a lecturer in zoology, and one third to the Aldrichian Praelector in Chemistry, now a university reader.[4]

Aldrich's chair in chemistry came at a critical point in the development of the subject. Throughout most of the 18th century chemistry had advanced in content but scarcely in understanding. By 1800, however, Lavoisier and his colleagues had settled the importance of weight in studying chemical transformations, had set out the pragmatic definition that an element was simply a body that could not be split into anything of smaller weight, had put paid to the phlogiston theory and so established the role of oxygen in combustion, and had introduced the binomial notation for inorganic compounds that we now use. A few years later Dalton was to propose his atomic hypothesis that was to set the scene for the great debates of the 19th century on the constitution of organic and inorganic compounds. All this, the new 'philosophic chemistry', was to lead to a clear intellectual and theoretical foundation of the subject; so that by the end of the century it could be truly said that chemistry had come of age. In Oxford it had then reached the point at which it was accepted within the University as a mature branch of study but its best days were still to come. However, Aldrich was an 18th century physician and it was chemistry's role in supporting medicine that was presumably his main interest. It was to be fifty years before the holder of the chair was not to be a medical man.

John Kidd[5] was the son of a captain in the merchant navy, a King's Scholar at Westminster, who took his BA at Christ Church in 1797 and then spent four years completing his medical education at Guy's Hospital. He returned to Oxford in 1801 to set up a medical practice and to succeed Robert Bourne in the unpaid and unestablished post of reader in chemistry. In 1803 the trustees released Aldrich's money and the House of Convocation proceeded to fill by election the two chairs of chemistry and of medicine. This House comprised all those who had taken their degree of Master of Arts (MA), whether they were now resident in Oxford or not, and it had then many of the powers that later in the century became the province of the more restricted House of Congregation, that is, those MAs within a few years of graduation, or living in Oxford, or holding resident posts within the University. Convocation met on 15 November 1803, when the Senior Proctor announced that John Kidd was the only man proposed for the post. Kidd was to say in 1809 that the Dean of Westminster had been his principal supporter.[6] The Dean was William Vincent who had been Kidd's headmaster at Westminster School and who had recently acquired a country rectory at Islip, near Oxford, where he lived in the summer.

The Vice-Chancellor "administered the oaths of allegiance and loyalty and admitted Mr Kidd as professor of chemistry with all the benefits and emoluments." The salary was only a little over £100 a year and Kidd continued in his medical practice, becoming a physician at the Radcliffe Infirmary in 1808 and Lee's Reader in Anatomy[ii] at Christ Church in 1816. A few years after his election the Crown added a further £100 a year to the chemistry chair, a sum for which Parliament soon took responsibility. Two days after Kidd's appointment, the professorship of medicine was filled by Robert Bourne in a contested election in which he beat George Williams, of the Radcliffe Infirmary and a Fellow of Corpus Christi College.[7]

Kidd was a conscientious man, much respected by his contemporaries. He gave a course of 26–30 lectures on chemistry in Michaelmas and Lent terms at 7 pm on Tuesdays, Thursdays, and Saturdays, and soon added also a course of lectures on anatomy. Chemistry was not then part of the recognised university curriculum; this was restricted to the classics, joined soon by mathematics. The examinations had been re-arranged in 1800 and honours classes, as we now know them, were introduced for the first time in 1807 in the two schools of Literæ Humaniores (or Classics and Philosophy, commonly called 'Greats'), and Mathematics and Physics.[iii] College tutors, anxious for their pupils to do well, often discouraged them from attending lectures, such as those in chemistry, that contributed nothing to their degree, but Kidd's lectures were well delivered and attracted an audience of both junior and senior members of the university and some intending medical men. He published a syllabus in 1808 from which it seems that he followed much the same course as Bourne had delivered ten years earlier, but with more time given to the metals and their compounds. He mentioned phlogiston but the emphasis was now on Lavoisier's role for oxygen.[8] His brief covered mineralogy also and for this course he wrote a two-volume treatise.[6] It shows little originality, being based, as he says, on the recent French texts of the Abbé Haüy (1801) and Alexandre Brongniart (1807), and on Richard Kirwan's geology. There are, however, some observations on what he had seen on trips around Oxford and elsewhere, and the whole was a substantial contribution to the field. His lecture courses eventually extended also to geology; on that he published a long 'Essay'.[9] This covered many of the facts but had also much rambling discussion on the theory of the subject which was then a matter of great dissension, not only on the merits or otherwise of what came to be called the 'uniformitarian' and 'catastrophic' hypotheses of earth history, but also on the increasingly worrying problem of reconciling geological findings with revealed religion. Kidd ended his 269-page

---

[ii] The holders of the Christ Church posts founded under the will of Matthew Lee were described indifferently as 'Lee's Reader' and as 'Dr. Lee's Reader' throughout the nineteenth century. Modern practice is to use the form 'Dr Lee's Reader'.

[iii] 'Mathematics and Physics' meant then what we should now call pure and applied mathematics. Physics, as we know it, was called experimental philosophy, a name that survives to this day as the title of the Dr Lee's Professor in the Department of Physics.

essay with the confession: "—the science of geology is so completely in its infancy as to render hopeless any attempt at successful generalization, and one may therefore be induced to persevere with patience in the accumulation of useful facts." He showed little regret at passing his course over to his successor as Reader in Geology and Mineralogy, William Buckland, who had attended his chemistry and mineralogy courses. Buckland was to enter the two linked fields with a gusto that surpassed anything that Kidd could muster. The intellectual attraction of these fields in the early nineteenth century was due in part to the perception that palaeontology, and beyond that, geology, mineralogy and chemistry, were sometimes seen as a natural backward extension of classical history into pre-historic times.[10]

Kidd's chemistry course was not intended to train professional chemists; that was not a field for which there was yet any call, except for apothecaries who prepared medicines and carried out many of the jobs that would today fall to a general practitioner. He never became deeply involved in the chemical disputes of the day but aimed only at an orthodox exposition of the subject as he saw it. He took the traditional view that the University should provide a liberal education that used the classics to teach by example the morals and literary skills that a gentleman should acquire, and used mathematics to teach him to reason exactly. Kidd believed that chemistry should be added to this programme to teach him how the world was constructed. Forty years later, Robert Walker, the Reader in Experimental Philosophy, was to complain that it was a disgrace that the University was still sending out graduates who believed that the four elements were earth, air, fire, and water.[11] One minor practical justification for the teaching of chemistry was the need for the future landowners among his auditors to know something of fertilizers and the science of agriculture. This was one of the first aims of the new Royal Institution in Albemarle Street, London, to satisfy which Humphry Davy published his *Elements of Agriculture* in 1813. Such a justification was not uppermost in Kidd's mind but it was to play a larger role in the career of his successor. Kidd had a chance to set out his philosophy in 1818 after William Brande of the Royal Institution criticised the teaching of chemistry at Oxford and Cambridge in an obscure footnote attached to a paragraph praising Boerhaave's contribution to 18th century chemistry.[12] Brande[iv] had written: "—for excepting in the Schools of London and Edinburgh, Chemistry, as a branch of education, is either entirely neglected, or, what is perhaps worse, superficially and imperfectly taught. This is especially the case at the English Universities, and the London Pharmacopœia is a record of the want of chemical knowledge—". It is clear that Brande was concerned particularly with the training of medical students in chemistry and this was not Kidd's prime intent for he defended his teaching of chemistry as an adjunct to a

---

[iv] Brande wrote the third of the Dissertations that preceded the regular articles in the Supplement to the 4th and 5th editions of Encyclopaedia Britannica, which was then edited and published from Edinburgh. The Dissertation appeared first as a tract of 79 pages that the British Library dates to 1817. The Supplement was issued in unbound half-volumes from 1808 onwards and this tract was apparently inserted into the half-volume that appeared in February 1818. When the Supplement was complete it was six whole volumes in 1824.

liberal education. He wrote; "— [while] the discipline of Classical and Mathematical studies is well calculated to form the groundwork of excellence in the Physical and Experimental Sciences, the converse of this is by no means true.— they are cultivated so far as is compatible with the views of a system of general education: hence the object—is, rather, to present a liberal illustration of their principles and practical application, than to run into the minutiæ of a technical, or even a philosophical detail of facts." He appended the syllabus of his courses given in 1815, 1816, and 1817. They differ little from that of 1808. He compared his 30 lectures with the 40 given at Guy's Hospital and claimed broad similarity except for the greater coverage of pharmacy for the medical students.

Kidd contributed little to the advance of chemistry. He published a long and inconclusive article on the efflorescence of saltpetre on the internal walls of his underground rooms in the Museum but was unable to explain the source of the potassium. He was not the first to notice this efflorescence.[13] Coal tar was a by-product of the distillation of coal to produce coke for iron- and steel-making and gas for illumination, but there was a sudden interest in a volatile solid that could be obtained by distilling the tar. On 19 December 1819 Alexander Garden, who ran a chemist's shop in Oxford Street, London, submitted a paper in which he reported that he had crystallised from the distillate a silvery solid that resembled camphor and benzoic acid.[14] This acid had been known from the 16th century when it was obtained by the dry distillation of the bark of some tropical trees. Two days after Garden's submission, Brande sent a paper to what was virtually the house journal of the Royal Institution in which he made similar observations.[15] Kidd followed with another paper, a longer and more thorough examination of the same solid, finding that it was composed of carbon and a little hydrogen but without making a quantitative analysis. He did, however, give the new material a name – naphthaline, or naphthalene as we now spell it.[16] In 1826 Faraday showed that two isomeric sulphonic acids could be made from it,[17] and many years later Auguste Laurent was to use a study of the isomers of naphthalene derivatives as part of the evidence for his theory of the constitution of organic compounds. These two papers of Kidd were submitted to the Royal Society by William Wollaston; Kidd became a Fellow only in March 1822. His last major publication was the second of the eight Treatises endowed by the Earl of Bridgewater to explain the manifestation of God in his Creation.[18] It was not one of the best of the group but both he and William Prout, who wrote the eighth Treatise on chemistry, emphasised the 'argument from design' in defending their theses.

Sir Christopher Pegge, the Regius Professor of Medicine, resigned in 1822 because of the ill health that had kept him away from Oxford for much of the year; he died in August. Kidd seized his chance and was elected in his stead to a post that carried a larger salary and more prestige than the chair of chemistry. His successor to the chemistry chair was at hand.

4.2 CHARLES DAUBENY AND REFORM

Charles Giles Bridle Daubeny was a Gloucestershire man, the son of a country rector.[19] He was educated at Winchester and Magdalen College, graduating with second-class honours in the classical school of Literae Humaniores in 1814. He had almost certainly attended also Kidd's lectures on chemistry and Buckland's on mineralogy. Oxford had then no provision for clinical medical training and Daubeny spent three years in Edinburgh before returning to Oxford to take his BM in 1818 and DM in 1821. In Edinburgh he attended Thomas Hope's lectures on chemistry and Robert Jameson's on mineralogy and geology. In 1819 he put his knowledge to good use on a tour of the extinct volcanoes of the Auvergne, which gave him material for several papers in Jameson's *Edinburgh Philosophical Journal*. He wrote also a more connected account of the geology of the region, and of other sites, which he published a few years later.[20] This work was dedicated to John Kidd, "who, by his lectures on geology, first called the attention of the University to this important branch of natural knowledge" and to William Buckland, "who has since so largely contributed by his exertions, both as a man of science, and as a lecturer, to open a wider field, and awaken increased interest in the same study." It proved to be one of Daubeny's more original contributions to science and contained much chemistry as well as geological fact and speculation. His contention was that volcanoes were caused by the influx of sea-water into the core of the mountain where the oxygen of the water reacted with an inflammable substance that Davy and Gay-Lussac had supposed to lurk underground, so generating the necessary heat to power an eruption. His evidence for this view was that most active volcanoes were found on or near sea-shores. The Continental vulcanists held that the heat came up from below and was not the consequence of a chemical reaction; they were to be proved right, although Daubeny was defending his 'chemical theory' before the British Association into the 1830s.[21]

His election by Convocation to the Aldrichian chair on 10 October 1822 came as no surprise to him for he gave his inaugural lecture only three weeks later, on 3 November.[22] He acknowledged that his proposed course would not be as thorough as those of Hope at Edinburgh and Brande at the Royal Institution, but these were aimed primarily at medical students. He thought that 30–40 lectures of one hour should suffice to give "a sufficiently comprehensive view both of the principles of chemistry, and of its more important applications, to satisfy the general student." He claimed, perhaps rather optimistically, that "chemistry has long ceased to be a mere empirical branch of knowledge, consisting of unconnected facts and obscure formulæ—." He moved a little beyond Kidd's defence of the subject but was not yet willing to advocate in Oxford an independent status for chemistry; it was still to be an adjunct to the liberal education of a gentleman. He was to hold this view in public for the rest of his life but it is interesting that when he gave a lecture in Bristol a few weeks after his inaugural lecture he was more willing to claim a practical value for the subject than is apparent from the tentative remarks in his Oxford lecture, or from the syllabus that he appended to it.[23] The course that he gave when first appointed was up-to-date but not intellectually venturous. Lavoisier and his theories had now displaced phlogiston, Davy's electrical experiments were discussed, but little was said at first about Dalton's atomic theory. The manner of the lectures seems to have been competent but not inspiring, and his experimental demonstrations were notorious for their failure, even when he had an assistant. Those attending the lectures could not try the experiments for themselves. Practical courses in chemistry came slowly into British universities; Thomas Thomson, after a tentative beginning at Edinburgh, introduced practical classes at Glasgow when he moved there in 1818, and Edinburgh followed in 1823, but in neither university was there established a research school in the way that occurred at Giessen under Justus Liebig from 1826.[24] In England practical classes became available only from 1837, under Thomas Graham at University College, and in the 1840s at King's College, London, and at the newly founded Royal College of Chemistry. On appointment Daubeny had obtained £200 from the Hebdomadal Board[v] for equipment but only on the condition that he provided for future needs himself and that he made over to the University the equipment that he already owned, of which a list was made[25] in 1823. He asked for a further £200 in 1826 but was refused. Nevertheless, in spite of these problems, his lectures attracted an audience of able men, many of whom were making their mark inside and outside Oxford. These included his present and future colleagues, the mineralogists William Buckland and Nevil Story Maskelyne, the mathematicians William Esson and Baden Powell,[vi] and Henry Liddell, the Dean of Christ Church. Among his early auditors were two future archbishops, Archibald Tait of Canterbury and William Thomson of York, and also Edward Pusey, the theologian, and William Froude, the mathematician and naval architect.[26] He proved a popular lecturer throughout the 1820s and into the 1830s. His salary remained the same as Kidd's, £221 a year after tax, but he added to this by continuing his medical practice for some years.

Daubeny's lectures were apparently designed initially to give his audience the principal facts of chemistry and not to go deeply into the many controversies of the day.

In 1831, however, he supplemented the course with a book on atomic theory, perhaps then the most contentious branch of chemistry.[27] He dedicated the book to Dalton whose atomic theory was the one that he generally supported, but noted with approval Gay-Lussac's law of combining volumes of gases, which Dalton never accepted. He assumed that water is HO, following Dalton's principle that a binary compound had the simplest possible formula since two atoms of the same kind were thought to have a repulsive force between them. This led to atomic weights of $H = 1$ and $O = 8$. He was more receptive to physical arguments than many chemists were at that time, drawing evidence for an atomic structure of matter from Dulong and Petit's rule for the specific heats of solids, from Mitscherlich's principle of isomorphism and, later in his career, from the existence of isomers that suggested that the atoms in a compound could be arranged in more than one spatial arrangement. Like almost all of his generation he apparently knew nothing of Avogadro's hypothesis or of its

---

[v] The Hebdomadal Board (later Council) then comprised the Heads of the Colleges, one of whom was the Vice-Chancellor, and the two Proctors. It was the executive body of the University.
[vi] The father of the founder of the Scout movement.

repetition by Ampère, and he had little time for Boscovich's "obscure and abstract speculations". The second edition in 1850 was a considerable revision and expansion of the first and of the supplement of 1840. Its contents and emphasis can be seen from the chapter headings:

> Opinions of ancient philosophers, and of moderns antecedently to the epoch of Dalton, with respect to the constitution of matter; Views of Dalton, and of other philosophers subsequently, with respect to the laws of combination betwixt matter, and its intimate constitution; On chemical symbols and notation; On the existence of atoms; On the mode of combination betwixt atoms; On isomorphous bodies; On isomeric bodies; On atomic volume; On the laws regulating the combination betwixt atoms; On the forces operating on living matter; On the applications of the atomic theory, and on its relation to other physical laws; Speculative inquiry into the elements of matter; On the knowledge of the ancients with respect the law of definite proportions *etc.*

This book became one of the important texts of the mid-century, being recommended to those taking the newly approved courses of the Natural Science School at Oxford and of the Natural Sciences Tripos at Cambridge in the 1850s.[28]

He was a short stout man, a kindly and somewhat Pickwickian figure, known to his contemporaries as 'little Dubs', who never married and lived at first in the basement of the Museum. Kidd had also lived in the Museum from 1811, although he was a married man with a family of daughters, but he moved first to Frewin Hall and then to 37 St Giles, on becoming the Regius Professor of Medicine. He left this house to Christ Church for the use of Dr Lee's Reader in Anatomy (a post he had held from 1816) and it was later occupied by successive Dr Lee's Readers in Chemistry, also a Christ Church post, until the retirement of Richard Wayne in 2006. The Museum can never have been a satisfactory home even after Daubeny had obtained £200 from the University for its improvement in 1829. He described it as "thoroughly unworthy of a great university", and moved to the Botanical Garden House opposite Magdalen College in about 1838. He was, however, not the last occupant, for Nevil Story Maskelyne moved in there after Daubeny had left.[29]

Daubeny's contribution to chemical knowledge was wider and more substantial than that of Kidd but not of the first rank if measured by what was being done at the time on the Continent. After his geological and chemical work on volcanoes he turned his attention to the analysis of mineral waters[30] but his interests soon moved in another direction when he acquired the additional chairs of Botany, in 1834, and Rural Economy in 1840. The action of light on plants, and of the plants on the composition of the atmosphere, were the subjects of a major paper in 1836 and of several communications to the new British Association for the Advancement of Science.[31] He found that it was only visible light and not the 'calorific' nor the 'chemical' rays that affected the plants, but was not able to take the matter much further. His interests moved then primarily into botanical areas and, following Davy's example, into the chemistry of agriculture. From 1842 he and his collaborators ran tests on the viability of stored seeds for the British Association, and eventually he bought a plot of land on the Iffley Road for field trials. This site was built over early in the 20th century, but there is still a Daubeny Road there. One of his pupils was (Sir) John Bennett Lawes, a wealthy landowner who turned his estate at Rothamsted into a field laboratory and, with J.H. Gilbert, a pupil of Liebig, created what became the finest agricultural research station in Europe. In 1843 Lawes invented and patented 'superphosphate', a mineral fertilizer obtained by treating bones with concentrated sulphuric acid. This proved more effective than the insoluble mineral fertilizer that Liebig tried to promote in Britain, much to the annoyance of the farmers who tried it. In spite of the later course of his interests, "Chemistry was, however, the thread which bound together all the researches of Dr. Daubeny", as John Phillips, the Keeper of the Museum, put it in his obituary for the Ashmolean Society and for the Royal Society.[19] Phillips felt, however, that he should add: "not that he was personally a dextrous manipulator of chemical instruments."

From 1830 there was a decline in the attendance at all science lectures. From 1822 to 1830 Daubeny had attracted an average of 28 undergraduates each year to his course. In the next decade this fell to 14, and in 1839 he suspended the course altogether for some time. The contrast with Yale must have been painful. He had travelled in North America the previous year and had seen there a chemistry class of 100 men and 30 women! His was not the only class to suffer; there were similar falls in the attendance at Kidd's course on anatomy, at Buckland's on mineralogy, and at Stephen Rigaud's on experimental philosophy. Part of the problem in Oxford was the sudden enthusiasm for theological disputes following the rise of the Tractarians, which culminated in the departure of John Henry Newman and his followers into the Roman Catholic Church, but this was not the whole story since there was a similar decline in attendance at the chemistry lectures at Cambridge.[28,32,33] One loss may have been of intending medical students since the Royal College of Physicians decided in 1835 that it would open its Fellowship to graduates from universities other than Oxford and Cambridge. From 1830 onwards there was increasing pressure for change. At the national level this resulted in the Parliamentary 'Reform Bill' of 1832; in the sciences it was manifest in Charles Babbage's blast directed principally at the dilettante Royal Society,[34] and in the moves inspired by David Brewster that led to the founding of the British Association for the Advancement of Science, the first meeting of which took place at York in 1831.[35] Daubeny was the only professor from Oxford or Cambridge to attend this successful meeting. He was made a member of the Chemistry and of the Zoology and Botany Committees. The first of these resolved that it was important that chemists re-determine the specific gravities of hydrogen, oxygen, and azote (nitrogen). Daubeny was asked "to undertake an investigation into the source from which organic bodies derive their fixed principles."

It was, perhaps, not difficult for Brewster and the Rev. William Vernon Harcourt (son of the Archbishop of York, and formerly an auditor of Kidd's and Buckland's lectures) to generate the enthusiasm that ensured the success of the first national meeting open to all the natural philosophers of metropolitan and provincial Britain, but the future of such enterprises often depends on the ability of the founders to see that the second meeting goes as well as the first. It was here that Daubeny was shown at his best. Without any authority from the University he publicly invited the new Association to meet the next year in Oxford. Roderick Murchison, the geologist, was to say "when—every difficulty hung around us, a Professor of this University, came forward and undertook, on his own responsibility, that Oxford would open its gates and receive us."[36] At York, Harcourt at once proposed Buckland as President for 1832, although he had not been present at the meeting. Daubeny and Powell, now the Savilian Professor of Geometry, were to undertake the duties of Local Secretaries, and by December 1831 Daubeny had secured the support of 43 professors and college tutors. In the event Daubeny's respected position and his popularity in the University won the day and Oxford rose magnificently to the challenge. The view that the University should be opened to those not members of the Church of England gained ground when, during the meeting, the University conferred honorary doctorates of civil law (D.C.L., there was no D.Sc. until 1900) on four dissenters of different persuasions: John Dalton, Michael Faraday, David Brewster, and Robert Brown, the botanist, none of whom would have been admitted to the degree of MA. The meeting was a success and Cambridge was quick to secure the attendance of the Association for the meeting in 1833. Daubeny tried to arrange for the fourth meeting in 1834 to be held in Bristol, in his home county of Gloucestershire, but was thwarted by James David Forbes who claimed the occasion for Edinburgh. After that it had to be Dublin in 1835, but then Bristol's turn came in 1836. Daubeny was again a Local Secretary and, since the President that year was a Bristol worthy, but no scientist, it fell to Daubeny to give the Address, an annual review of the recent advances in all fields of science.[37] One tricky point arose from the earlier discussion in Edinburgh and Dublin on atomic symbols. These discussions had been chaired and guided by Edward Turner, the Professor of Chemistry at what came to be called University College, London.[38] Dalton, who claimed the rights of the propagator of the atomic theory, naturally wished to retain his diagrammatic symbols, but these were rejected in favour of Berzelius's 'algebraic' symbols, essentially those that we now use, based on the first letter or two letters of the element's name. Not only were these simpler for the printer, but their number was capable of indefinite expansion as more and more elements were discovered, and they did not carry any implications of the possible structure of the compounds. The increasingly positivistic attitude of the chemists and other natural philosophers of the day held this to be a fault of Dalton's system. At Bristol, Daubeny, who had rated Dalton's work so highly in his recent book, had to steer a hard path between support of the majority and avoiding distress to Dalton. He tentatively suggested that Dalton's symbols might be retained for elementary instruction. In 1856 Daubeny was President of the Association when it met in Cheltenham in Gloucestershire, where he then gave a second address and was honoured with the award of a medal. He then noted that in the twenty years since his first address organic chemistry had been added to inorganic, and he ended with the comment that the phenomenon of the isomerism of organic compounds, which he had not fully accepted in his book of 1831, implied the correctness of the atomic theory.[39]

Reform within Oxford itself proved difficult and a battle was fought throughout the 1830s and 1840s.[33,40] Perhaps the first indication that the scientists were organising themselves came with the foundation of the Ashmolean Society in 1828, a more formal successor to a dining club in which Kidd, Daubeny, Powell and others had discussed their concerns. Daubeny was one of the initial members of the Society, which was, perhaps, inspired by the foundation of the Cambridge Philosophical Society in 1819, but it was never to rival its Cambridge counterpart in importance and, after a period of decline was to merge with the Natural History Society in 1901, in which form it still survives. Meanwhile, the dining club, whose early history and relation to the Society is obscure, continued to flourish, and today it provides a common ground on which Heads of Science Departments and Heads of Houses can meet to discuss university problems.[41] Daubeny was also a founder member of the (national) Chemical Society in 1841, and its President in 1853, but the principal movers there were Thomas Graham and the London and Scots chemists. The call in Oxford for more attention to the sciences and less concentration on the classical languages came from Kidd, Daubeny, Powell, Philip Duncan, Keeper of the Museum from 1831, Robert Walker, Rigaud's successor as Reader in Experimental Philosophy from 1839, and Henry Acland, after he came back to Oxford as Lee's Reader in Anatomy in 1846. Powell opened with a pamphlet based on the first lecture of his course of 1832, in which he noted how "Scientific knowledge is rapidly spreading *among all classes* EXCEPT THE HIGHER, and the consequence must be, that that class *will not long remain* THE HIGHER"; he was immediately supported by Walker.[42] In 1839 Convocation rejected a statute that would have required the professors to give courses of lectures from which all students would have been compelled to make a choice. For those taking the Mathematics and Physics School it was suggested that the eligible lectures would have included chemistry.[43] Francis Jeune, the Master of Pembroke College, and several of his colleagues put forward a scheme for improving the examination statutes in 1843 but this got nowhere. The question was reopened in 1845 when the question of the reform of the statutes and teaching at Oxford and Cambridge was raised in Parliament. A motion to set up a commission of enquiry was defeated there by 143 votes to 82, but the increasing concern of the outside world was clear. The cause of reform was again reinforced locally at the meeting of the British Association in Oxford in 1847, when the young Benjamin Brodie, then living in London, was one of the secretaries of the Chemistry Committee. On 12 July, immediately after the meeting, Daubeny, Walker, Duncan, and the newly arrived Acland circulated a memorial calling for a new Museum for the proper housing of the University's now considerable scientific collections and for the better accommodation of the professors and their classes. Buckland, by now living in London as Dean of

Westminster, refused to sign their memorial believing that the promotion of science in Oxford was a lost cause.[44] Even Acland had his doubts; he wrote,

> The science studies of the University were for various causes almost extinct, notwithstanding the efforts of Buckland, Kidd and Daubeny: Kidd, a man of truly scientific spirit: Daubeny, a sincere, high-minded able chemist, both had failed. I felt the work before me desperate or hopeless. The intellect of the University was wholly given to ecclesiastical and theological questions. All physical science was discountenanced.[45]

Without Buckland's support, this memorial also got nowhere, but the climate of opinion soon changed; the University began to tire of the endless theological disputes and some of the non-scientific Fellows began to think more seriously about reform of its statutes and syllabus. Among these was Goldwin Smith,[vii] a historian at University College, who later wrote: "A few of us, Mark Pattison and [Benjamin] Jowett among us, began to meet in the rooms of Arthur Stanley at University College, and addressed to Lord John Russell, the head of the Liberal Government" a request for a commission of enquiry.[46] Daubeny was not a signatory. Russell accepted the case put to him by the Oxford fellows and, possibly more effectively, by distinguished scientists outside Oxford, including Charles Babbage, Charles Lyell, Charles Wheatstone and Charles Darwin. After some political arguments – "the wasps at once swarmed out upon him" – he set up two Royal Commissions, one for each university. That for Oxford was chaired by Samuel Hinds, Bishop of Norwich and a Whig prelate, and contained Baden Powell, Archibald Tait, Francis Jeune, and Henry Liddell, now headmaster of Westminster School. Arthur Stanley became the Secretary and Goldwin Smith, the Assistant Secretary. The membership was clearly loaded in favour of the reformers. The remit of the Commission was "to inquire into the State, Discipline, Studies, and Revenues" of the University and Colleges. The Prime Minister announced his decision to the Duke of Wellington, the Chancellor at Oxford, on 8 May 1850, saying that "—the discoveries of physical and chymical science have rendered changes of the course of study of our national universities highly expedient." The Duke at once passed this hot potato to Frederick Plumptre, the Vice-Chancellor who replied to Russell on 15 May to say that the University was reluctant to cooperate with the Commission, calling it "an unconstitutional proceeding"; most of the colleges were also hostile. Nevertheless, Russell went ahead and the Commission was set up on 31 August.[47] Since it was a Royal Commission, not a Parliamentary Commission, it had no power to insist on the production of papers or the giving of evidence. Thus, Daubeny told them all about his university chemistry classes, but since the Bishop of Winchester, the Visitor of Magdalen College, had told the fellows not to cooperate, he refused to say anything about his college activities. This was unfortunate since he had, in 1848, finally lost patience with the poor

---

[vii] He later had a varied career in England, the United States, and Canada, and the Goldwin Smith Professorship of Chemistry at Cornell is named after him.

facilities that the University provided in the Ashmolean Museum and had erected, partly at his own expense and partly at the College's, a new laboratory in the Botanic Garden, across the High Street from Magdalen.

Meanwhile, both Oxford and Cambridge had anticipated the likely criticisms of the outside commissions and had at last admitted science and other modern studies as proper subjects for education, examination and graduation. A pamphlet by Daubeny in 1848 was probably the most important single influence at Oxford.[32] He noted that 45 of the 64 resident college tutors had recently signed a memorial[33] asking for new examination statutes and then set out what he himself thought was needed, since although the memorial included mathematics and the physical sciences, "treated mathematically, as at Cambridge", it made no mention of chemistry. He divided the primary sciences into three branches: mechanical or natural philosophy, chemistry, and physiology. Every educated man should have some knowledge of these. The first included gravity and its consequences, but optics, acoustics and electricity were "not to be insisted on". For chemistry, he required a knowledge of "the properties and composition of the atmosphere, of water, and of some of the more important elements but not that of the entire catalogue of simple substances". Physiology included some knowledge of the whole kingdom of organic nature. The secondary sciences, which he thought should not be compulsory, included botany, astronomy, geology and mineralogy. New examination statutes were drawn up in 1849 and finally approved by Convocation on 14 May 1850 although it was another two years before a chemistry syllabus could be agreed. The statutes recognised at last that the natural sciences had a part to play in the formal education provided at Oxford but fell short of what some of Daubeny's colleagues would have liked. The Gospels, the classical languages and some mathematics formed the subjects of 'Responsions', taken between terms 3 and 7, and in the First Public Examination, in terms 8 to 12, and it was only then, after up to three years of study (each of four terms), that a candidate then had a choice of courses in the Second Public Examination, in terms 13 to 18. This examination comprised a choice of one from pure and applied mathematics, or law and modern history, or the natural sciences.[48] The syllabus of the last was clearly influenced by Daubeny's pamphlet of two years previously, as he himself claimed in the introduction to the last edition of his *Introduction to Atomic Theory*[27], but the new School of Natural Science did not go as far as Daubeny would have liked. The 'Greats' men objected to the requirement that they had to take also one of the 'modern' subjects in their final honours examination and this provision was dropped in 1865.

The new statutes were condemned as inadequate by the Commissioners when they reported in 1852 because of the inordinate amount of time that had to be spent on the classical subjects before a candidate could get to grips with some science.[49] Moreover, the aim of most of the reformers, including Acland and Daubeny, that all graduates should have some knowledge of natural science was never seriously addressed, and, indeed, the degree of specialisation that a university course would soon require meant that such general knowledge passed beyond what could be expected from a course of three or four years. This aim gradually dropped out of the reformers' demands. The first graduates of the Natural Science School emerged in 1853, one received a third-class degree and two passed, but the next year there was one man with first-class honours and two with second. There were three examiners in Natural Science, of whom Daubeny was the chemist. In all, the number of candidates graduating in the different honours schools, between 1853 and 1860 was: Literae Humaniores, 577; Law and Modern History, 253; Mathematics, 117; Natural Science, 71. Thereafter, the numbers of natural scientists rose steadily, with an average of 23 a year in the 1870s, 26 in the 1880s and 44 in the 1890s. A change from the aspiration of Acland and Daubeny that all men should have a knowledge of the three basic sciences, mechanical philosophy, chemistry and physiology, was marked by new statutes in 1864 and 1872 that allowed a candidate to choose one subject for his final examination. From this change stemmed the divergence of the Oxford and Cambridge courses in chemistry. The former has become predominantly an honour school based firmly on a single subject, with ancillary subjects, such as mathematics, now supplied as required, while the Cambridge Tripos was more broadly based from the outset.

An assistant to the Professor of Chemistry was agreed in 1853 and Nevil Story Maskelyne[viii] was appointed; from 1850 he had also been deputy to Buckland, the now ailing Reader in Mineralogy. Daubeny offered demonstrations in practical chemistry in his laboratory at Magdalen in 1850, and some instruction was available to Balliol men in a cellar of Staircase 16 from 1853. Story Maskelyne went further that year and admitted undergraduate chemists to 'hands-on' practical classes for a fee of £5 a term; perhaps following the example of St John's College, Cambridge, which had offered a similar facility the previous year, although it was 1865 before there was any university instruction in practical chemistry there.[50] In Oxford it was only in 1857 that the examiners gave notice that they required candidates in Natural Science to have demonstrated practical skills.[51] Daubeny was by now persuaded of the importance of practical work and defended it in a lecture at the Royal Institution in 1855.[52] He was again obliged also to defend the new school against outside criticism, arguing, as before, that the physical sciences are not to be valued principally for their practical use, nor was it the case that they could only be studied successfully in large industrial towns. The evidence from Germany was decisive on that point, although few Oxford and Cambridge graduates had taken part in the trek of British chemists to Liebig at Giessen or later to Bunsen at Marburg and Heidelberg. Daubeny took the same view as Auguste Comte[53] in urging that instruction in chemistry should not require any mathematical treatment; indeed, his choice of words suggests that his view was directly inspired by Comte's writings that had a strong influence on scientific thought in Britain in the early part and middle of the century. He was, no doubt, also influenced by a desire to attract to the study of chemistry those with little ability or interest in mathematics. He confessed that Oxford's record in research was so far rather weak: "It must be admitted that the

---

[viii] Story Maskelyne was born Story, and, similarly, Vernon Harcourt (see below) came from a family whose name was Vernon. Both second names were added for domestic reasons and it was common in the 19th century to leave such doubled surnames without a hyphen.

inmates of our Colleges, notwithstanding the learned leisure that the munificence of their founders secures to them, have seldom taken part in those great discoveries by which the active spirits of our age have so entirely changed the face of science."[54] Cambridge was just two years ahead of Oxford. In 1851 six candidates sat there the examination of Natural Sciences Tripos, which was also at first essentially a 'graduate' addition to the classical and mathematical syllabus.[55]

The principal criticisms of the Commissioners had related to the restriction of most college fellowships to particular classes of candidates and to the fact that all modern subjects were excluded. Moreover, few fellows were required to reside in Oxford or to teach but were expected only to use their salary to support them while they sought ordination in the Established Church or other professional instruction elsewhere. The Commission also addressed the methods used to choose new professors when a vacancy occurred. Election by Convocation, the method required for the Aldrichian chairs, was condemned as "the worst mode of appointment" and the change to election by a small 'Delegacy' was recommended. The Commission's endorsement of the need for more science was expressed by a call for more professors, for college fellowships for all professors, and for the creation of "a subordinate class of professors, under the name of lecturers or readers".[ix] Christ Church responded some years later by founding the Lee's Readerships in Chemistry in 1859, and in Physics in 1869.[56] An Act of Parliament of 1854 set up a second, or Executive Commission, to carry out the necessary reforms. These included the removal of the need for all graduates to be members of the Established Church, but still leaving the colleges the power to reject such men, and increasing the powers of the resident MAs, as expressed in the House of Congregation, although the wider body of Convocation still had more power than many would have wished. Daubeny's last service to the cause of reform came in 1864 when the Hebdomadal Council and the resident fellows at last agreed that the sciences could be studied after only two years of classical studies. This contentious proposal was likely to fail in Convocation but Daubeny wrote a letter to *The Times*[57] on 24 February 1864 urging the non-resident MAs not to descend on Oxford and vote down what the University clearly wanted. A few days later the motion was narrowly carried in Convocation by 281 to 243. After 1886 Natural Science candidates required the classics only to the level of the Preliminary Examination.

The support of the 1850 Commission was also valued by the campaign for a new Museum that was gathering steam in Oxford. Here, it was hoped, the sciences would find a worthy home.

4.3 THE MUSEUM

The Ashmolean Museum in Broad Street had served chemistry and other sciences since 1683, although some rooms in the neighbouring Clarendon Building

---

[ix] These non-professorial university teachers were called 'demonstrators' in the laboratory sciences until, in 1964, Oxford fell into line with the rest of the country and called them 'university lecturers'.

had also been occupied when required in the 19th century. If the sciences were to take the place in the University that the reformers envisaged then clearly something bigger and better was needed. We have seen that a memorial drawn up after the meeting in Oxford of the British Association in 1847 had proved ineffective. It had called for a new Museum with space for the increasing collections of mineralogical and biological specimens, for lecture rooms, a meeting room, and a library but, perhaps significantly, there was no mention of laboratories or of research. A second attempt started with a meeting on 9 May 1849 in the Lodgings of the Warden of New College, David Williams. About twenty men were present and a committee was formed with the Warden as chairman and Daubeny, Stanley, Powell, and Jeune as members. Even Buckland was sufficiently impressed to join later. Their attempt to raise money privately for the new building failed, although there were generous donations of £100 or more promised by some of the proposers.[58] It was then discovered that the University Press had accumulated over the years a profit of £60,000, mainly from the printing and sale of Bibles and Books of Common Prayer. The reformers at once saw a solution to their problem, but there was naturally opposition both from those who thought that the sciences were an ungodly subject to benefit from the sale of God's word, and, perhaps more rationally, from those who thought that a field that attracted so few devotees should not be the main beneficiary of this unexpected bounty. Convocation had defeated in 1851 a proposal to spend money on a museum but after the report of the Commission, which called strongly for a new museum "with proper laboratories", it was agreed in principle that up to £30,000 of the Press money could be so used, and on 24 June 1853 the University bought 4 acres of land for £4000 from Merton College. The site was north of Wadham College gardens, in the University Parks. A Delegacy was set up in January 1854 to supervise the building. The secretary was Story Maskelyne and the members included Daubeny and Acland; it was the last who quickly became the driving force behind the enterprise.[44,59] For chemistry, the Delegacy allocated, on paper, a museum for chemical products ($20' \times 30'$), a lecture room for 100 men with an adjoining apparatus room, a professorial laboratory ($20' \times 20'$) with a sitting room, a laboratory for 11 to 20 men ($35' \times 30'$, or more), a furnace room, a balance room, and an outhouse.[x] Some of the remainder of the Press money was spent on the assistant for the Professor of Chemistry and on the general improvement of professorial salaries. The crucial vote in Convocation that gave the Delegacy power to appoint the architect and to sign contracts passed on 12 December 1854, by 70 votes to 64 after Acland had persuaded Edward Pusey, the theologian, to drop his opposition.[60]

But what form should the new building take and who should be the architect? The Delegates had called for a competition and the two leading anonymous entries, of the 32 submitted, were very different in their external form. One was a classical building in Palladian style and one in was what was described as 'Rhenish Gothic'. In Anglican Oxford, still in part under the influence of the

---

[x] $1' = 1$ foot $= 0.3048$ m.

Chemistry Comes of Age: The 19th Century

Tractarian movement, the latter was sure to prevail. In a second vote on 12 December[60] it was chosen by 81 votes to 38. One of the arguments that carried weight in Oxford was that the Museum should reflect in its function and style the manifestation of God's purpose through the study of the natural world; this was a theme of Acland's prayer written for the occasion of the laying of the foundation stone.[61] A classical and therefore secular style would not fit this aim. Moreover, the architects of the gothic building were revealed to be Benjamin Woodward and Thomas Deane, the former a disciple of John Ruskin and a friend of Acland. The internal form is striking with a central court with a glass roof supported on cast-iron pillars and decorated with columns representing the different rocks found in the British Isles; Daubeny was one of those who contributed £5 or more to pay for a column. The only space in the Museum that was described specifically as a laboratory was that for chemistry. It seems that neither the architect nor Acland ever considered the designs used for recent laboratories on the Continent or in London; Oxford would do things its own way. Laboratories were known to be sources of noxious fumes and liable to catch fire. The same could be said of the kitchens in medieval houses and monasteries and so the pattern chosen was that of the Abbot's kitchen of the famous abbey at Glastonbury. The laboratory was semi-detached from the rest of the museum to prevent its being a fire hazard. It features prominently in the engraving of the Museum that was chosen to illustrate the University Almanack of 1860 (Figure 4.1). The building was almost square at ground level but octagonal in the upper part, $32' \times 32'$ internally. It was inadequately lit by

Figure 4.1 Engraving of the Museum, from the University Almanack of 1860, showing the Abbot's Kitchen as the extension to the right of the main building and the Keeper's residence, Museum House, standing alone to the right of that.

gothic windows and ventilated by four chimneys, one at each corner, and by open louvres at the top of the conical roof. The chemists who had to work there never came to share Acland and Ruskin's optimistic belief in its convenience. Daubeny had not intended to move to the new laboratory,[62] since he had his own at Magdalen, and he seems to have had little or no influence on its design. He resigned the chemistry chair in 1854, shortly after completing his service as President of the Chemical Society, in order to concentrate more on his botanical and agricultural interests, although he continued to teach the Magdalen chemists, when his health permitted, until his death in 1867.

Once the design of the Museum was settled, things moved quickly and the new Chancellor, the Earl of Derby, laid the foundation stone on 20 June 1855. The building was sufficiently far advanced for the first chemistry lectures to be given in 1858 although gas had not yet been laid on and the doors and windows were still unpainted. The move into the Museum of the mineral and biological collections, of Radcliffe's library of scientific works, and of the professors and their equipment followed soon after, and the new Museum was substantially complete in time for the British Association to meet there in 1860, when Bishop Wilberforce and Thomas Huxley had their clash over Darwin's theory of evolution. As well as the laboratory in the Abbott's Kitchen, chemistry was allocated a range of ground-floor rooms in the south front of the main building and a yard with stores in the space between the main building and the Kitchen. The Kitchen soon proved to be too small; indeed it was no larger than the space that had been provided for chemistry in the old Museum in 1683. An extension to the east was built in 1877–9 that comprised a large single room on the first floor that served as an inorganic and analytical laboratory, with smaller rooms below (shown in the plan of 1893, Figure 4.2,) for organic chemistry. The smoothness of the moves and the efficient running of the building in its first years owed much to the good sense of the first Keeper, John Phillips, the geologist, who was the oldest of the professors who moved into the Museum, and who ran the building until his death in 1874.[63] Had he lived longer science in Oxford might have flourished more strongly in the 1880s and 1890s. The administrative and financial affairs of the Museum were placed in the hands of a Delegacy, which comprised the Vice-Chancellor, the Proctors, and six members elected by Congregation, but none of the science professors, although they could attend its meetings and speak.

4.4 BENJAMIN BRODIE

Daubeny's retirement from the chair of chemistry in 1854, while retaining those of botany and rural economy, left uncertain the future of the chemistry laboratory in the new Museum. One man who had a strong claim to the chair was Benjamin Collins Brodie,[64] the son of the baronet Sir Benjamin Brodie, a surgeon at St George's Hospital and soon to be President of the Royal Society. The younger Brodie, who was to inherit the baronetcy in 1862, had attended Balliol College where he studied mathematics under Powell, in which he obtained second-class honours in 1838, and chemistry under Daubeny, before

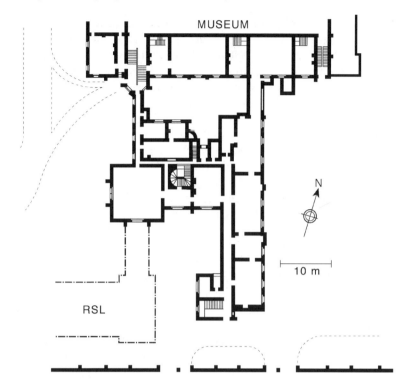

Figure 4.2 Ground floor of the Chemistry Laboratories in 1893, adapted from the plan in the edition of that year of H.W. Acland and J. Ruskin, *The Oxford Museum*. The Abbot's Kitchen is the square building in the centre and the laboratories to the east of it are the additions of 1877–79. The dashed lines in the south-west corner mark the site of the Radcliffe [Science] Library, built in 1901–02. The large open yard between the laboratories and the main part of the Museum was later filled with a lecture theatre.

going to work in 1844 with Liebig, a friend of his father's. Here he studied beeswax and found that it contained what we now call a homologous series of solid alcohols. J.B. Dumas had broached the idea of such series and it had been discussed in Charles Gerhardt's first book, published in 1844–45. This laboratory work by Brodie, and no doubt his father's position, led to his election to the Royal Society in 1849 and to the award of a Royal Medal in 1850. He set up a private laboratory on his return to London where he continued his work on waxes, alcohols and the related fatty acids. He worked also in the Royal Institution as an assistant to William Brande, another of his father's friends. When Brande retired in 1852 Brodie had hoped to succeed him and to set up there a research school of chemistry similar to that of Liebig at Giessen. The Managers of the Royal Institution thought, however, that his lectures lacked popular appeal and appointed John Tyndall instead, with the title of Professor of Natural Philosophy. Brodie was better fitted to be a candidate for the Oxford post when Daubeny retired two years later. He was, however, not without his detractors, partly because of his forceful way of expressing his views but principally because his lack of faith had led to his refusal to sign the Thirty-Nine Articles of the Church of England. There was another possible candidate in Story Maskelyne, who was a more popular and more orthodox man. He divided his time between Oxford, where he was Daubeny's assistant in the old Museum and deputy to Buckland, and London where he had become friendly with Brodie, six years his senior, and where he worked in Brodie's private laboratory. He was one of the most radical, if junior, of the Oxford reformers and a considerable authority on the new subject of photography. He did not have Brodie's private means and so the prospect of a permanent post was enticing. His friends urged him to stand but he was restrained by his feeling of obligation to Brodie who had helped him so much with both personal and laboratory accommodation in London. It seems he did all he could to support Brodie's election, which was still at the mercy of an unpredictable Convocation, but his name was nevertheless put forward. Brodie won the poorly attended election[65] by 33 votes to 10 on 13 November, 1855. The choice of Brodie marked, for the moment, the end of the traditional link between chemistry and medicine, and the field that he was to choose for his research ran counter to Daubeny's view that chemists need not study mathematics.

Brodie faced an immediate problem on taking up the Oxford post in 1855; where was he to work? Daubeny had moved to his own laboratory at Magdalen, but Story Maskelyne had won a battle with Powell for Daubeny's former space in the old Museum, where he was now running his practical chemistry classes. The construction of the new Museum was just beginning and would not be finished for another five years. Meanwhile Brodie's old college, Balliol, came to his aid and in December 1855 the Master offered him two rooms as a laboratory and lecture room on Staircase 16 of their new Salvin Building, and a grant of £77/10/0[xi] for equipment. He shared the space with his Balliol colleague, the Irish mathematician Henry John Stephen Smith who had recently taken on the teaching of chemistry to Balliol men, after a crash course in the subject from Story Maskelyne and from A.W. Hofmann at the Royal College of Chemistry. He had apparently also been a possible candidate for the chemistry chair, but was elected to the more suitable Savilian Professorship of Geometry in 1860. Brodie had to manage with this space in Balliol until he could move to the still incomplete new Museum in 1858. There he lectured, at first twice and later three times a week during two of the four terms.[66] The colleges were slow to appoint tutors for this new subject and Brodie supplemented his formal lectures by what we should call tutorial classes, and he called "Catechetical Lectures—intended especially for candidates for honours in the School of Natural Science." These were held each week on Thursdays. Story Maskelyne followed him to the Parks a little later, thus bringing to an end the teaching of chemistry in the old Museum. Brodie soon discovered the shortcomings of the Kitchen. The open louvres at the top and the four chimneys

---

[xi] £77.50 in modern notation.

generated intolerable draughts and he persuaded the Delegacy to glaze the louvres. The light was inadequate and the windows had to be enlarged. In his appeals to Convocation for money for improvements and for equipment Brodie had to be represented by Smith since his atheism had precluded him from taking the degree of MA, and so had deprived of him of the right of addressing Convocation in person. His research had moved on to the study of allotropy, particularly of phosphorus, sulfur, and carbon, and to the preparation of graphitic oxide or acid, made by the oxidation of graphite by a mixture of nitric and sulfuric acids. Brodie introduced practical classes as soon as the state of the Museum laboratory allowed, for which he charged £3 a term. The position of 'research student' was one not to be entertained until many years later but soon after Brodie had retired his successor, William Odling, announced in 1875 that "The Professor will be glad to hear from gentlemen who, having passed the necessary examinations for the degree of Bachelor of Arts, wish to carry on original work in the University Laboratory." It is not known how many takers of this offer there were until the research degree of B.Sc. was instituted in 1895. This was taken by 79 candidates from that date up to the start of the First World War, of whom 28 were chemists, some from outside Oxford.[67]

Organic chemistry, in particular, passed through a time of chaotic development in the 1840s and 1850s. There was total confusion about the concepts of atoms, molecules, and equivalents.[68] William Odling[69] went to Paris in 1854 to meet J.B. Biot, a visit that led to his translating the most up-to-date work on the subject, Auguste Laurent's posthumous *Méthode de Chimie*, which appeared in English as *Chemical Method: Notation, Classification & Nomenclature* in 1855. There, Odling listed the different formulae to which the principal chemical schools of the first half of the century had inclined.[70] For some of key protagonists and molecules, he chose as his examples: Dalton [HO, HCl, NaCl], Berzelius [H_2O, H_2Cl_2, $NaCl_2$], Gerhardt [H_2O, H_2Cl_2, Na_2Cl_2], and Gerhardt and Laurent [H_2O, HCl, NaCl]. It was the 'theory of types' of the last pair of chemists that was to prevail, particularly after Alexander Williamson's decisive paper on the 'water type' in which he showed that there were two hydrogen atoms in water that could be successively and independently replaced by hydrocarbon radicals to prepare alcohols and ethers, and that the latter could be simple or mixed, as in methyl ethyl ether. Similarly, when Edward Frankland reacted ethyl iodide with zinc to produce what he thought initially was the ethyl radical it eventually became clear that the 'radical' had dimerised and produced what we call butane. Brodie had attempted in 1850 to extend to such reactions the same idea that Williamson had applied to the ethers, and to prepare 'mixed' hydrocarbons with odd numbers of carbon atoms by trying to react a mixture of two aliphatic iodides with zinc. He failed,[71] and it was left to Adolphe Wurtz to succeed with this synthesis in 1855 by using sodium instead of zinc. Many of the controversies could have been settled if the chemists of the day had been more willing to take into account the evidence provided by the densities of the vapours, many of which were known, but few organic chemists were then prepared to rely on physical facts and their often still disputed interpretation. Odling was never as influential as Williamson or Hofmann in converting British chemists to the modern form of organic chemistry, or as Kekulé and Frankland in the subsequent development of the ideas of molecular structure, which Odling regarded with suspicion, and of valency, but his work actively supported these pioneers. Brodie was generally of the same persuasion as Odling. He tried to reconcile the old electrochemical 'dualism' with the increasingly conflicting facts revealed by the rapidly advancing field of organic chemistry. Brodie used some of the modern atomic weights ($C=12$, $O=16$, $S=32$, $Se=79.2$, $Te=128$) in his first lecture course[72] of 1856, but in 1859 an Oxford examination paper[43] was still using the atomic weights of $O=8$, $S=16$, and $Ca=20$. The candidates must have been more confused than the examiner!

Brodie had asked Odling to produce a textbook suitable for his class at Oxford and 'arranged in accordance with his own methods of teaching'. In 1861 Odling published the first part of his *Manual of Chemistry: Descriptive and Theoretical*,[73] which dealt with general principles and with some inorganic compounds. (The second part never appeared.) Here, he used most of the new atomic weights: $C=12$, $O=16$, $Cl=35.5$, and for the metals, $Na=23$, $K=39$, but also $Ca=20$ and $Al=13.7$, that is, half the current values. He had some doubts about the low weights that he had ascribed to some metals, for he suggested that instead of writing them as 'double atoms', (Fe_2), (Al_2) *etc.*, as gravimetric analyses required, it might be better to denote them by symbols in which a letter was repeated, Ffe, All, *etc.*. By 1864 he was convinced that the doubling of these weights was the right thing to do, and William Crookes came to the same conclusion that year and used them in his journal, *Chemical News*. Odling's book is important for it describes how he and Brodie saw chemical theory, and so how this was taught to the Oxford chemists in the third quarter of the century. Brodie and Odling distrusted the idea of a corpuscular atom and of the idea that complex molecules were formed of groups of such atoms arranged in determinable geometric structures. Odling wrote:

> To every element is assigned a particular number, termed its proportional number, which expresses the least indivisible portion of the element that is found to enter into combination.—The determination of what is the smallest indivisible combining proportion, or *chemical atom*, of a body involves considerations of very many circumstances—.'[73]

Such careful positivist language avoids any commitment to a corpuscular atom. It precluded speculation about structures of molecules, and indeed his use of the words 'atom' and 'molecule' is not always as clear as we should now wish; he wrote, for example, of an atom of chlorhydric acid, HCl, of an atom of chlorine, Cl, and of a molecule of chlorine, Cl_2. His scheme of things cannot explain the existence of isomers. He noted the existence of these, giving as an example, propionic acid, methyl acetate, and ethyl formate, all of which are $C_3H_6O_2$, but was content to say only that "In many cases we can at present only speculate on the nature of the isomerism." On the more positive side we can see in his groupings of the elements the beginnings of the ideas that led to their periodic classification, and, in his denoting of 'equivalency' by dashes, as in Bi''', the modern idea of valency.

Early in 1860 Kekulé had put forward the proposal that there should be a major chemical congress to resolve the conflicting notations, structural formulae, and atomic weights, but Brodie had been unenthusiastic. He replied:

> If you consider that the objects which you have in view are at all likely to be promoted by the addition of my name to the names of the distinguished Chemists who have expressed their willingness to sign the document, which I have received from you, I should certainly not withhold it—It appears to me however, that we should be on our guard against even the apparent attempt to force the opinion of others or to attempt the premature settlement of questions, which are by no means ripe for it, and on which opinion is so unformed.[74]

The meeting went ahead at Karlsruhe and was attended by about 140 chemists on 3–5 September 1860. Brodie was there in his capacity as President of the Chemical Society. The meeting did not pass formal judgments of the kind that he and Williamson would have deplored. Most of those present came soon to the modern view of atomic weights and of determinable molecular structures, prompted by the circulation of Stanislao Cannizzaro's pamphlet of 1858 in which he persuaded the more traditional chemists to take seriously the evidence of physical research, and particularly of the import of Avogadro's hypothesis. Many chemists had already come to the belief that the simple gases had diatomic molecules but Cannizzaro's manifesto was only fully accepted when they agreed also to double the atomic weights of what we see as the di- and tri-valent metals. Henceforth most scientists accepted that the physicists' molecules, evident in the new kinetic theory of gases, and the chemists' multi-atomic molecules, O_2, N_2, *etc.*, were the same things. Not all accepted this rationalisation, and much later in the century there was a reaction against atoms by Wilhelm Ostwald and the 'energeticists', but they never had a strong following among British chemists.

Brodie's research soon took an unorthodox direction. The long arguments over the atomic formulae of organic compounds, the positivistic views he shared with Williamson, and perhaps Faraday's doubts about corpuscular atoms which he had met when working in the Royal Institution, had made him increasingly sceptical about the new enthusiasm for the structural implications of the modern formulae. Williamson, however, managed to retain his support of the ideas of Comte, under whom he had studied in France, with a firm conviction in the reality of atoms. Brodie argued that all we could know was what happened when two or more substances reacted together – they produced one or more new substances. He therefore invented a calculus of operations in which the elements were not the chemical atoms but the operations needed to bring certain compounds into being.[53] He was a competent mathematician and could draw for advice on his friends Henry Smith and William Donkin, sen. He had worked previously on the organic peroxides and, in a paper entitled 'The organic peroxides theoretically considered'[75] he noted the parallelism of the equations,

$$2H_2O = H_2 + H_2O_2 \quad \text{and} \quad 2HCl = H_2 + Cl_2,$$

so that hydrogen peroxide is to water as chlorine is to hydrogen chloride, and such parallel equations described the operations of the creation of the compounds on the right-hand side. In developing his ideas he was encouraged by Odling[76] in his Presidential Address to the Chemistry Section at the meeting of the British Association at Dundee in 1864, although at that time Odling did not fully understand them. The first part of Brodie's 'Calculus of chemical operations' was read to the Royal Society on 3 May 1866 and published later in the volume for that year; Odling was the referee, calling it the "first application of algebra to chemistry" and saying that "its importance could scarcely be exaggerated."[77] Brodie reasoned that if α is the symbol for the creation of a litre of hydrogen, then water, which contains oxygen, requires two symbols, α and ξ, and can be expressed as $\alpha^m \xi^n$, where m and n are integers to be determined. Since two volumes of steam can be decomposed into two of hydrogen and one of oxygen, the simplest solution in prime integers is to write the operation of the creation of steam as $\alpha\xi$, when oxygen becomes ξ^2, in order to satisfy the equation of decomposition: $2\alpha\xi = 2\alpha + \xi^2$. (In his calculus, $x + y = xy$.) Hydrogen peroxide becomes $\alpha\xi^2$. Similarly, hydrogen chloride is $\alpha\chi$ and chlorine is $\alpha\chi^2$, ammonia is $\alpha^2 v$ and nitrogen αv^2. Carbon is not so simply characterised since the element cannot be volatilised, but by less direct routes he concludes that carbon may be κ or κ^2, and, if the former, then methane is $\alpha\kappa^2$, carbon monoxide is $\kappa\xi$, and carbon dioxide is $\kappa\xi^2$. He divided the elements into three classes: those created by one operation, such as hydrogen, those by two, such as oxygen (and carbon?), and those by three, such as nitrogen and chlorine. To those who asked if the symbols ξ, χ, and v represented real or imaginary entities he indulged in a verbal sleight-of-hand and said that he thought of them as 'ideal'. He was invited to explain his ideas at a meeting of the Chemical Society on 6 June 1867. Because of the interest aroused by his work and the large size of the audience, the meeting was held in the Royal Society's lecture theatre and attended also by Fellows of that Society who were not chemists. His lecture there was entitled 'On the Mode of Representation afforded by the Chemical Calculus, as contrasted with the Atomic Theory'; it received long reports in the *Chemical News* and in a new journal called *The Laboratory*.[78] It was reprinted as a separate pamphlet in 1880, under the title 'Ideal Chemistry'. As well as promoting his own system he also attacked the notion that chemists could determine the three-dimensional structure of molecules. He was particularly severe on the so-called 'glyptic formulae'. These were models in which atoms were represented as solid bodies with small holes into which there were inserted rods to join them to the appropriate number of other atoms.[xii] Williamson chaired the meeting, but was soon to join Brodie's critics, who included some of the physicists who had attended. Clerk Maxwell said that on entering the meeting he was shocked to find that space was a chemical substance and that hydrogen was an operation. He commented, as Brodie himself had done, on the relation of the new calculus to George Boole's algebra, and repeated his belief in the kinetic theory of gases and the validity of its evidence for the real existence of atoms. George Stokes pointed out that nitric oxide and nitrogen dioxide presented anomalies that needed to be resolved. Odling was perhaps the most receptive of the chemists and was prepared to give the new theory a fuller hearing.

Brodie's calculus received a second airing at the British Association in Dundee in 1867. Thomas Anderson, the President of the Chemical Section, discussed it favourably, but noted that if hydrogen had been denoted by α^2, rather than α, then no element (in the conventional sense) would have needed multiple symbols, a variant of Kekulé's earlier criticism that the whole construction depended on the arbitrary choice of hydrogen as its base.[80] Brodie returned to the fray in 1872, after a period of illness, with a second part of his scheme[81] but by then the chemists had lost interest, realising that his calculus lacked any credible basis, or a predictive power to match the rapidly moving theories of structural organic chemistry. It failed, for example, to differentiate between isomers, it contributed nothing to the emerging ideas on valence nor to Mendeleev's periodic classification of 1869. Meanwhile, the new journal, *The Laboratory*, had folded, after only five months' publication, and Crookes's *Chemical News* ignored the second paper. In 1872 Brodie resigned his chair because of his ill-health and retired to his house on Box Hill in Surrey. He returned there briefly to his work on ozone, a subject that had long interested him, commenting on Odling's conjecture and Soret's proof that it was triatomic oxygen. His apparatus for its preparation, 'Brodie's ozoniser', was invented there. He died of a rheumatic fever in Torquay on 24 November 1880.

4.5 WILLIAM ODLING AND HIS DEMONSTRATORS

In 1865 Magdalen College had, in accordance with the recommendations of the 1854 Executive Commission, taken over the Aldrichian chair and had renamed it the Waynflete professorship after their fifteenth-century founder, William of Waynflete, who had stipulated that his college should have praelectors in natural philosophy.[82] Magdalen was not, however, required to offer the professor a fellowship; that was a reform that came only in 1877, after Brodie had retired. The Aldrichian trust money thus released might have escaped from the chemists had Brodie not intervened and ensured that it was used first to

---

[xii] Glyptic = carved. There is a boxed set of these models in the History of Science Museum in Oxford in which carbon is represented by a black cube with four holes arranged in a plane.[79] This presumably dates from between 1867, when such models were first made commercially, and some time after 1874, when the tetrahedral carbon atom was announced. It is not clear who used these as teaching aids since neither Brodie nor Odling, his successor, believed in them.

purchase apparatus and, in 1870 to set up the Aldrichian Demonstratorship in Chemistry.[83] This post was the successor to a demonstratorship held from 1863 to 1869 by Henry George Madan.[84] From 1871 to 1873 the new Aldrichian post was held for two years by Heathcote Wyndham,[85] who was followed by Walter William Fisher.[86] Initially, this post, like Madan's, was for an assistant to the professor, but a second form of the title, the Aldrichian Praelectorship, a name that first appears in the University Calendar in 1930, was later bestowed on the senior lecturer in chemistry, by appointment, and now it goes to a lecturer chosen on merit, when it carries with it a promotion to a readership.

Another reform had come into force by the time that Brodie retired; his successor was no longer to be elected by an open vote in Convocation. A delegacy of five electors was appointed: the Chancellor of the University, who was then the Marquis of Salisbury (although presumably his deputy, the Vice-Chancellor, the Rev. Henry Liddell, acted in his place); the Visitor of Magdalen College, who was the Bishop of Winchester; the President of Magdalen, the Rev. Frederic Bulley; the President of the Royal Society, Sir George Airy; and the President of the Royal College of Physicians, Sir George Burrows.[xiii] None was a chemist, but the committee could seek advice and was clearly better qualified to choose a professor than the largely non-resident body of MAs that comprised Convocation. The strongest local candidate was Augustus Vernon Harcourt, an early pupil of Smith and Brodie at Balliol, and later Brodie's assistant in the new laboratory in the Museum.[87] In 1859 he was elected to be the first Lee's Reader in Chemistry at Christ Church and, a few years later, on the departure of the anatomical specimens to the new Museum, he converted the Christ Church building into a chemical laboratory, a role it had played for a time in the 18th century. He wrote there his *Exercises in Practical Chemistry* (1869) with Henry Madan; the book uses modern atomic weights.[87] In the Christ Church laboratory Vernon Harcourt embarked on a novel study of the factors that determined the rates of chemical reactions, an innovation that, as we shall see in the next section of this chapter, and in later chapters, was to have profound effects on the course of chemical research in Oxford for the next hundred years. He had been elected to the Royal Society in 1868, was a Secretary of the Chemical Society from 1865 to 1873, and was to be its President from 1895 to 1897. He was, however, five years younger than Odling, the obvious external candidate, and had not the same reputation, nor had he yet received the same recognition. Odling had been educated at Guy's Hospital, had acquired a medical degree, had been elected to a Fellowship of the Royal Society at the age of 29, had taught practical chemistry at Guy's, and had succeeded Faraday as Fullerian Professor at the Royal Institution in 1868. He had been a Secretary at the Chemical Society from 1856 to 1869 (thus overlapping Vernon Harcourt's tenure by four years), and was to be its President from 1873. It was therefore no surprise that Odling was preferred to Vernon Harcourt, although Ray Lankester, in a letter to Thomas Huxley, said that the choice was influenced by the fact

---

[xiii] A Statute of 1881 added the Professor of Experimental Philosophy, the Professor of Chemistry at Cambridge, and one other, thus giving the board a form closer to its present composition.

that internal candidates were at a disadvantage at Oxford when compared with those, like Brodie and Odling, who had established their reputations in laboratories with better facilities than were available there.[88] Odling was, however, the first man to be elected to the Aldrich/Waynflete chair without having been educated at Oxford, and so became the first of a succession of such appointments that was not broken until the 21st century. The election was on 11 July 1872, and it can have come as no surprise to the University since he was elected a fellow of Worcester College the next day.

Odling married on his appointment and set up house at 15 Norham Gardens, just north of the University Parks, and now the home of the Department of Education. He had no great record of experimental research, although he had run one of the earliest practical laboratories at Guy's Hospital. He had similarly showed little interest in medical research, although he had given a course of lectures on 'animal chemistry' to the Royal College of Physicians and was, for a time, the Medical Officer of Health for the London Borough of Lambeth. His reputation rested on his part in advancing the theoretical ideas behind the chemical reforms of the 1850s and 1860s. By the time of his move to Oxford these were becoming complete, with Kekulé's hexagonal ring for the aromatic compounds in 1865, Mendeleev's periodic table in 1869 (following Odling's preliminary ideas along similar lines ten years earlier), and van 't Hoff and Le Bel's tetrahedral carbon atom in 1874 for the aliphatic compounds. What then was Odling to do in Oxford? The short answer, as far as research went, was *nothing*. There was a joint paper with James Marsh, a demonstrator from 1885, on phenol derivatives,[89,90] but otherwise he published nothing that can be described as original work between his appointment in 1872 and his retirement forty years later. Nor did he contribute to the practical instruction of the undergraduates; that he left to the demonstrators whom he appointed. It was said that he considered that it was not proper etiquette for the professor to set foot in the Laboratory.[xiv] This had been enlarged in 1877–1879 to accommodate the growing numbers of undergraduates reading chemistry, after Convocation had voted by 62 to 22 to allocate £7000. Odling certainly deserves the credit for raising this considerable sum from an often unsympathetic University. The upper floor of the two-storey building was to be used, and indeed still is, for classes in inorganic chemistry, and the rooms below were for organic chemistry. An intermediate floor was inserted into the Kitchen in 1902, but this space only compensated for the loss of the ground floor that became the entrance hall of the new Radcliffe [Science] Library (Figure 4.3).

Soon after his arrival in Oxford, Odling was elected President of the Chemical Society for two years, 1873–1875. His time in office coincided with disputes on how far the Society should be restricted to qualified professional chemists and how far it should function as a learned society open to all who wished to contribute to its aims. The issue was settled by establishing the Institute of Chemistry as a separate professional body that restricted its membership to

---

[xiv] The evidence for this often-quoted belief is not clear; it possibly originated with a remark by Bertram Lambert, one of the young demonstrators appointed in 1905.

Figure 4.3 Ground floor of the Abbot's Kitchen in 2007 when it was being refurbished by the Radcliffe Science Library, from a photograph by Susan Green, Chief Cataloguer of the Library. The door is in the north wall and led to a passage connecting the Kitchen and the Museum. The four iron pillars were inserted in 1901–02 to carry the weight of the new upper floor to which the chemists then moved.

those who could pass its examinations or obtain equivalent training elsewhere. Odling was on the organising committee of the Institute and became a founding Vice-President when it was set up in 1877. He became the third President from 1883 to 1888 and it was during his term of office that the body obtained a Royal Charter. In spite of this connection with the outside world of practical chemistry he did not persuade most Oxford chemists to follow industrial careers. An analysis of the records of seven colleges shows that 20 out the 89 chemical graduates whose careers are known went into industry between 1900 and 1914.[67] The proportion in other English universities seems to be similar between 1880 and 1910 although the figures are again rendered uncertain by the high numbers whose careers are not known.[91]

The sciences slowly but gradually gained independent status at Oxford. Boards of Studies, in 1872, were succeeded by Boards of Faculties in 1877, which had a majority of professors and readers and were empowered to appoint examiners. Hitherto, these had been appointed by the Vice-Chancellor and Proctors. Candidates in the sciences had been exempted from Honours Moderations in 1886, but Odling did not support a move to exempt them completely from showing competence in Greek, since he thought that chemists who knew no Greek would lose 'face' within the university. This requirement was also held sacred by some of the younger classics fellows who were responsible for a feeling in many Colleges towards the end of the century that the scientists were becoming arrogant in their demands and had to be kept in their subservient place.[92] These classics fellows often opposed the scientists in Congregation, 'the Museum vote', by allying themselves with the Oxford graduates among the city clergy and professional men, 'the black dragoons', who were members of Congregation from 1854 until 1913. So compulsory Greek survived until after the First World War, and compulsory Latin until after the Second. A move to establish separate degrees of Bachelor and Master of Natural Science (BNS and MNS) failed when the University obtained counsel's opinion that an Act of Parliament would be needed if the holders of these degrees were to have the same privileges as the existing BAs and MAs.

A synoptic view of the teaching of chemistry in Oxford at the time of Odling's election is to be found in the report of the 'Devonshire' Commission in 1873. This Commission[93] was set up, at the urging of the British Association, to enquire into the state of scientific education and research throughout Britain. The chairman was the Duke of Devonshire, the founder of the Cavendish Laboratory in Cambridge, and Oxford science was represented by Henry Smith, the Savilian Professor of Geometry. The chemist on the commission was W. Allen Miller from King's College, London, and it was he who put most of the questions to Brodie, then in almost his last days in post. Unfortunately, Miller was taken ill on his way to the British Association meeting at Liverpool in 1870 and died soon after, so he never signed the Report. Moreover, Brodie had retired before it was complete and so it is Odling's lectures that feature in the Report as issued in 1873. He lectured twice a week in Michaelmas and Lent terms on 'The succession of chemical ideas,' in which presumably he summarised the position reached at the end of the arguments of the last twenty years. Walter William Fisher, now the Aldrichian Demonstrator, lectured twice a week in Easter and Lent terms on 'Elementary organic chemistry', and William Frederick Donkin,[94] in the same terms, on 'Elementary inorganic chemistry'. These lectures were supplemented by practical classes. There were also lectures and classes held for men from different groups of the Colleges, notably Vernon Harcourt on 'Inorganic chemistry', and Heathcote Wyndham, briefly the Aldrichian Demonstrator and Natural Science tutor at Merton, on 'The facts on which the modern studies of chemistry are founded' and on 'Rock specimens', with a practical class on mineralogy held in the Mineralogy Laboratory in the Museum. These lectures may have justified the appearance of the occasional examination question on what we would now recognise as physical chemistry from 1871 onwards. The course as a whole seems to have been a light and not well-balanced diet of instruction even for a 'supplementary' subject such as chemistry was still perceived to be by the traditional scholars. Certainly the examination papers were not very demanding, as Odling found when he compared the London papers that he was used to with the norm at Oxford. Thus, when Henry Roscoe discovered the new element vanadium in 1869, there was a question on it from Odling in the London examinations the next year, an innovation that would have been unacceptable at Oxford.[95] Odling took his turn as an examiner in the Honour School from 1874 onwards, and in 1880–81 was joined by William Tilden, newly appointed as Professor of Chemistry at Mason College (now the University) in Birmingham. He was followed in later years by external examiners from London, such as Henry Armstrong and William Ramsay. At Oxford the external examiner played a full part in the setting and marking of papers and it is only in recent years that the role has changed to that of a 'moderating' examiner, as in other universities.

The Devonshire Commission of 1873 had been preceded by the financial fact-finding Cleveland Commission of 1871 and was soon followed by one chaired by the Earl of Selborne that took evidence in 1877, and reported four years later. In his evidence Odling said that his problem of lack of space would be overcome when the new laboratory was complete but he still needed more staff in order to cover adequately all the branches of chemistry. He was adamant that these men should be subservient to the one professor and was clearly not happy with the independence of the Christ Church laboratory under the college reader, Vernon Harcourt, calling it "really a rival establishment". He asked also for what would now be called research studentships since there was no way of supporting any of the best men who might wish to start a career in research. He defended his policy of leaving the running of the practical laboratory to the senior demonstrator and made no comment on his own dropping of research.[96] This Commission was responsible for the setting up of the Common University Fund (CUF), which was to transfer money from the Colleges to the University so as to strengthen its core activities and the professors' position. It was never as effective as its proponents hoped since the agricultural depression of the end of the 19th century considerably reduced the colleges' incomes.

From 1886 the published lists of those obtaining honours in the natural sciences showed the single subject in which each candidate had chosen to be examined and so it is possible to follow the changing numbers of honours chemists. There were an average of 18 from 1890 to 1894, and in the successive periods of five years this number remained fairly steady: 21 from 1894 to 1899, 24 from 1900 to 1904, 27 from 1905 to 1909, and 25 from 1910 to 1914. In the earlier years the chemists formed about half the total number of natural scientists but this proportion had fallen to about a third by the end of the period.[97]

The arrangement of the lectures in the Museum changed little with the years,[98] and their overall quality was probably never high. In 1888 Odling lectured in Michaelmas, Hilary, and Easter terms on organic chemistry, Fisher on elementary inorganic chemistry, and John Watts[99] gave a more advanced course on organic chemistry. In the Summer term Victor Herbert Veley,[100] a demonstrator from 1887, gave a course on physical chemistry, and Marsh a course on 'Some points of chemical theory', in which he presumably sorted out the reforms of notation, formulae, and atomic weights of the 1860s. About 60 attended the laboratory each term, and rather more came to the lectures. There were also lectures and practical work for the medical students. Soddy found in 1896, when he was an undergraduate, that the lecturers gave an adequate account of the atomic and molecular doctrines of the time since Odling, even if he did no research himself, kept up with what was being done elsewhere, but, Soddy claimed, the teaching in the Museum of practical inorganic chemistry "was almost incredibly bad, or rather, non-existent".[101] Odling had by now abandoned his positivistic objections to atoms and structural formulae.

By 1902 Odling claimed that there were now eight or more lectures a week in the Michaelmas and Hilary terms and fewer in the Easter and Trinity terms. The syllabus was again little changed, and Odling's lectures had retained some of their historical aspect. Henry Tizard, a pupil of Nevil Sidgwick's at Magdalen in 1905, said of Odling's lectures that he "went to them for a time, more in order to improve my perspective of history than to learn any chemistry. I got the vague impression that he and Dalton had been boys together."[102] A course of lectures from Allan Frederick Walden[103] strengthened the instruction in inorganic chemistry, Marsh also lectured on stereochemistry, and Veley's lectures on physical chemistry now occupied two terms. This branch of the subject was at last becoming recognised as an integral part of chemistry. Odling reported that "In Michaelmas term, a temporary Laboratory for Experimental work in Physical Chemistry was organized, and practical instruction given by Dr. Veley." Twenty seven students attended "and this number seems likely to increase in the future."[98] One senses, perhaps, Odling's surprise that this hitherto unrecognised branch of chemistry was here to stay. He asked, in vain, for a CUF readership for Veley in physical chemistry, even claiming that this subject "has of late become perhaps the most important of all branches of chemistry".[66] There was some research by "Demonstrators, B.Sc. candidates and others" and there were now about 70 undergraduates working in the laboratory of whom about 4 were women.

The Association for the Higher Education of Women was set up in Oxford in June 1878, and Somerville College and Lady Margaret Hall opened their doors in October 1879. The Association undertook the provision of extra lectures for women by members of the University and the necessary provision of chaperones. It was another four years before 122 members of Congregation successfully petitioned that body to allow women to be admitted to the regular university examinations. By a statute of 1 March 1884 they were admitted to the Final Honour Schools of Mathematics, Modern History and Natural Science, but were not required to show proficiency in Latin or Greek. In 1885 Margaret Seward was awarded a certificate for a first-class performance in chemistry, but it was many years before women were permitted to take the degree itself. She was at once appointed tutor (and chaperone) at Somerville, but stayed for only two years, when she went to Royal Holloway College, and then on to King's College in London. The chemists were early advocates of the admission of women to lectures and laboratories. Harold Dixon[104] tried to admit them to his lectures in the laboratory run by Balliol and Trinity Colleges, but this attempt was at first vetoed by Benjamin Jowett and John Percival, the heads of the two colleges. The issue was forced by Vernon Harcourt in 1880 when he threatened to stop his lectures arranged solely for women in Christ Church unless Dixon had his way. Chaperones were required until 1893, although this proved no problem for Dixon's class since his young wife attended his lectures. The University laboratories were opened to women in 1885. The first committee that set up Somerville College had included

Mrs Harcourt and Mrs Esson, the wife of Vernon Harcourt's mathematical collaborator at Merton. Odling had long been a supporter of women's education and, in 1893, became the first representative of the Hebdomadal Council on the Council of Somerville.[105]

Odling's demonstrators were, on the whole, a conscientious body of men who helped him to mount a course of inorganic and organic chemistry. John Watts had been a London graduate who had worked under Odling at the Royal Institution. When Odling was elected to the Waynflete chair, Watts moved briefly to Edinburgh University, but soon followed Odling to Oxford. There, in spite of having already a London D.Sc., he matriculated as an undergraduate, took a first-class degree from Balliol in 1876, and moved to Merton, after the early death of Wyndham, where he spent the rest of his life as the Tutor in Natural Science. He lectured on organic chemistry and taught quantitative analysis on the ground floor of the then undivided Kitchen. His strength was as a college tutor where he had a string of successful pupils that included, among the later Oxford chemists, Tom Moore, Bertram Lambert and Frederick Soddy, the last of whom ascribed Watts's success principally to his ability to choose the best men and draw them out rather than to any particular skill at teaching. Walter Fisher was in post when Odling arrived. He was a Merton graduate of 1870 who had worked briefly as an assistant to Vernon Harcourt, and was appointed by Brodie to the Aldrichian Demonstratorship. He held this post for the whole of Odling's tenure of the chair, and for six years from 1874 he taught chemistry at Balliol also. He was as much a 'professional' chemist as an academic, being for forty years the Public Analyst for Oxfordshire and the surrounding counties, and serving on the council and as an examiner for the Institute of Chemistry. His duties as Public Analyst sometimes clashed with those as a demonstrator and, Soddy alleged, he often left instruction in qualitative analysis to the laboratory steward.[101] His contributions to academic research were few. William Donkin was the son of William Fishburn Donkin, the Savilian Professor of Astronomy, and after taking his degree from Magdalen in 1868, became the science tutor at Keble and a demonstrator. After the death of his wife in 1876 he moved to London as professor at St George's Hospital, but lost his life while climbing on Dychtau in the Caucasus in 1888. Of the early demonstrators, James Marsh made the greatest reputation for his academic research. He graduated in 1882 from Balliol, spent a year in Kekulé's laboratory at Bonn, a year in France, and a short time as a school master, before becoming a demonstrator in 1885, and publishing his first paper, with Odling,[89] in 1887. He was elected to the Royal Society in 1906, and to a fellowship at Merton the same year. He taught also at Exeter College. His research centred on the terpenes, and particularly on camphor and its derivatives, although he never did sort out the three-dimensional structure of this difficult molecule, a problem that was solved by Julius Bredt in 1893. This work led him to an interest in stereochemistry in general, on which he lectured, and to his translating and expanding a book by van 't Hoff that the University Press published in 1891 as *Chemistry in space*. He was assisted in some of his work on camphor derivatives by John Gardner[106] of Magdalen, who, after a year in

Heidelberg, was a demonstrator in Oxford for two years from 1901, when he went as a lecturer to St George's Hospital. Gardner's departure and the need to teach a growing number of undergraduates led to the appointment of further demonstrators in and about 1905: Bertram Lambert,[107] Allan Walden, and Nevil Sidgwick,[108] who had been a fellow at Lincoln since 1901 and a lecturer at Magdalen since 1903. Lambert was one of John Watts's successful pupils at Merton, who was appointed a demonstrator soon after he graduated and remained in Oxford for the whole of his career. His first research was on the rusting of iron that, he correctly concluded, was an electrolytic process, but later he turned to the measurement of the density and adsorption of gases. In the course of this work he taught himself glassblowing, becoming perhaps the best amateur glassblower in the country, and teaching the elements of this art to numerous research students (including the writer of this chapter). Tom Moore,[109] another Merton man who taught at Magdalen, is remembered principally for a paper of 1912 with Thomas Field Winmill.[110] In a section of this, written by Moore, he showed that the degree of ionisation of aqueous solutions of amines could best be understood if there was a weak bond, or attraction, between the hydrogen atom of the water molecule and the nitrogen of the amine, the likely origin of which was a partial positive charge on the H atom and a negative one on the N. The bond can be expressed, O–H···N. Similar work soon afterwards in Germany and in the United States, notably by Latimer and Rodebush in 1920, led to this weak bond being called the 'hydrogen bond' and to an appreciation of its importance particularly in biological systems. Moore became a fellow of Magdalen in 1907, but left Oxford in 1914 for the chair of chemistry at Royal Holloway College in London University. Sidgwick, a nephew by marriage of Brodie, was one of Vernon Harcourt's pupils at Christ Church, who took a 'first' in chemistry in 1895, and then a 'first' in Literae Humaniores in 1897, a feat that is unlikely ever to be repeated. He became the best known of the group but did little research of note until after the first World War. His choice of research topics before then was described by Henry Tizard, his Royal Society obituarist, as 'haphazard'. He did, however, write a notable monograph, *The organic chemistry of nitrogen* (1910), which rescued that field from its rather boring reputation, and throughout the 1920s and 1930s he became one of the stalwarts of Oxford chemistry. Walden's reputation was as a tutor and fellow at New College; his nickname of 'Teacher' was well-bestowed. The most productive of the organic chemists in the years before the War was Frederick Daniel Chattaway[111] whose work was mainly on organic and inorganic compounds of nitrogen, often of an explosive nature. He had had an unusually prolonged education, through Birmingham, London, Aberystwyth and Christ Church, Oxford, where he was also one of Vernon Harcourt's pupils and where he graduated in 1891. His early work was in the Christ Church laboratory, then Odling allowed him to work at the Museum in one of the rooms in the extension of 1879, and, after a spell as Professor of Chemistry at St Bartholomew's Hospital, he returned to Oxford in 1907 and later took charge of the laboratory at Queen's College. The overall numbers of staff was, however, still small, with only 13 college tutors or lecturers for all the natural sciences in 1892, although many of these taught for more than one college and they were supplemented by those holding less formal appointments.[67]

J.C. Smith, himself a demonstrator in organic chemistry from 1931, summarised the years during which Odling and his demonstrators were in charge of the laboratory at the Museum by writing:

> Thus while the great chapters of Organic Chemistry were being written by von Baeyer, Victor Meyer, Kekulé, Emil Fischer, Ladenburg, and Willstätter abroad, and in England by Frankland, H.E. Armstrong and the Perkins, Oxford's contribution had been elegant and intelligent but never dominating.[112]

Henry Tizard and William Perkin put it less charitably. The former wrote that most of the staff were "content to live like gentlemen, passing to the younger generation the knowledge that been amassed by others".[102] Perkin said in his Presidential Address[113] to the Chemical Society in 1915, shortly after he had taken up the Waynflete chair:

> If the record of our universities is examined, it is at once obvious that many of these famous places, and more particularly the Universities of Oxford and Cambridge and the Scottish Universities, contributed practically nothing to the advancement of organic chemistry during the latter part of the last century, and their output of research in this subject is still far less than it ought to be.

A blistering attack on Oxford science came in the form of a lecture there in 1903 by John Perry, a Professor of Engineering in London, but its impact was diminished by its strong language, and, perhaps, by the memory that he himself had previously been a failed candidate for a post in physics in Oxford.[114]

Some feeling that matters might be improved had come to the undergraduates in 1882 when they founded the Junior Scientific Club.[115] Here, they met every two weeks during term to hear papers from their own members and from senior speakers from Oxford and beyond. From 1897 these were collected in a termly journal. The subjects ranged across a wide field, with a surprising number on biological and mechanical subjects although about half the undergraduates were chemists. Among the chemical papers, physical topics were more popular than the more traditional organic and inorganic fields, and H. Brereton Baker, Soddy, and Sidgwick were prominent members in their time as undergraduates. But there was also felt to be a need for a more specialised forum. There had, for some years, been an informal chemical club based on the Balliol laboratory, but by 1901 a determined effort had been made to encourage chemical discussion on a wider basis with two more formal clubs, the Chemical Club of 1899 and the Alembic Club of December 1900, which absorbed the Chemical during the first World War. The leader here was W.M. Hooton, a Christ Church undergraduate, and the first President G.W.F. Holroyd,[116] also of Christ Church, who was an assistant to Vernon Harcourt from 1897 to 1899, and then a college lecturer until 1903. He was supported by Watts, Walden (who chose the name), and some of the younger men, Sidgwick, Lambert, Moore and others. The club included both staff and undergraduates (Soddy was an early member), and it was divided into senior and junior branches. These met both separately and together in joint meetings, when they gathered to hear outside speakers. The social highlight of the year was the annual dinner, usually held in the Randolph Hotel. The best days of the club were in the 1920s and 1930s when many former members of the University returned for the dinner and to discuss problems in the laboratories with their former tutors and research supervisors. For many years ICI paid the expenses of members of its staff who wished to attend. Women were admitted only in 1950. The senior branch of the club survived until May 1965 but the increasing specialisation of chemistry and the spread of departmental seminars then made subject-wide lectures less attractive. The junior part of the club survived for another decade.

4.6 THE COLLEGE LABORATORIES AND THE GROWTH OF PHYSICAL CHEMISTRY

The account given above of the work of Brodie, Odling and the demonstrators deals only with part of the teaching and research in chemistry in Oxford in the second half of the century. A parallel organisation had developed, based on the Colleges, and essentially independent of University support and of the work of the professor. There had been a laboratory in Christ Church, used at first both for anatomy and for chemistry, since the benefactions of John Freind and Matthew Lee in the previous century, and this returned to being a chemistry laboratory when anatomy moved to the Museum in 1860. Daubeny, dissatisfied with the inadequate provision for chemistry in the old Ashmolean Museum, had established his own laboratory at Magdalen, in the Botanic Garden, in 1848, and chemistry was taught and practised there even after he had acquired other chairs. Balliol College had set up a small laboratory in the basement of its Salvin Building to teach its own men and, as it turned out, to accommodate Brodie when he found that Story Maskelyne had appropriated Daubeny's laboratory space in the old Ashmolean for mineralogy and for practical classes in chemistry. By the middle of the century there were, therefore, centres of instruction and of occasional research in some of the Colleges, but these did not amount to a coherent or deliberate plan to complement, or perhaps even to compete with the centralised course under the professor at the Museum. Matters took a decisive step when Brodie retired in 1872. Had Vernon Harcourt been elected to the chair, he would presumably have moved to the Museum and the Christ Church laboratory might have been abandoned. He had, however, been elected to the Lee's Readership, a college post, in 1859 and remained at Christ Church until his retirement in 1902. These three laboratories, Balliol (joined later by its neighbour, Trinity), Magdalen and Christ Church then formed the nucleus of college-based teaching of chemistry, at first to their own undergraduates but later more widely, and were to see the development of research that complemented and soon surpassed that emanating from the Museum. The division of effort and resources between the Museum and the colleges was apparently unplanned. Thus in 1892, half-way through Odling's tenure of the chair, there were four demonstrators: Fisher, the Aldrichian Demonstrator, and Marsh, Watts and Veley. Of these Watts taught also in Merton and Magdalen, and Veley in Queen's and to the non-collegiate students. The University Calendar for that year shows also that Vernon Harcourt taught in Christ Church, Conroy and Nagel (see below) taught in both Balliol and Trinity, and Percy Elford, another Christ Church man who had taken a first-class degree in 1889, taught in and was elected a fellow of St John's in 1892. There were other less formal arrangements for the tutorial teaching of undergraduates of other Colleges that were not recorded in the Calendar, and also a group of 'coaches', unattached to the University, who were often employed privately by the keener honours candidates.

It was probably more by chance than design that much of the research undertaken in the college laboratories was in the newly distinguishable field of physical chemistry, although the teaching of both physics and chemistry had always been a function of college laboratories. Many of the tutors in natural science took both subjects as their responsibility although the level of physics attained by those who were really chemists was probably never very advanced. The attitude of chemists towards physics had often been uneasy. Early in the 18th century Newton had regarded the two subjects as closely linked, as he set out in his *Opticks*, but Herman Boerhaave in Leiden soon emphasised the autonomy of chemistry. Both views were to be found throughout the century but early in the 19th the French chemists, notably Berthollet, Gay-Lussac, and Dumas, embraced physical ideas with enthusiasm. The two generalisations that had the greatest influence on chemical thought were Dulong and Petit's discovery of the (approximate) constancy of atomic specific heats and Mitscherlich's law of isomorphism, since these threw light on the vexed problem of atomic weights. Avogadro's (and Ampère's) hypothesis was the missed opportunity of the time. Most British chemists in the 19th century held to the view that theirs was an autonomous subject and they regarded physical evidence on the constitution of molecules with scepticism. Daubeny was one of those most inclined to accept such evidence. Meanwhile, the laws of solution were slowly evolving; Faraday's laws of electrolysis, Daniell and Hittorf's interpretation of them in terms of transport numbers, Raoult's study of the depression of freezing points and elevation of boiling points, and Arrhenius's interpretation of the properties of solutions of electrolytes in terms of ionic dissociation, all led to the view that physical chemistry was essentially the chemistry of solutions. This view was confirmed in 1887 when Wilhelm Ostwald, supported by J.H. van 't Hoff and Svante Arrhenius, founded the journal, *Zeitschrift für physikalische Chemie*, which, although not entirely given to this field, emphasised the chemistry of solutions. This work came not from Germany, where organic chemistry reigned supreme, but from its neighbouring countries. This field, and the related development of chemical thermodynamics by Willard Gibbs some years earlier, largely passed by the Oxford chemists who started to develop their own ideas on the physical aspects of chemistry, and to follow it in the college laboratories.[117] In Oxford, as on the Continent, it was at the centre that organic chemistry dominated the scene and physical chemistry grew up in the outlying areas.

The leader of the college-based chemists was undoubtedly Vernon Harcourt. He continued to work in the Museum with Brodie for a few years after his election to the Lee's Readership at Christ Church until he could refit the laboratory there for chemical teaching and research in 1865. His dissatisfaction with the mere accumulation of chemical facts was later made clear in his Presidential Address[118] to the Chemistry Section of the British Association in 1875, but it is not obvious what led him first to enquire into the rates of chemical reactions. It was not then a recognised branch of chemistry although some chemists and physicists on the Continent had made a start in the field, notably Ludwig Wilhelmy who had studied the rate of inversion of sucrose in 1850, and Marcellin Berthelot who studied the hydrolysis of esters in the early 1860s. None of this work, however, made much impression on the chemical community at the time. Later, Leopold Pfaundler at Innsbruck began to speculate that such rates were governed by the frequency of collisions between rapidly moving molecules. Most inorganic reactions, with which Vernon Harcourt was more familiar than with organic, proceed so rapidly that the limiting speeds are determined more by the rates of mixing of the reactants than by chemical constraints, but he found one reaction that went at a measurable rate, the oxidation of oxalic acid by potassium permanganate and sulfuric acid.[119] This proved an unfortunate choice since the reaction occurs in more than one step and the rate is affected by the manganous sulfate produced as the reaction proceeds. The partial disentangling of these complications was beyond Vernon Harcourt's limited mathematical skill and he called in William Esson, the mathematical fellow at Merton, to help him out.[xv] They were able to deduce that "the total amount of change occurring at any moment is proportional to the quantity of substance remaining". In an appendix, Esson used for the first time the phrase "velocity of chemical change" for what we now write as $-d[A]/dt$, where $[A]$ is the amount of one of the reactants. He showed that this velocity is usually proportional to $[A]$. They were more successful with their second reaction, the oxidation of hydrogen iodide to iodine by hydrogen peroxide, for it proved to be simpler and was first-order in the concentration of the iodide. They published their first paper[120] on this reaction in 1867, but then continued their work, off and on, for nearly thirty years before publishing a massive paper of 78 pages as the Royal Society's Bakerian Lecture[121] of 1895. This included a discussion of the effect of temperature on the reaction rate but they ignored the important work of Svante Arrhenius on

---

[xv] It is not known why he did not call on Charles Lutwidge Dodgson (or 'Lewis Carroll'), the mathematical tutor of his own college, with whom he was friendly, and, it has been suggested, who later took Harcourt as the model for the White Knight in *Alice through the looking-glass*. But Dodgson took less interest in science than Esson.

this subject in 1889, preferring to express the rate as a power of the absolute temperature. The University backed both horses when, in 1908, it conferred honorary doctorates on both Harcourt and Arrhenius at the celebrations of the jubilee of the Museum.

Berthollet had discussed many years earlier the effect of amounts of substance on the position of chemical equilibrium, but without any quantitative conclusions. Wilhelmy had, however, deduced essentially the same expression as Harcourt and Esson for the dependence of rate on amount of substance, although it was some years before his work was noticed. The year before Harcourt and Esson published their paper, another pair, also one a chemist and one a mathematician, Peter Waage and Cato Guldberg, had published what we now call this 'law of mass action'. But their paper, in the Norwegian language, and published in a Norwegian journal, was certainly unknown to the two men in Oxford.[122] The importance of Harcourt and Esson's work was, therefore, mainly local since others were also advancing the subject abroad. Within Oxford, however, it became the source of inspiration for a wider range of research on the kinetics of chemical reactions.

Such work soon took on a second aspect. Harold Dixon[104] came up to Christ Church from Westminster School to study classics, but he spent much of his time on athletics and was in danger of being sent down from the university when Vernon Harcourt aroused in him a latent interest in science. He took first-class honours in chemistry in 1875 and, at his tutor's suggestion, took up a study of gaseous explosions the next year. Dixon's father's house, and others, had been badly damaged by a gunpowder explosion in London in 1874, when Harcourt had been called in to investigate.[xvi] This explosion may have prompted the proposed line of research. Dixon worked for some years in the Christ Church laboratory and his first discovery was that a mixture of carbon monoxide and oxygen could not be ignited by an electric spark if the components were rigorously dried. He then proceeded to show that some of Bunsen's work in this field was not correct, and, in particular, that the speed of explosive waves was generally much higher than Bunsen had reported. In 1879 he was appointed Millard Lecturer at Trinity College, and in 1881 Bedford Lecturer at Balliol. These neighbouring colleges had each received benefactions that enabled them to do more to support teaching in the sciences, and were now collaborating. Dixon's work was carried in the Balliol laboratory and, when he needed more space for the pipes along which his explosive waves travelled, he apparently set them up in passages under the college hall itself. He worked also

---

[xvi] Harcourt was appointed a Gas Referee in 1872 and served for 45 years. The three Referees had to oversee, on behalf of the consumers, the testing of house gas for its calorific and illuminating power. It was in conjunction with this work that he devised a means of removing carbon disulfide by catalytic hydrogenation, and designed his photometer based on a pentane burner that served as the British standard of intensity of illumination until the 1930s. Tyndall, one of his colleagues as a Referee, took a poor view of it, describing it a "trumpery thing that has occupied him to the exclusion of better work for the last ten years."[123] His other contribution to applied science was a chloroform inhaler that continued in use until safer anaesthetics were introduced.

in the Daubeny Laboratory at Magdalen under circumstances that he described, as follows:

> I think that it was in 1881 that my friend C[harles] J[ohn] F[rancis] Yule, Fellow of Magdalen, showed me round the Magdalen Laboratory, and explained the electric chronograph with which he had been doing physiological experiments. The instrument seemed to be adaptable to my wants, so as he was no longer using it, I asked him to let me try it on my gases. This he allowed me to do, and I soon found that it would accurately record the very rapid rate of explosions. Accordingly I got leave from the College to work in the Laboratory, and I worked there for about two years. In 1883 or 1884 I bought the chronograph from Magdalen and moved it to Balliol.[124]

Yule's chronograph had been built in 1874, but Frederick John Smith, [Jervis-Smith from 1897], the University Lecturer in Mechanics, developed a superior instrument based on similar principles in 1888. His 'tram chronograph' had a trolley whose speed was regulated reproducibly by a freely falling weight; this he exhibited in the summer of 1889 at the Royal Society's Conversazione. Dixon soon adapted this also to his work on explosions.[125] He remained at Balliol until he succeeded Henry Roscoe at Manchester in 1887, in what was then the leading chemistry department in the country, but returned briefly to Oxford to be the President of the Chemistry Section at the meeting there of the British Association in 1894. His address[126] on that occasion was a strong plea for Oxford not simply to load its students with the facts of chemistry but to teach them how to undertake research in which, he said, the University was still weak. His address must surely have been seen as a criticism of the regime of Odling and his colleagues at the Museum. Dixon's later work on gaseous reactions and explosions established him as the leader of this field; he gave the Bakerian Lecture to the Royal Society in 1893 and was President of the Chemical Society from 1909 to 1911.

During Dixon's time at Balliol he was assisted for two years by Herbert Brereton Baker,[127] a Brackenbury Scholar of the College. Baker became fascinated by Dixon's observation that reactions could be inhibited by intensive drying, and continued research in this direction during the twenty years he spent teaching at Dulwich College. For his work there he was elected to the Royal Society in 1902, a rare instance of the work of a schoolmaster being so recognised. His enthusiasm exceeded his caution, and his explanation of some of his results has not stood the test of time, but he did ample to earn himself the nickname of 'Dry Baker'. He was appointed to the Dr Lee's Readership at Christ Church in 1902, when Vernon Harcourt retired, and at once set about reorganising the haphazard arrangements for college instruction. He arranged that the Christ Church laboratory should specialise in teaching inorganic chemistry and that its course should be supplemented by one in quantitative analysis at Magdalen by John Job Manley. In 1888 Manley had succeeded John Harris, Daubeny's assistant who had moved with him from the old Museum.

Manley had had no university education but became a competent instructor and, in his research, an authority on the chemical balance. The University rewarded him with an honorary MA in 1903 and he obtained a D.Sc. in 1931. Under Baker's scheme, practical physical chemistry was taught in the Balliol-Trinity laboratory and, from 1907, also in the laboratory that Jesus College established that year. Organic chemistry was taught in the Museum by Odling's Demonstrators and in Queen's College, which had set up a laboratory in 1900. Perhaps the most valuable aspect of Baker's reform was that all College lectures and laboratory classes were open to all students, rather than being available without charge to those of the college concerned, and those of other colleges with which it had negotiated collaborative agreements, or who had agreed to pay the small fees.

It was the Balliol-Trinity Laboratory that became eventually the strongest of the colleges' research institutions in chemistry.[128] After a faltering start under Henry Smith and Benjamin Brodie little was done until 1874 when Balliol decided to build a new hall, and to incorporate into the design a laboratory for the natural sciences. This was built below ground level to the north-east of the hall and was ready for use in 1879. Trinity then proposed a collaboration in which Balliol provided the laboratory and Trinity the lecturer; Harold Dixon. When he left for Manchester in 1887 the Colleges each elected a lecturer. Balliol chose Sir John Conroy, another pupil of Vernon Harcourt at Christ Church, and tutor at Keble College, who became their Bedford Lecturer. His degree was in chemistry but he was basically a physicist who did a little research in optics but who taught both physics and chemistry. In 1888 Trinity chose David Henry Nagel, a Scot, a graduate first of Aberdeen, and later of Oxford as a pupil of Dixon's at Trinity. Like Dixon he was an excellent teacher – H.G.J. Moseley[xvii] was one of his pupils – but he did no research. Trinity supplemented the Balliol laboratory with one of its own in 1897 and a breach was made in the dividing wall between the Colleges to enable the combined laboratories to be run as one institution. Conroy was originally in charge but after his early death in 1900 Nagel took over, with Harold Hartley as his assistant.[130] Hartley had been taught as a boy at Dulwich by H.B. Baker, and, after some earlier work with Veley in the Museum and at Magdalen, became one of the few early Oxford chemists to take up the subject of the physical chemistry of solutions, although this field was no longer the dominant theme on the Continent that it had been twenty years earlier. It had been first introduced into Oxford when Magdalen College sent Duncan Randolph Wilson, later a college lecturer, to work with

---

[xvii] Henry Moseley[129] came up to Trinity College in 1906 to read physics but his studies spread also to chemistry, as is shown by his presence in a group photograph of the Alembic Club in 1907 and by his presence in the Balliol-Trinity laboratory (Figure 4.4). When he applied to work with Rutherford at Manchester in 1910 his referees claimed for him a considerable knowledge of chemistry although he demurred. His discovery in the Electrical Laboratory at Oxford of the linear relation between the atomic number of an element and the square root of the frequency of the main line of its X-ray spectrum was a landmark in physics and a solution to a whole set of chemical anomalies over atomic weights and the identification of the rare earths. His early death at Gallipoli in August 1915 was, as Rutherford said, "a striking example of the misuse of scientific talent".

Chemistry Comes of Age: The 19th Century 119

Figure 4.4 Henry Moseley in the Balliol-Trinity Laboratory in 1910. The photograph was provided by the Museum of the History of Science.

Ostwald and to bring back a set of notes on the course at Leipzig. From 1900 to 1904 these notes formed the basis of a course in practical physical chemistry that ran for a few years at Magdalen, before this subject moved to the Balliol-Trinity laboratory under Baker's rationalisation. The course that Hartley then ran there was again heavily biased towards the properties of solutions, apparently ignoring the research work carried out by other Oxford chemists on reaction kinetics.[128] But Vernon Harcourt never accepted the idea that solutions of electrolytes were ionised and there was probably little meeting of minds between the two groups. The Oxford chemists never became deeply involved in the field of chemical thermodynamics, and still less in its union with kinetic theory, or the field of statistical thermodynamics. This was a major subject for discussion at the British Association when it met in Oxford in 1894. This meeting was attended by the leader of the field, Ludwig Boltzmann, but the only local contributor to the ensuing correspondence in *Nature* was Robert Edward Baynes, Dr Lee's Reader in Physics at Christ Church. Oxford chemists played only a minor role in the meeting. Hartley became an efficient head of laboratory and the foundations that he laid were to support research achievements after the first World War that were to surpass anything that Oxford had previously achieved in physical chemistry.

The laboratory at Queen's College was run at first by George Bernard Cronshaw and Allan Frederick Walden, neither of whom contributed much to research in organic chemistry. That flourished at Queen's only after Chattaway returned there from St Bartholomew's Hospital and joined Walden in 1907. In the same year Jesus College set up its own laboratory, a fine three-storied building in Ship Street named after Sir Leoline Jenkins, a seventeenth century Principal and benefactor.[131] The director of this laboratory for almost all of the forty years of its existence was David Leonard Chapman,[132] another pupil of Vernon Harcourt, who later worked also with Dixon at Manchester. With this parentage it is not surprising that he also turned to the field of explosive reactions. He was a better mathematician than most Oxford chemists of his time and studied the gas dynamics as well as the chemistry of the reactions. This work, published in his first paper[133] in 1899, gave his name to the Chapman–Jouguet layer of the gas immediately behind a detonation front, a layer whose properties he deduced from the fact that the detonation 'wave' is one of steady state, minimum speed, and of maximum entropy. The theory was developed independently by Émile Jouguet six years later. The second phenomenon to which his name is given is the approximate theory for the distribution of electric charge outside a planar electrode; this is called the Gouy-Chapman layer, since in this case another Frenchman, Georges Gouy, was the first in the field.[134] Most of his work, however, was in the now established Oxford field of the rates of chemical reactions. His first work in the Jesus laboratory was on the rate of the homogeneous decomposition of ozone to oxygen, the first gas reaction shown to be truly homogeneous when carried out in a vessel of sufficiently large volume/surface ratio. He soon returned, however, to a photochemical reaction that he had studied first at Manchester. This was the notoriously tricky reaction of hydrogen and chlorine. Chapman, sometimes with his wife, Muriel, resolved many of the problems that caused the rate to be so irreproducible, but the mechanism was not understood until after the suggestion of Walther Nernst in 1918 that the first step was the dissociation of chlorine into atoms and the subsequent propagation of a chain of reactions between an atom and a molecule. This was a mechanism that Chapman was slow to accept.

The related but, in the end, largely independent efforts of Vernon Harcourt, Dixon, Baker, and Chapman amounted to the creation of a 'school' of reaction kinetics that went on to even greater things as the twentieth century advanced. For the first time chemistry in Oxford had become a major player on the international scene. Nothing that had been done at the centre, in the Museum, or the Old Chemistry Department as it was later called, was of the same calibre. In a lecture to mark the jubilee of the erection of the Museum in 1908 Vernon Harcourt put it more tactfully:

> In the Museum, sometimes one study advances more rapidly and attracts more pupils, sometimes another—but when "the tired waves vainly breaking, seem here no painful inch to gain", the College Laboratories, with young and eager teachers are prepared to supply any deficiency that may occur.[135]

The college laboratories grew haphazardly to meet the needs of increasing numbers of students of chemistry, for this was the most popular science subject throughout the era. The laboratories were not planned centrally and they cooperated little until the reforms instigated by Baker in 1904. As with many features of Oxford, their independence and lack of coordination was both a strength, in that it allowed the growth of a research school in a distinctive branch of physical chemistry, and also a weakness later in the twentieth century when matters had to be done differently. The autonomy of the colleges allowed them to choose their own tutors without reference to the professor who controlled only the appointment of the demonstrators. Some men held joint appointments but there was no formal mechanism for cooperation. It was therefore difficult to plan a coherent course of instruction and, still less, a coherent school of research. What collaboration that did occur depended little on the discussions of the Faculty Board but more on personal contacts and the ability of the tutors to move their research work from one college laboratory to another, or to or from the Old Chemistry Department, as seemed best to them, and as could be agreed in man-to-man negotiations. The enthusiasm generated by the college laboratories was reflected in the examination results, for the colleges that were most successful in getting first-class degrees for their candidates between 1880 and 1914 were those with laboratories: Christ Church, Balliol, and Magdalen, with New College (but with no laboratory), Trinity, and Jesus also doing well.[136]

The college laboratories were, however, not the only centres of research in subjects at least cognate with chemistry that grew up outside the Museum laboratory. Early in the 19th century chemistry was closely allied with mineralogy and geology; these links were later maintained through Nevil Story Maskelyne, who taught the earliest course of practical chemistry when he was the assistant to Daubeny. He had been a reluctant candidate for the Aldrichian chair of chemistry in 1855, and became the Professor of Mineralogy the next year after the death of Buckland. Two years later, needing a larger salary on which to get married, he accepted the post of Keeper of Minerals at the British Museum, but he retained the Oxford chair, with the approval of the University. In 1894 Magdalen College took it over and it became another Waynflete professorship, whereupon Story Maskelyne, satisfied that its future was secure,[xviii] soon relinquished it, to be succeeded by Henry Alexander Miers, a former pupil and a colleague at the British Museum.[137] Miers's forte was crystallography and this interest was soon to give birth in Oxford to another field, closely related to chemistry, that was to flourish in the next century. He had been a classical and mathematical scholar at Trinity College who, after graduation in 1881, spent a year partly with Story Maskelyne, partly at Cambridge and partly with Paul von Groth in Strassburg (as it was then called). He joined the British Museum in 1882, where he designed new goniometers and applied them to study the growth of crystals from solution. When he took up the Oxford chair he found that there were no proper laboratories for teaching or for mineralogical and crystallographic research. He set about establishing these with enthusiasm and soon attracted a few keen students to work with him on crystal growth.[138] These included Ernald Hartley[139] (no relation of Harold

---

[xviii] In this assumption he was not wholly correct; for, unusually for such a chair, the Waynflete Professorship of Mineralogy was abolished in 1941.

Hartley) from Christ Church and Thomas Barker[140] who had graduated in chemistry in 1904 with a distinction in mineralogy and crystallography. In 1908 Barker worked for a year in the laboratory of E.F. Federov in St Petersburg whose aim was to find a systematic way of classifying crystals. Barker shared this aim and his work led nearly fifty years later to the *Barker Index of Crystals* of which the principal editor was Mary Winnearls Porter, who had also been one of Miers's pupils, and was the first woman to receive an Oxford degree when she was awarded her B.Sc., a research degree, in 1920. Miers left Oxford in 1908 to become Principal of London University, and then later Vice-Chancellor of Manchester. The crystallography group that he had established veered more to chemistry than to geology. This bias proved its worth to the University many years after he had left when the treatment of the subject in Oxford was transformed by the introduction of the determination of the molecular structure of crystals by means of X-ray diffraction patterns.

The eighth Earl of Berkeley[141] started life in the Navy but resigned his commission at the age of 22 to study chemistry at the Royal College of Science and then crystallography with Henry Miers at Oxford. He bought a house at Foxcombe on Boars Hill, to the west of Oxford. This was outside of the limit of $1\frac{1}{2}$ miles from Carfax within which undergraduates were required to live. The University would not waive the rule so Berkeley withdrew his application to become an undergraduate at Christ Church, and had no further formal connection with the University. In 1897 he had worked for a short time in the old Balliol laboratory ('Brodie's laboratory' in the Salvin Building) but the next year he built his own laboratory at Foxcombe and all his later work was done there, much of it with Ernald Hartley, whom he had met in his short time at Christ Church and most of whose research was to be carried out as a partner with Berkeley in their work at Foxcombe on osmotic pressure. This pressure is one of the classical measures of the activity of a solute, whether an electrolyte or not. It was, however, notoriously difficult to determine accurately since the high pressures, generally into the tens or sometimes hundreds of atmospheres, had to be carried by a membrane that was permeable to water, but impermeable to the solute. Berkeley and Hartley overcame these difficulties and their work in this field remains a classic piece of research.[142] For simple solutions, such as cane sugar in water, there is now nothing more to be done, and the field has been abandoned by modern physical chemists, but the technique still has its uses in the polymer field where the pressures are much lower and membranes easier to construct. Hartley meanwhile had embarked on a more systematic study of the ferrocyanides and their derivatives; this led to some worthwhile new chemistry but no new membranes suitable for osmotic pressure measurements. After service in the First World War, in which he collaborated with Bertram Lambert, he returned to Oxford where he became a demonstrator in the Old Chemistry Department.

Thus by the early years of the twentieth century chemistry had come of age in Oxford; it was mature, but the best days of its adulthood were still to come. It was the most popular of the natural sciences, and its research effort, although scattered over several college laboratories, was at last of international importance. Whereas the state of physics in Oxford was now clearly well behind that of Cambridge, chemistry had held its own. Both universities were to be strengthened by recruits from the flourishing departments at Manchester: Pope to Cambridge in 1908, and Perkin to Oxford on Odling's retirement in 1912. Several weaknesses remained; there was still only the one professor, at a time when other universities were moving to two or more. Odling had not established a school of research, although long before he retired he had become a respected member of the University and of the community of British chemists. There was still the unresolved separation between the professor with his demonstrators, and the autonomous colleges with their fellows and tutors and, in some cases, their own laboratories. The level of teaching seems to have been patchy but improving, but those who most prided themselves on their teaching often felt no obligation to do any research. There was little connection between the formal teaching of chemistry and cognate fields (with the exception of mineralogy and crystallography), so that Oxford chemists were often sent out from the University with a scanty knowledge of mathematics, physics or of any biological science. In the twentieth century the University was to resolve most but not all of these difficulties.

REFERENCES

1. R. A. Austen-Leigh, *Eton College Register, 1698–1752*, Spottiswoode Ballantyne, Eton, 1927, p. 2; J. Foster, *Alumni Oxoniensis, 1715–1886*, Foster, London, 1887, p. 13.
2. T. H. Levere, *Ambix*, 1981, **28**, 61.
3. *OU Archives*, Will of George Aldrich, W. P.y25β.4.
4. University Statutes; Trusts, Schedule B, *OU Gazette*, 25 July 2001, **131**, Suppl. 1 to No. 4593, p. 1311.
5. John Kidd (1775–1851), *ODNB*; *DSB*; A. G. Gibson, *The Radcliffe Infirmary*, London, 1926, pp. 109–12.
6. J. Kidd, *Outlines of Mineralogy*, Oxford and London, 1809, 2 vol., see the dedication in vol. 1.
7. *OU Archives*, Register of Convocation, 1802–1809, pp. 81–3.
8. [J. Kidd], *Syllabus of a course of lectures on chemistry at Oxford in 1808*, Oxford, 1808.
9. J. Kidd, *A Geological Essay on the imperfect evidence in support of a theory of the Earth—*, Oxford, 1815.
10. N. A. Rupke, 'Oxford's scientific awakening and the role of geology' in M. G. Brock and M. C. Curthoys, ed., *History of the University of Oxford, Nineteenth-Century, Part 1*, Oxford University Press, 1997, v. 6, pp. 543–62; M. Curthoys, *Oxford Magazine*, 2007, No. 269, p. 8.
11. R. Walker, *A letter to the Rev. the Vice-Chancellor, on improvements in the present examination statutes and in the studies of the University of Oxford*, Oxford, 1848.
12. W. T. Brande, *Dissertation third: Exhibiting a general view of chemical philosophy, from the early ages to the end of the eighteenth century*, n.p.,

n.d.; idem., *ibid.*, *Supplement to the 4th and 5th Editions of Encyclopaedia Britannica*, Edinburgh, 1824, v. 3, pp. 1–79; J. Kidd, *An answer to a charge against the English Universities contained in the supplement to Edinburgh Encyclopædia*, Oxford, 1818. Kidd's syllabus was reprinted by P. W. Kent, *Some scientists in the life of Christ Church, Oxford*, printed privately, Oxford, 2001, App. 4.
13. See Chapter 3, and A. V. Simcock, *The Ashmolean Museum and Oxford Science, 1683–1983*, Museum of the History of Science, Oxford, 1984, p. 5; J. Kidd, *Philos. Trans. Roy. Soc.*, 1814, **104**, 508.
14. A. Garden, *Annals Philos.*, 1820, **15**, 74.
15. W. T. Brande, *Quart. J. Sci.*, 1820, **8**, 287.
16. J. Kidd, *Philos. Trans. Roy. Soc.*, 1821, **111**, 209.
17. M. Faraday, *Philos. Trans. Roy. Soc.*, 1826, **116**, 140.
18. J. Kidd, *On the adaptation of external nature to the physical condition of man*, London, 1833.
19. C. G. B. Daubeny (1795–1867), *ODNB*; *DSB*; J. Phillips, *Proc. Ashmolean Soc.*, 1868, **5** [new series], 8; *Proc. Roy. Soc.*, 1868–1869, **17**, lxxiv; R. T. Günther, *A history of the Daubeny Laboratory, Magdalen College, Oxford—*, printed privately, Oxford, 1904, Appendix D; F. S. Taylor, *Ann. Sci.*, 1952, **8**, 82; D. R. Oldroyd and D. W. Hutchings, *Notes Rec. Roy. Soc.*, 1979, **33**, 217; A. P. Willsher, *Daubeny and the development of the chemical school in Oxford*, Chemistry Part II thesis, Oxford, 1961; N. I. Miller, *Chemistry for gentlemen: Charles Daubeny and the role of chemical education at Oxford, 1800–1867*, Chemistry Part II thesis, Oxford, 1986.
20. C. Daubeny, *A description of active and extinct volcanos* [sic]—, London, 1826.
21. C. Daubeny, *Rep. Brit. Assoc. Adv. Sci.*, 1831, **1**, 92; *ibid.*, 1836, **6**, *Trans. Sec.*, 81.
22. C. Daubeny, *Inaugural lecture on the study of chemistry, read at the Ashmolean Museum, November 2, 1822*, Oxford, 1823.
23. C. Daubeny, *On the application of chemistry to the arts*, a lecture at Bristol, 6 January, 1823, printed with ref. 22; *Outline of a course of lectures on chemistry, to be delivered at the Museum, Oxford in the winter 1822, 23*, printed with ref. 22.
24. J. B. Morrell, *Ambix*, 1972, **19**, 1.
25. C. R. Hill, *Catalogue 1: Chemical Apparatus*, History of Science Museum, Oxford, 1971, pp. 69–75.
26. Günther, ref. 19, Appendix E.
27. C. Daubeny, *An introduction to the atomic theory: comprising a sketch of the opinions entertained by the most distinguished ancient and modern philosophers with respect to the constitution of matter*, London, 1831; *Supplement—*, London, 1840; 2nd edn, London, 1850.
28. Oldroyd and Hutchings, ref. 19; W. H. Brock, 'Coming and Going: the fitful career of James Cumming', Chap. 6 of M. D. Archer and C. D. Haley, eds, *The 1702 Chair of Chemistry at Cambridge: Transformation and Change*, Cambridge University Press, 2005.

29. V. Morton, *Oxford Rebels: the life and friends of Nevil Story Maskelyne, 1823–1911, pioneer Oxford scientist, photographer and politician*, Sutton, Gloucester, 1987, p. 74.
30. C. Daubeny, *Philos. Trans. Roy. Soc.*, 1830, **120**, 223; *Rep. Brit. Assoc. Adv. Sci.*, 1836, **6**, 1; *Miscellanies: being a collection of Memoirs and Essays on Scientific and Literary Subjects, published at various times, vol. 1, Experimental and Geological Memoirs*, Oxford, 1867, pp. 195–205.
31. C. Daubeny, *Philos. Trans. Roy. Soc.* 1836, **126**, 149; *Rep. Brit. Assoc. Adv. Sci.*, 1831, **1**, 436; 1837, **7**, 505; *Miscellanies*, ref. 30, pp. 3–92.
32. C. Daubeny, *Brief remarks on the correlation of the natural sciences: drawn up with reference to the scheme for the extension and better management of the studies of the University, now in agitation*, Oxford, 1848; Brock, ref. 28.
33. [B. Jowett and A. P. Stanley], *Suggestions for an improvement of the Examination Statute*, Oxford, 1848.
34. C. Babbage, *Reflections on the decline of science in England, and on some of its causes*, London, 1830.
35. J. B. Morrell and A. Thackray, *Gentlemen of Science: Early years of the British Association for the Advancement of Science*, Oxford University Press, 1981; *Gentlemen of Science: Early Correspondence of the British Association for the Advancement of Science*, Royal Historical Society, London, 1984.
36. R. Murchison, *Rep. Brit. Assoc. Adv. Sci.*, 1832, **2**, 108.
37. C. Daubeny, *Rep. Brit. Assoc. Adv. Sci.*, 1836, **6**, xxi; *Miscellanies*, ref. 30, v. 3, pp. 136–179.
38. E. Turner, *Philos. Trans. Roy. Soc.*, 1833, **123**, 523; A. Thackray, *John Dalton: Critical assessments of his life and science*, Harvard University Press, Cambridge MA, 1972; T. L. Alborn, *Ann. Sci*, 1989, **46**, 437.
39. C. Daubeny, *Rep. Brit. Assoc. Adv. Soc.*, 1856, **26**, lvi; *Miscellanies*, ref. 30, v. 2, pp. 136–79.
40. W. R. Ward, 'From the Tractarians to the Executive Commission, 1845–1854', in Brock and Curthoys, ref. 10, pp. 306–336.
41. [Anon.], *A hundred years of the Ashmolean Club, 1869–1969*, printed privately, Oxford, n.d.
42. B. Powell, *The present state and future prospects of Mathematical and Physical Studies in the University of Oxford—*, Oxford, 1832; [R. Walker], *A few words in favour of Professor Powell and the Sciences, as connected with certain educational remarks: by Philomath. Oxoniensis*, Oxford, 1832.
43. Taylor, ref. 19.
44. H. M. Vernon and K. D. Vernon, *A history of the Oxford Museum*, Clarendon Press, Oxford, 1909, pp. 40–1.
45. J. B. Atlay, *Sir Henry Wentworth Acland, Bart., K.C.B., F.R.S., Regius Professor of Medicine in the University of Oxford. A Memoir*, Smith Elder, London, 1903, p. 33.
46. Goldwin Smith, *Reminiscences*, Macmillan, New York, 1910, pp. 98–115.
47. *Report of Her Majesty's Commissioners appointed to inquire into the State, Discipline, Studies, and Revenues of the University and Colleges of the*

University of Oxford: together with the evidence and appendix, C.1452, HMSO, London, 1852, pp. 60–67, 79, 104, 124, evidence, p. 13; G. M. Young and W. D. Hancock, ed., *English Historical Documents*, Eyre and Spottiswoode, London, 1956, v. 12, pt. 1, pp. 869–891.
48. 'New Examination Statutes', in *Papers relating to the Proceedings of the University*, Oxford, 1850, Bodleian Library, GA Oxf. c.66, item 120; *New examination statutes. Abstracts of their principal provisions, with a catalogue of books, either expressly mentioned, or treating of the subjects required*, Oxford, 1851.
49. *Report—*, ref. 47, p. 79.
50. J. Shorter, 'Chemistry at Cambridge under George Downing Liveing', in Archer and Haley, ref. 28, chap. 7.
51. R. Fox and G. Gooday, ed. *Physics in Oxford, 1839–1939: laboratory, learning, and college life*, Oxford University Press, 2005, p. 33.
52. C. Daubeny, 'On the importance of the study of chemistry as a branch of education for all classes', in E. R. Lankester, ed., *Science and Education*, London, 1855.
53. [A. Comte], *The Positive Philosophy of Auguste Comte*, trans. Harriet Martineau, London, 1853, v. 1, pp. 298–9; W. H. Brock, ed., *The Atomic Debates: Brodie and the rejection of the atomic theory*, Leicester University Press, 1967, Appendix, 'Comte, Williamson and Brodie', pp. 145–52.
54. C. Daubeny, *Can physical science obtain a home in an English University? An inquiry, suggested by some recent remarks contained in a late number* [June, 1853] *of the Quarterly Review*, Oxford, 1853; *Miscellanies*, ref. 30, v. 2, pp. 41–59.
55. Brock, ref. 28.
56. Kent, ref. 12, p. 30.
57. C. Daubeny, 'A new examination statute at Oxford', *The Times*, 24 February 1864, p. 12, col. 5.
58. 'Oxford University Museum', item 161 of *Papers—*, ref. 48.
59. Atlay, ref. 44, chap. 8; R. Fox, 'The University Museum and Oxford Science, 1850–1880', in Brock and Curtoys, ref. 10, pp. 641–693; M. Crosland, *Ann. Sci.*, 2003, **60**, 399.
60. *OU Archives*, Register of Convocation, 1854–1871, p. 31.
61. Vernon and Vernon, ref. 44, p. 59.
62. C. Daubeny, *Brief remarks on the statute De Lectoribus Publicis to be submitted to Congregation on Tuesday, June 17*, Oxford, 1851.
63. J. B. Morrell, *John Phillips and the business of Victorian science*, Ashgate, Aldershot, 2005.
64. Benjamin Collins Brodie (1817–1880), *ODNB*; *DSB*; W. H. Brock, *HYLE*, 2002, **8**, 49–54.
65. *OU Archives*, Register of Convocation, 1854–1871, p. 82; Morton, ref. 29, pp. 118–120.
66. B. C. B[rodie], *Syllabus of a Course of Lectures on Chemistry, to be delivered in the Museum in Michaelmas and Hilary Terms, 1864 and 1865,*

on Tuesdays and Saturdays, at one o'clock, Oxford, 1864; W. Odling and W. W. Fisher, 'From the Waynflete Professor of Chemistry', in *Statements of the Needs of the University, being Replies to—the Vice-Chancellor*, Oxford, 1902, pp. 44–47.

67. J. Howarth, '"Oxford for Arts': the Natural Sciences, 1880–1914', in Brock and Curthoys, ref. 10, Part 2, 2000, v. 7, pp. 457–497.
68. A. J. Rocke, *Chemical atomism in the nineteenth century: from Dalton to Cannizzaro*, Ohio State University Press, Columbus OH, 1984, chap. 7–9; *The Quiet Revolution: Hermann Kolbe and the science of organic chemistry*, University of California Press, Berkeley CA, 1993, chap. 4.
69. William Odling (1829–1921), *ODNB*; *DSB*; H. B. D[ixon], *Proc. Roy. Soc. A*, 1921–1922, **100**, i; J. E. Marsh, *J. Chem. Soc.*, 1921, **119**, 553; J. L. Thornton and A. Wiles, *Ann. Sci.*, 1956, **12**, 288; J. Freeman, *The life and times of William Odling, 1829–1921, Waynflete Professor of Chemistry*, Chemistry Part II/B.Sc. thesis, Oxford, 1962/1963.
70. A. Laurent, *Chemical Method: Notation, Classification, & Nomenclature*, trans. W. Odling, Cavendish Society, London, 1855, 'Translator's Preface'; reprinted in facsimile in D. Knight, ed., *The Development of Chemistry, 1789–1914*, Routledge, London, 1998, v. 7.
71. B. C. Brodie, *J. Chem. Soc.*, 1851, **3**, 405–11.
72. T. W. M. Smith, 'The Balliol-Trinity Laboratories', in J. M. Prest, *Balliol Studies*, Leopard's Head Press, London, 1982, p. 192.
73. W. Odling, *A Manual of Chemistry: Descriptive and Theoretical, Part 1*, London, 1861, 'Advertisement' and 'General Considerations'.
74. Letter from B. C. Brodie to A. Kekulé, 27 May 1860, cited by Rocke, ref. 68, 1984, p. 292.
75. B. C. Brodie, *J. Chem. Soc.*, 1864, **17**, 281.
76. W. Odling, *Rep. Brit. Assoc. Adv. Sci.*, 1864, **34**, 21; *Chemical News*, 1864, **10**, 149.
77. B. C. Brodie, *Philos. Trans. Roy. Soc.*, 1866, **140**, 781; Royal Society Referees' Reports, 1866–1869, item 6.30, 23/10/1866.
78. *Chemical News*, 1867, **15**, 295, **16**, 2; *The Laboratory*, 1867, **1**, 78; B. C. Brodie, *Ideal Chemistry, a lecture*, Macmillan, London, 1880.
79. Hill, ref. 25, p. 63.
80. T. Anderson, *Rep. Brit. Assoc. Adv. Sci., Trans. Sec.*, 1867, **37**, 28; A. Kekulé, *The Laboratory*, 1867, **1**, 333.
81. B. C. Brodie, *Philos. Trans. Roy. Soc.*, 1877, **167**, 35.
82. R. T. Günther, *The Daubeny Laboratory Register, 1904–1915—*, published privately, Oxford, 1916, pp. 141–42.
83. *Chemical News*, 1866, **13**, 106, 252.
84. Henry George Madan (1838–1901), H. A. M[iers], *J. Chem. Soc.*, 1902, 628.
85. Thomas Heathcote Gerald Wyndham (1842–1876), [Anon.], *J. Chem. Soc.*, 1877, 510.
86. Walter William Fisher (1842–1920), B. Dyer, *J. Chem. Soc.*, 1920, 456.
87. Augustus George Vernon Harcourt (1834–1919), *ODNB*; *DSB*; H. B. D[ixon], *Proc. Roy. Soc. A*, 1920, **97**, vii; J. Shorter, *J. Chem. Educ.* 1980,

57, 411; M. C. King, *ibid.*, 1983, **60**, 177; *Ambix*, **31**, 1984, 16; A. Vernon Harcourt and H. G. Madan, *Exercises in Practical Chemistry*, Oxford University Press, 1869.
88. J. Howarth, *Eng. Hist. Rev.*, 1987, **102**, 334.
89. W. Odling and J. E. Marsh, *Rep. Brit. Assoc. Adv. Sci.*, 1887, **57**, 646.
90. James Ernest Marsh (1860–1938), F. Soddy, *Obit. Not. Fell. Roy. Soc.*, 1936–1938, **2**, 549; J. A. Gardner, *J. Chem. Soc.*, 1938, 1130.
91. J. Donnelly, *Brit. J. Hist. Sci.*, 1991, **24**, 3.
92. A. J. Engel, *From Clergyman to Don: the rise of the academic profession in nineteenth-century Oxford*, Oxford University Press, 1983, pp. 217–230.
93. Third Report of the Royal Commission on Scientific Instruction and the Advancement of Science, C.868, HMSO, London, 1873, §§ 29–31.
94. William Frederick Donkin (1845–1888), [Anon.], *J. Chem. Soc.*, 1889, 292.
95. Freeman, ref. 69, p. 69.
96. *The Oxford University Commission*, C.2868, HMSO, London, 1881, Evidence of Odling, pp. 171–176, Qu. 2916–3006, esp. 2958.
97. OU Calendars; Fox, ref. 59; J. Howarth, ref. 67.
98. 'Report of the Waynflete Professor of Chemistry' in *Report of University Institutions for the Year 1888 [—1902]*, Oxford University Press.
99. John Watts (1843–1933), F. S[oddy], *J. Chem. Soc.*, 1933, 1652.
100. Victor Herbert Veley (1856–1933), J. A. Gardner, *Obit. Not. Fell. Roy. Soc.*, 1932–1935, **1**, 229; *J. Chem. Soc.*, 1934, 570.
101. M. Howorth, *Atomic Transformation: the greatest discovery ever made. From the Memoirs of Frederick Soddy*, New World Publications, London, 1953, pp. 20–22.
102. R. W. Clark, *Tizard*, Methuen, London, 1965, pp. 10–12.
103. Allan Frederick Walden (1871–1956), H.R. Ing, *Proc. Chem. Soc.*, 1957, 237.
104. Harold Baily Dixon (1852–1930), *DSB*; H.B.B[aker] and W.B[one], *Proc. Roy. Soc. A*, 1932, **134**, i; *J. Chem. Soc*, 1931, 3349.
105. A. M. A. H. Rogers, *Degrees by degrees: the story of the admission of Oxford women students to membership of the University*, Oxford University Press, 1938; P. Adams, *Somerville for women: an Oxford College, 1879–1993*, Oxford University Press, 1996.
106. John Addeyman Gardner (1867?–1946), [Anon.], *J. Roy. Inst. Chem.*, 1946, 196.
107. Bertram Lambert (1881–1963), E. J. Bowen, *Proc. Chem. Soc.*, 1964, 94; C. S. G. P[hillips], *Postmaster* (Merton College Magazine), 1963, 14.
108. Nevil Vincent Sidgwick (1873–1952), *ODNB*; *DSB*; H. T. Tizard, *Obit. Not. Fell. Roy. Soc.*, 1954, **9**, 237.
109. Tom Sydney Moore (1881–1966), R. Augustin, *Chem. Brit.*, 1967, 494.
110. T. S. Moore and T. F. Winmill, *J. Chem. Soc.*, 1912, **101**, 1635.
111. Frederick Daniel Chattaway (1860–1944), G. R. Clemo, *Obit. Not. Fell. Roy. Soc.*, 1942–1944, **4**, 713.

112. J. C. Smith, *The development of organic chemistry in Oxford, Parts 1 and 2*, unpublished mimeographed manuscript, no date, Radcliffe Science Library, Oxford, 1933d.302.
113. W. H. Perkin, *Trans. Chem. Soc.*, 1915, **107**, 557.
114. J. Perry, *Nature*, 1903, **69**, 209.
115. P. J. Rowlinson, *Oxf. Rev. Educ.*, 1983, **9**, 133.
116. E. J. B[owen], *The Alembic Club, The first fifty years*, 1967, Manuscript 167, History of Science Museum, Oxford; George William Fraser Holroyd (1871–1934), R. H. Pickard, *J. Chem. Soc.*, 1935, 407.
117. K. J. Laidler, *Arch. Hist. Exact Sci.*, 1985, **32**, 43; 1988, **38**, 197; A. V. Simcock, 'Laboratories and Physics in Oxford Colleges, 1848–1947', chap. 4 of Fox and Gooday, ref. 51.
118. A. Vernon Harcourt, *Rep. Brit. Assoc. Adv. Sci.*, 1875, **45**, 32.
119. A. Vernon Harcourt, *Chem. News.*, 1864, **10**, 171; A. Vernon Harcourt and W. Esson, *Philos. Trans. Roy. Soc.*, 1866, **156**, 193.
120. A. Vernon Harcourt and W. Esson, *Philos. Trans. Roy. Soc.*, 1867, **157**, 117.
121. A. Vernon Harcourt and W. Esson, *Philos. Trans. Roy. Soc. A*, 1895, **186**, 817.
122. [Anon.], *The Law of Mass Action: a centenary volume, 1864–1964*, Universitetsforlaget, Oslo, 1964.
123. A. S. Eve and C. H. Creasey, *Life and Work of John Tyndall*, Macmillan, London, 1945, pp. 257–8.
124. Günther, ref. 19, p. 24.
125. F. J. Smith, *Proc. Roy. Soc. A*, 1889, **45**, 451; F. Jervis-Smith, *The tram chronograph*, published privately, Oxford, 1904; A. V. Simcock, ref. 117, pp. 195–9.
126. H. B. Dixon, *Rep. Brit. Assoc. Adv. Sci.*, 1894, **64**, 594.
127. Herbert Brereton Baker (1862–1935), *ODNB*; J. F. Thorpe, *Obit. Not. Roy. Soc.* 1932–1935, **1**, 523; J. C. Philip, *J. Chem. Soc.*, 1935, 893.
128. E. J. Bowen, *Notes Rec. Roy. Soc.*, 1970, **25**, 227; Smith, ref. 72.
129. Henry Gwyn Jeffreys Moseley (1887–1915), *ODNB*; *DSB*; J. L. Heilbron, *H.G.J. Moseley: The Life and Letters of an English Physicist, 1887–1915*, University of California Press, 1974.
130. Harold Brewer Hartley (1878–1972), *ODNB*; A. G. Ogston, *Biog. Mem. Roy. Soc.*, 1973, **19**, 349.
131. D. A. Long, *Jesus College Record.*, 1989, 17.
132. David Leonard Chapman (1869–1958), *ODNB*; *DSB*; E. J. Bowen, *Biog. Mem. Fell. Roy. Soc.*, 1958, **4**, 35; D. L. Hammick, *Proc. Chem. Soc.*, 1959, 101.
133. D. L. Chapman, *Philos. Mag.*, 1899, **47**, 90.
134. D. L. Chapman, *Philos. Mag.*, 1913, **25**, 475.
135. Vernon and Vernon, ref. 44; H. Hartley, *Studies in the History of Chemistry*, Oxford University Press, 1971, p. 231.
136. R. T. Günther, ref. 82, pp. 160–1.
137. Henry Alexander Miers (1858–1942), *ODNB*; *DSB*; T. H. Holland, *Obit. Not. Fell. Roy. Soc.*, 1942–1944, **4**, 369.

138. A. Vincent, *Geology and Mineralogy at Oxford, 1860–1986: History and Reminiscence*, published privately, n.p., 1994.
139. Ernald George Justinian Hartley (1875–1947), M. P. Applebey, *J. Chem. Soc.*, 1948, 899.
140. Thomas Vipond Barker (1881–1931), H. A. Miers, *J. Chem. Soc.*, 1931, 3344.
141. Randal Thomas Mowbray Rawdon Berkeley, Earl of Berkeley (1865–1942), *ODNB*; H. Hartley, *Obit. Not. Fell. Roy. Soc.*, 1942, **4**, 167.
142. Earl of Berkeley and E. G. J. Hartley, *Proc. Roy. Soc. A*, 1904, **73**, 436, and later papers.

CHAPTER 5
Research as the Thing: Oxford Chemistry 1912–1939

JACK MORRELL

Division of History and Philosophy of Science, School of Philosophy, University of Leeds, Leeds LS2 9JT

5.1 INTRODUCTION

The transformation of Oxford chemistry between the two world wars was not revealed in undergraduate numbers, which grew slowly from just less than forty to just over forty finalists per year, of whom two to three were women. This chemical cohort was small compared with those in arts subjects and social studies that attracted 80% of undergraduates between the wars: at the end of the 1930s the most popular subject was history, with about 250 finalists per year, followed by Greats, law, PPE (philosophy, politics, economics), English, and modern languages with about 130 on average in each. Within the physical sciences, however, chemistry was the largest final honours school, well ahead of mathematics and physics (about 30 and 20 finalists per year).[1] In the sciences three-year degree courses were the norm at Oxford, except for chemistry: in 1916 a year of research (Part 2) was added to the existing three-year degree course (Part 1). No other British university had a statutory year of research in its undergraduate chemistry degree. The expanded chemistry course at Oxford created problems of extra laboratory accommodation and generated about 160 undergraduates at any one time. Thus the cause of chemistry could be promoted *via* the argument from increased or substantial undergraduate numbers. Compared with chemistry departments elsewhere, Oxford's was big: by 1939 it was the largest in any British university. Between the wars the old adage

---

Chemistry at Oxford: A History from 1600 to 2005
Edited by Robert J.P. Williams, John S. Rowlinson and Allan Chapman
© Royal Society of Chemistry 2009
Published by the Royal Society of Chemistry, www.rsc.org

'Cambridge for science, Oxford for arts' applied to many subjects but not to chemistry, in which Cambridge produced no more than thirty finalists per year.

Chemistry was indeed the science in which Oxford felt decidedly superior to Cambridge, not just in size but also in reputation. In Fenland the subject drifted in the 1930s under William Pope, the senior professor, who was an ill man from 1927. He had overall charge of the laboratories so that when his papacy lost impetus it was not easy to recharge his department as a whole. At Oxford, where chemistry was pursued in a variety of locations, research prospered *via* outstanding individuals and productive research groups. It was increasingly accepted among Oxford's chemical dons that research was not a personal luxury but an institutional imperative as important as teaching and scholarship. Though the idea of research as the thing and the associated development of research groups were long familiar from German examples and newly powerful in the USA, they were not pervasive in Oxford chemistry until the inter-war period. The increased research salience of Oxford's chemists in these years was shown in office-holding in two national chemical societies. The key Oxonians in the Chemical Society of London were Nevil Sidgwick and Robert Robinson. The former was a member of its council for six years, a vice-president for another four years, and president 1935–7. He was also the influential first chairman of the Society's publication committee for six years. Robinson was especially prominent after 1935, being vice-president for four years and then president for two.[2] The Faraday Society, founded in 1902 to promote physical chemistry, enjoyed Sidgwick as president 1932–4 and visited Oxford five times between the wars to conduct thematic meetings that lasted two days and usually attracted leading figures from home and abroad.[3]

Oxford's chemists became more visible in the national and international worlds of science. In 1939 there were in post six Oxford chemists who were FRSs, namely, Cyril Hinshelwood and Robinson, the two professors, and David Chapman, Sidgwick, Edward Bowen, and William Hume-Rothery, who were college fellows. By 1954 another eight of those in post in 1939 had been elected FRS. They were Ronald Bell, Harold Thompson, Wilson Baker, Dorothy Hodgkin, Leslie Sutton, and Dalziel Hammick (all college fellows) and Frederick King and Herbert Powell (University demonstrators). Two Oxford chemists, Robinson and Hinshelwood, occupied the presidency of the Royal Society of London for ten years between 1945 and 1960, thus disturbing the previous monopoly by Cantabrigians. Three of those in post in 1939 were to receive the highest international accolade, a Nobel prize, the recipients being Robinson 1947, Hinshelwood 1956, and Hodgkin 1964. In analysing Oxford chemistry between 1912 and 1939 the main problem is to explain how research prospered at a time when the place of chemistry in undergraduate education remained relatively unchanged.

5.2 THE IMPACT OF PERKIN

The arrival in Oxford of W. H. Perkin junior, as Waynflete professor of chemistry in 1913, was a decisive event in the University's history. For the first time Oxford had head-hunted a provincial university professor of science, with no previous connection with Oxford, who was renowned as a researcher himself, as a research school leader, as a fund-raiser for no fewer than four new laboratories, which he helped to design and plan, and as a disciple of Adolf von Baeyer with whom he had worked in Munich from 1882 to 1886. As the leading British organic chemist of his generation, he was lured aged 52 to Oxford from the University of Manchester, where he had enjoyed 20 years as professor, specifically to boost research. He was so successful at Oxford that his successor, Robert Robinson, was his chemical son. Given Oxford's inbuilt conservatism, this policy of poaching established non-Oxonian researchers from provincial universities, though subsequently so common-place as to be notorious, was not only novel before the First World War but also prospectively disruptive.[4]

It was first implemented through Perkin for several reasons. Firstly, in 1900 the University had shown that it was prepared to outflank, though not replace or dismiss, an embarrassingly weak science professor when Townsend, first occupant of the new Wykeham chair of experimental physics, was appointed to remedy the failings of Clifton, the professor of experimental philosophy. Secondly, the bringing together of the college chemistry laboratories in 1904 under H. B. Baker's scheme was a public confession of the deficiencies of the University's chemistry laboratory run by William Odling, the aging Waynflete professor of chemistry. Thirdly, in 1908–9 the Hebdomadal Council, the chief administrative body of the University, was so alarmed by Odling's performance that it took the unprecedented step of consulting metropolitan and provincial university professors about how to organise a subject spread between one University laboratory and five college ones. Though the replies were not always harmonious, most of the advisors were trenchant about Oxford's weakness in research. From Manchester Perkin's professorial colleague, H. B. Dixon, advocated the Mancunian model of two professors, a centralised department, and the aim of advanced research. From the Central Technical College, London, H. E. Armstrong denounced as a national disgrace Oxford's failure to count as 'a school of chemical research'. From University College, London, William Ramsay (a Nobel prize winner) agreed with Dixon that a professor of organic chemistry was required. Having consulted Oxford's college fellows, who praised the Balliol-Trinity Laboratory to the skies, the Council decided that when Odling retired his chair should be filled by an organic chemist, that at some time in the future a chair for inorganic chemistry should be established from the ancient endowment of Dr Lee, and that a new University laboratory, which would work harmoniously with the college laboratories, was required. Meanwhile, Edward Poulton, professor of entomology, had been urging that Oxford should give more attention to being an imperial and international university *via* research and in 1909 Lord Curzon, Chancellor of the University, called on it to give greater attention to research and to postgraduate training.[5]

The result of such sustained pressure was that when Odling retired in 1912, aged 83, the University found £15,000 and a site for a new chemistry laboratory by 19 November 1912. Meanwhile the advert for the Waynflete chair of chemistry stressed that the occupant would be required to 'assist the pursuit of knowledge and contribute to the advancement of it'. Three electors to the chair had strong reasons for seeing Perkin as the man to fulfill these desiderata. Herbert Warren, president of Magdalen, was a strong supporter of science and research at Oxford and presumably wanted what Odling had not provided, *i.e.* impressive research from the occupant of the chair, the salary of which cost Magdalen £900 pa. Dixon, an Oxonian who had taught in the Balliol-Trinity laboratory, had pleaded publicly in 1894 for greater encouragement of chemical research at Oxford and of teaching the art of research; having been a colleague of Perkin at Manchester for twenty years, he had seen from close quarters what Perkin could do as a researcher and research school leader. Pope, who had replaced Liveing at Cambridge in 1908, was a moderniser who had no previous Oxbridge connection; and, as a professor of applied chemistry at the Manchester Municipal School of Technology from 1901, he knew Perkin's work at the University not least through collaboration in research with him. Presumably Pope thought Perkin capable of doing at Oxford what he was doing at Cambridge. Two other electors would have approved the choice of Perkin. Townsend, a new broom in Oxford's physics, presumably welcomed the prospect of another vigorous sweeper. Henry Miers had been the first full-time professor of mineralogy at Oxford from 1895 to 1908. As a Waynflete professor, productive researcher, outstanding University administrator, and teacher of crystallography to chemistry undergraduates, he was presumably embarrassed by the failings of Odling's regime and saw Perkin as a redeemer. Driven by such views, the electors offered the chair to Perkin on 23 November, he accepted it formally on 10 December 1912, and became automatically a fellow of Magdalen.[6]

Perkin accepted the Waynflete chair, which he occupied until his death, on the understanding that his research time would not be interrupted by low-level teaching or shredded in other ways, his aim being to build up a school of research as rapidly as possible. For Perkin research, not teaching, was the basis of the reputation of any university school of science. Unlike Odling, he set a grand personal example, appearing and working in the laboratory six or seven hours a day and maintaining a flow of important publications. In 1916 his famous paper on the alkaloids cryptopine and protopine, which ran to 214 pages, took up a whole number of the *Journal of the Chemical Society*. Even his undergraduate lectures were research-orientated, with their stress on original sources, and intended mainly for research students; but his top priorities remained research at the laboratory bench and building up a research school. Well aware of the danger of a British university professor becoming 'an academic fossil and unproductive,' he emphasised the centrality of 'output of research', in phraseology that Odling would have found alien; and soon Oxford's publications in organic chemistry began to surge at Manchester's expense.[7]

As soon as Perkin appeared in Oxford in January 1913, he attended to the question of the promised new laboratory, which was to be devoted entirely to organic chemistry. He convinced the University that the architect should be Paul Waterhouse, the trusted designer of the 1909 Morley Laboratories at

Manchester: he would be cheaper than anyone else because the various measurements made for the Morley Laboratories could be used in Oxford. Perkin even invited Oxonians to Manchester in February 1913 to gain a 'pretty accurate idea of what the suggested accommodation really amounts to.' As before, Perkin collaborated with his architect on the design. The new Oxford laboratory copied the Morley Laboratories in several respects: in height, in the professorial eyrie on the top floor from which Perkin could look down into the main teaching laboratories, and in the public-lavatory style of the walls, lined with hard-glazed brown and cream bricks, which Perkin regarded as a necessity, not a luxury.

As at Manchester, Perkin was able to attract external funding for the new building. Strongly supported by Warren, by summer 1915 he had secured from C. W. Dyson Perrins, after whom the Oxford laboratory was named, a total of £30,000 to supplement the £15,000 available from the University. Dyson Perrins provided £5,000 for the building, £5,000 for equipment, and a permanent endowment of £20,000, the interest on which was to be used for promoting research under the direction of the Waynflete professor only for as long as organic chemistry remained the chief subject of his chair. The Dyson Perrins Trust, used to cover maintenance costs, was of great value to Perkin because it made his laboratory less dependent for income on the University and on laboratory fees. Dyson Perrins, an Oxford graduate in law and partner in the famous Lea and Perrins Worcester Sauce firm, was a very wealthy philanthropist who was presumably impressed by Perkin's general approach and in particular by his research on the alkaloid berberine, of which Dyson Perrins's father had established the empirical formula in 1862.

The Dyson Perrins Laboratory, Figure 5.1, was built in two stages, the first ending with the completion of the central block and western wing in 1916, which enabled Perkin and his co-workers to move from the old chemistry building. The second stage was completed in 1922 at a final cost almost twice that of the first, which had required £20,000 for the building and £5,000 for apparatus. Perkin pressurised the University to such an extent that much of the total cost of the second stage (£45,600) was met by it, the rest (£7,500) being secured by Perkin as gifts (mainly £5,000 from the British Dyestuffs Corporation and £1,000 from Barclays Bank). The University resorted to two financial devices. First, in a remarkable act of financial obfuscation it borrowed the capital of £20,000 from the Dyson Perrins Trust with his permission and undertook to replace it with annual instalments of £1,000 payable for twenty years, without affecting at all the interest payable to the laboratory! This loan doubled, at one stroke, the University's total borrowing. Second, it borrowed £19,000 from the Special Reserve Fund. It was forced into these desperate financial moves by Perkin who was keen to alleviate the congestion in his palace of chemistry produced by the post-war increase in research students. His methods provoked dismay and wrath. He gave the University optimistic estimates about his power to raise money from chemical firms, persuaded the University to commit itself to the ambitious scheme for extending the Dyson Perrins Laboratory, and when external money failed to materialise he openly

Figure 5.1 The front of the Dyson Perrins laboratory for organic chemistry. (Reproduced with permission of the University Estates Directorate.)

advocated that the University should have no financial reserves, borrow what it could, and 'in forme pauperis' trust to appeals. With relentless opportunism and entrepreneurial wiliness bordering on deception, Perkin exploited the administrative confusion between the University Chest and the Hebdomadal Council.[8]

There is some truth in the old adage that the First World War was above all a chemists' war. Most departments of chemistry in British universities were employed in dealing with the scientific problems posed by the war, beginning in 1915 when the Ministry of Munitions was established. By 1917 most academic chemists were engaged in war-related research. The major foci, which often involved devising reactions that could be scaled up with a good yield, were explosives, poisonous gases, drugs, and dyestuffs, in all of which university chemists enabled the chemical industry in Britain to rival that in Germany by late 1918. Four examples must suffice. At Cambridge, Pope showed that under specific conditions mustard gas could be made by direct synthesis from sulphur monochloride and ethylene. By the end of the war British output of this gas was thirty times that of the Germans at one thirtieth of the cost. At Imperial College, London, the three professors of chemistry, J. F. Thorpe, H. B. Baker, and J. C. Philip gave much attention to the production and detection of poison gases and defences against them. Sometimes this research had unintended consequences. One of their students, Christopher Ingold, gassed himself with phosgene (carbon oxychloride), from which parlous state he was rescued by another research student, Hilda Usherwood, whom he later married. At Manchester, Weizmann, who had collaborated there with Perkin on industrial research, devised a successful pilot plant for making acetone, essential for making cordite, by fermenting starch. Even at the small University of St Andrews, Irvine led successful research on the synthesis of three drugs, dulcitol, inulin, and novocaine.[9]

At Oxford, Perkin quickly turned his department to war work, first in the Old Chemistry Laboratory and from spring 1916 in the new Dyson Perrins Laboratory. His deeds matched his words in his second address as president of the Chemical Society 1913–15. In March 1915 he attacked the organic chemical industry in Britain as decadent and lamented that the grip and power of the Germans were total. His remedies were that at a time of national emergency university departments should not only devote themselves to applied research but also cooperate with industry, however distasteful these practices might be to some Oxonian academics.[10]

The war work that Perkin led or supervised took two forms. The first was research for government bodies such as the Department of Explosive Supplies, the Air Board, and the Chemical Warfare Committee. One major focus was devising new commercial preparations of substances useful in the war, such as acetone from alcohol, toluene from petroleum, a substitute for acetone in aeroplane 'dope', and a non-inflammable rubber coating for airships. Another focus was improving existing processes such as the sulphonation of benzene and the manufacture of synthetic phenol and picric acid. A minor focus was research on mustard gas, absorbents for carbon monoxide, and the preparation of arsenic compounds. Such work was carried out mainly by groups, varying in size from just two people to a maximum of nine. Thus, Perkin introduced into Oxford the practice of team research under a supervisor. His teams drew on college fellows, laboratory demonstrators, half a dozen undergraduates, and the odd graduate, but on only few outsiders. It was ironic that the most enduring researcher in Perkin's various groups was Sidgwick, a renowned individualist who scorned Perkin's style of organic chemistry as limited and passé. The largest team, which worked on acetone, included for a time around nine people of varied status and experience: Perkin as professor was leader, supported by Sidgwick (a demonstrator and college fellow), Bertram Lambert (a demonstrator), and E. G. J. Hartley, who for 14 years had been full-time assistant in Lord Berkeley's laboratory at Foxcombe Hall near Oxford. Supporting them in turn were a couple of undergraduates and F. C. Hall who was appointed in 1916 by Perkin as steward of the Dyson Perrins Laboratory where he was an effective substitute demonstrator and a skilled quantitative analyst. One temporary member of the acetone team was Hammick, an Oxford graduate who was previously a schoolmaster. Later in the war Hammick became a captain in the Chemical Warfare Brigade of the Royal Engineers. In contrast to the sizable teams organised by Perkin, two of his demonstrators worked in a more traditional way. Frederick Chattaway focused solo on drugs, antiseptics, and the preparation of offensive organic compounds; while James Marsh, fellow of Merton, attacked problems suggested by the War Office such as the cheap production of pyrogallol, absorbents for poisonous gases, and drugs for military hospitals.[11]

The second form of war work that Perkin organised at Oxford was 'industrial research', a term and practice that he introduced into Oxford in 1916. As a leading supporter of the Department of Scientific and Industrial Research, established by government in 1916, Perkin drew on his previous experience at Manchester by promoting in the Dyson Perrins Laboratory research of value to industry carried out there by workers paid by fine-chemical firms. W. J. Bush and Company and Boake, Roberts and Company seconded two and three people respectively but British Dyes supported seven while the war lasted. British Dyes had been formed in 1915 and backed by the government to strengthen and rationalise the industry. By early 1916 Perkin was chairman of British Dyes' advisory council and supervised its research department, which was composed of colonies of organic chemists working in several universities including Oxford. The members of his Oxford colony wrote confidential research reports for their employer and helped it to take out patents. Of the leading or enduring members of the colony, only F. A. Mason and T. V. Barker, demonstrator in mineralogy and research fellow of Brasenose, were Oxonians. The remainder were outsiders. Edward Hope was lured from Manchester University where he had researched under Perkin. Joseph Kenyon, formerly a teacher at Blackburn Technical College, came from the Medical Research Council where he was a research chemist; George Clemo from a Cornish school; and William Kermack, a recent Aberdeen graduate, from the RAF's experimental station at Martlesham Heath. Only Hope found permanent employment at Oxford, the other three becoming professors elsewhere and FRSs. Though Perkin controlled or oversaw the chemical war work done in Oxford, one contribution was made independently. In the Jesus Laboratory Chapman tested many of the proposed inventions sent to the Inventions Committee of which he was a member.[12]

Away from the University and often in the armed forces, Oxford chemists contributed significantly to the war effort through their expertise in applied chemistry. One of them, Andrea Angel, tutor at Christ Church, made the ultimate sacrifice. During the war he was chief chemist at the Silvertown works of Brunner Mond that from early 1915 were used for purifying trinitrotoluene. Early in 1917 the works were wrecked by huge explosion which left 69 people dead. While trying to extinguish a fire and rescue his injured staff, Angel was killed and for his bravery was posthumously awarded the Albert medal. Three other Oxonian chemists were concerned with explosives. H. J. George, who graduated in 1914 from Jesus, served at Gallipoli and subsequently worked in the Explosives Department of the Ministry of Munitions where one of his colleagues was M. P. Applebey, who from 1914 had been lecturer at Jesus. The war prevented one young chemist from taking up his scholarship at Balliol: instead in 1916 Cyril Hinshelwood joined the Department of Explosives, Queensferry Royal Ordnance Factory, where he became deputy chief chemist. There he studied the slow decomposition of solid explosives by measuring the gases evolved. This research stimulated his interest in chemical kinetics, which became the main focus of his subsequent work. He returned to Oxford early in 1919, took the shortened chemistry degree that allowed servicemen to graduate in five terms, was immediately elected a research fellow of Balliol, and next year was elected a tutorial fellow of Trinity. The war thus provided a research field and an accelerated Oxford career for the future Nobel laureate.[13]

Three Oxford chemist-soldiers contributed greatly to gas-warfare research. Harold Hartley, fellow of Balliol, became in 1915 chemical adviser in France to the Third Army, interpreting intelligence about German capacity to wage chemical warfare, planning chemical warfare operations, and devising means of defence against gas attacks. By 1917 he had been promoted to assistant director of gas services at the Army's general headquarters and to the rank of lt-colonel. From 1918 to spring 1919 he was in charge of the chemical warfare department of the Ministry of Munitions with the rank of brigadier-general. As a soldier Hartley was awarded the MC, the OBE, and the CBE. After the war he led a government enquiry that inspected German chemical factories concerned with chemical warfare. The resulting report led Parliament to promote by legislation the British chemical industry. Hartley's initiation into military chemical research proved so agreeable that on his return to Oxford he styled himself General Hartley, an appellation that provoked a caustic comment from Hinshelwood:

When the war was over
General Hartley
Returned to civil life again,
Partly.

The more modest Bertram Lambert and E. G. J. Hartley worked on respirators, sometimes collaboratively. The former devised in 1915 a new type of anti-gas respirator which in modified form became by 1916 the British service gas mask. He ended the war as a captain, doing chemical research with the British Expeditionary Force. He received not just an OBE but also a huge reward of £12,500 from the War Inventions Board. It seems that E. G. J. Hartley served in France as a chemical adviser to the First Army and from 1917 was posted to Gas Services (Home) Research in London where he worked on respirators.[14]

Two Oxford chemists, both pupils of Hartley, showed versatility by working as experimental researchers for the Royal Flying Corps. R. B. Bourdillon, fellow of University College, worked on bomb sights and flying in clouds, first at the Corps' Flying School at Upavon and then at the Armament Experimental Unit at Orfordness. H. R. Raikes, who had graduated in 1914, volunteered for military service in France where he was severely wounded in 1915. When recovered he joined the Corps, at Upavon and Orfordness, working on bombing techniques and rising to be chief experimental officer and a major. Both men were early recipients of the Air Force Cross. No such honour was conferred on E. J. Bowen, another Hartley pupil, who spent one year at Balliol and then as a volunteer went to France in 1917, serving as a lieutenant in a howitzer unit.[15]

In spite of a widespread feeling that controversial legislation should not be introduced during the war behind the backs of Oxonian soldiers, two important changes were adopted. In 1916 the extended chemistry degree was approved, with the new doctorate of philosophy (DPhil) following the next year.

The former was mainly the brain-child of Perkin for whom research at the bench, building up a research school, and what he called output of research were top priorities. He had long believed that the normal three-year BSc in British universities was tediously stereotyped and glaringly deficient in that the undergraduates were not required to do any research. His early experience at Oxford quickly convinced him that in order to master organic chemistry students needed to spend more time in the laboratory. The outbreak of war in 1914 and the government's scheme of 1915 for the organisation of scientific and industrial research provided Perkin with an opportunity to lead a campaign to add a fourth year of research (Part 2) to the existing three-year degree course (Part 1) in chemistry. He convinced his colleagues that research training was essential even if employers of graduate chemists were currently indifferent to it. He stressed that research experience was necessary for all honours chemistry undergraduates and that it should not be viewed as a luxury restricted to those destined for a first-class degree. His own contribution to war research gave credibility to his claims that research chemists would produce rapid changes in some parts of the chemical industry and that in future there would be a greater demand for research chemists. At a time of national crisis the new examination regulations in chemistry, which required each candidate to present records of experimental investigations, were quickly and unanimously approved by the University in May 1916 on 'national grounds', even though there was no research year in any other undergraduate degree course in science. Practical questions, such as the extra expense of an extra year for chemistry undergraduates and problems of extra laboratory accommodation, seem to have been ignored. For decades the Part 2 research year in the undergraduate chemistry course was unique: no other undergraduate degree course in science at Oxford and no other chemistry degree course elsewhere in Britain required a year's supervised research for a classified honours degree. Effectively operative from 1919, the Part 2 soon became a valuable method of training and a means of contributing to the advancement of knowledge. It was popular with undergraduates, about three quarters of whom took it. The others were rewarded with the strange distinction of an honours degree without a class.[16]

Before the war the bulk of Americans, and many Australians and Canadians, who wished to acquire a doctorate went to German universities. Comparatively few Americans came to Oxford for this purpose because there was no postgraduate degree that was sufficiently attractive to them. As Lord Curzon had stressed in 1909, Oxford needed a larger cohort of advanced students who could be attracted by greater encouragement and rewards. The Oxford DPhil degree was introduced to induce graduates to come to Oxford from universities in Britain, the dominions, and especially the USA, after the end of the war. It was also a move to encourage the claims and practice of research at Oxford. Not surprisingly Perkin was the leading spokesman for the scientists, ably abetted by Poulton. Perkin saw the DPhil as a means of persuading good graduates to do advanced work for two to three years under a supervisor and of then rewarding them, thus remedying what he regarded as a conspicuous defect of Oxford. He believed that the two existing research degrees in science at Oxford did not promote high-level supervised research: the one- or two-year BSc, introduced in 1895, was not sufficiently testing so it did not entice the best graduates; whereas the DSc, given for published solo work done over several years, was too exclusive, not a regular career option, and lacked the element of research training. The DPhil was approved in May 1917 without serious opposition, though there was considerable debate about its relation to other research degrees available at Oxford.[17] One consequence of its introduction was neglected. Postgraduate work in a collegiate university, with an elaborate and hallowed system of undergraduate teaching based in colleges, was likely to be an execrescence. Large-scale expansion of post-graduate supervision could only be met properly by appointing more professors and readers, and by increasing the teaching burden of college tutors. Thus the introduction of the DPhil was bound to exacerbate, eventually but not immediately, the old thorny problem of the relation between the University and the colleges, because the latter lacked facilities for postgraduates, a deficiency that endured for decades. The introduction of the DPhil and the Part 2 put pressure on the colleges in another way. From their perspective it had been possible to view specialised research as a boorish German notion and postgraduate supervision as a Yankee device for inserting into a patrician university plebeians who could not help themselves. From 1917 it was no longer easy for colleges and their tutorial fellows in chemistry to ignore or deprecate specialist research and the related supervision of research carried out by final-year undergraduates and advanced students. The colleges had to face a future, created for them mainly by Perkin, in which the traditional aim of liberal education *via* chemistry was supplemented by the new one of training specialist chemists.

5.3 THE CONTRIBUTIONS OF THE COLLEGES

The considerable input of the colleges into chemistry was shown by the number of tutorial fellows in the subject, which was greater than that in any other science. In the 1920s there were fifteen in the men's colleges but none in the women's. By 1939 there were eighteen, with new tutorial fellowships in three colleges, Keble who elected Parkes in 1930, Somerville with Hodgkin in 1935, and St Edmund Hall with Irving in 1938. All three were chemists who acted as physical science tutors. Those colleges without a tutorial fellow in chemistry employed a college lecturer to tutor their undergraduates or farmed them out to a suitable demonstrator or to a tutorial fellow in another college. A few colleges with a tutorial fellow also employed on short-term contracts a lecturer to teach aspects of chemistry which the fellow felt unable to cover adequately. In the late 1930s both Balliol and Magdalen, who had physical chemists as fellows (Bell and Sutton), appointed King as lecturer in organic chemistry. It was not unknown for a tutor to send his pupils to a fellow of another college for specialised teaching for a term. Thus both Sutton and Kenneth Hutchison were sent by Sidgwick at Lincoln and Hammick at Oriel to study traditional organic chemistry under Hope at Magdalen. At Queen's both E. P. Abraham and E. F. Caldin were dispatched by their tutor, Wilson Baker, another traditional organic chemist, for tutorials with J. H. Wolfenden, a physical chemist and fellow of Exeter.[18]

The colleges were universities in miniature in which undergraduates reading different subjects studied and lived together. They had a corporate life and ethos different from those of a university department or a hall of residence. Informally they gave a general education. Formally they provided the tutorial at which an undergraduate, often solo but sometimes paired, met his or her college tutor for an hour or so every week in term time. The basic aim of the tutorial was to compel undergraduates to read, think, write, and argue. The standard procedure was that the tutor set an essay title, gave a reading list, and a week later the undergraduate returned with an essay or some prepared work which was read out by its author and criticised by the tutor. The tutorial system developed the abilities to find out from reading, to write coherent prose, and to defend one's position while accepting criticism. It encouraged the cult of oral and dialectical confidence and enlarged the powers of the mind, but was not designed to impart factual information or technique. In chemistry it was supplemented by lectures, at which attendance was optional, and by laboratory work at which attendance was expected. For undergraduate chemists there were three sources of teaching: the college that organised tutorials, the Physical Sciences Faculty Board that was responsible for the lecture programme, and the University and college laboratories that offered laboratory work under the aegis of this board. This arrangement meant that college chemistry tutors had to dovetail their tutorials, in terms of timing and content, with the lectures and laboratory sessions to avoid overlap and to give overall coherence. This complicated organisational problem was not faced by their arts colleagues whose jobs were simpler and less onerous in that they gave tutorials and directed the reading of their charges. This difference in the teaching responsibilities of arts and chemistry college fellows was reflected in their daily lives: the chemists split their time between their laboratory and their college; arts tutors spent a greater proportion of their working time in college and were therefore more likely than the chemists to contribute to college life.

Stephen Leacock's famous sketch of an Oxford tutorial depicted it as a social, smoky, and educative occasion: 'what an Oxford tutor does is to get a little group of students together and smoke at them. Men who have been systematically smoked at for four years turn into ripe scholars . . . . A well-smoked man speaks and writes English with a grace that can be acquired in no other way.' At St John's undergraduate chemists were well-smoked by Applebey who required them to expound their essays at the blackboard and to read original research papers not just in English but also in French and German.[19] In practice, of course, tutorials varied according to the tutor. Chemistry sported some well-known tutors as well as eccentrics. Harold Hartley took a great deal of trouble with his Balliol pupils: he never missed a tutorial though he was a busy man. He steered them into jobs, laid great emphasis on exact expression on paper, and guided but did not force his pupils in their Part 2 to work on his favourite topics. In his tutorials he avoided formal instruction: using penetrating questions and remarks, he gained the confidence of his pupils while making them do all the work. He challenged his pupils from the start by making them read and write about original papers and not text books.[20] At Trinity Hinshelwood was held in awe as an all-round tutor: though he was a physical chemist he covered inorganic and organic chemistry. He also set demanding vacation reading on which pupils were questioned. At the end of K. J. Laidler's first term at Oxford, Hinshelwood asked him to read Sommerfeld's *Atombau und Spektrallinien* in the original, though there was an English translation of earlier editions available.[21] At Lincoln Sidgwick made no attempt to be a forcing tutor. He read essays casually and uncritically, but with odd illuminating comments. He discussed topics that seemed irrelevant or non-chemical. He did not require students to attend his own lectures. He offered a liberal scientific education on which the pupil could build as he chose: Sidgwick's trademark was wide-ranging and discursive discussion in an informal atmosphere.[22] At Magdalen Sutton, a pupil of Sidgwick's, continued this non-forcing approach. He set essays on very broad topics and asked his pupils to take them as far as they could. During the tutorial he would read the essay, sometimes asking for clarification, disputing a point, or expressing surprise at a novelty. Sutton gave a good training in working on a subject and getting a grip on it. He encouraged his pupils by giving the impression that they could illuminate him on important questions. Consequently his pupils did not leave his tutorial exhausted through having been told what they should have known: they left with some sense of achievement.[23] This idea that the tutorial involved a creative partnership between tutor and tutee was developed further by Hammick who covered all branches of chemistry at Oriel: several of his published papers developed from discussions of queries raised at tutorials. Hammick preferred discussion on a previously arranged topic instead of submitting an essay. As a tutor he put the onus onto his pupils and gave them their heads, frequently quoting Gibbon's view that the powers of instruction are rarely effective save in those fortunate cases where they are almost superfluous.[24] At St John's Thompson was a driving and demanding tutor who required an essay of 4,000 words minimum each week from each pupil. His tough approach was revealed to the ambitious Fred Dainton at their first tutorial. Dainton's essay was so severely criticised by Thompson that he was told to re-write it for the following week. Though Dainton learned from this experience, less-resilient pupils were crushed by Thompson. After four years of Thompson as tutor, Dainton retaliated by going to Cambridge for his doctorate.[25]

Older tutors in chemistry had some difficulties in maintaining the vaunted advantages of the tutorial system. At Queen's Chattaway did not spoon-feed his pupils and made no attempt to transform second-class men into examination firsts. But his bibliographies were out-of-date; and as a preparative organic chemist, he could not cope with physical chemistry which his pupils studied through lectures and private reading. But they liked him, affectionately called him 'Poisoner', and were interested in his accounts of great chemists he had known.[26] At Jesus, Chapman was an Oxford character, whose eccentricities, but not his tutorial arrangements, impressed his pupils. Chapman did not set essays regularly and gave no individual tuition so he knew little about the academic progress his charges were making. He seems to have given tutorials to four or five students together. Sometimes these were classes on topics not covered in lectures; occasionally an individual student addressed the others on the contents of his essay. Though Chapman directed his pupils to original research papers, it is clear that there was not the regular essay writing on a wide range of topics and the educative discussion that were the norm in other colleges. Other rituals appealed more to Chapman. The most choice, witnessed inadvertently by his pupils, involved golf. In his college laboratory Chapman would rise from his desk, take up a golfing stance, grasp an imaginary club, practice a few swings, and drive an imaginary ball through a window. Finally, with one hand over his eyes, he would peer into the distance to locate his ball.[27]

College tutors in chemistry enjoyed or suffered a responsibility denied to those in all other subjects: they supervised Part 2. For tutors who were not researchers the devising of feasible projects, which were worthwhile but completeable in a year, could be taxing. For tutors active in research Part 2 was a godsend because it was a fertile and steady source of research pupils and of publications. Undergraduates also benefited from Part 2. They learned under supervision how to do research at the bench, how to finish a project, and how to write a thesis embodying their results. They developed practical skills to supplement the literary and dialectical ones they had acquired *via* tutorials. For ambitious and able undergraduates working under stimulating tutors, such as Hartley, Hinshelwood, and Thompson, Part 2 could lead to joint publications, sometimes even before graduation. Generally, Part 2 made college tutors and their charges research-orientated.

The system of college tutorials in chemistry had weaknesses. In comparison with physical chemistry, inorganic chemistry was neglected because many tutors were physical chemists. There was little regular formal instruction in mathematics and physics. These lacunae existed because there was no detailed syllabus for the final degree examination that might have prescribed them. Weak students could be floored by challenging tutors; and at times a few tutors merely coached for success in examinations. Yet overall the combination of tutorials and Part 2 made Oxford's chemistry unique. Other universities had teaching laboratories and effective lecturers but only Oxford and Cambridge offered personal tuition *via* tutorials in the former and supervisions in the latter. In four respects, however, Oxford's procedures were more fruitful. At Cambridge, supervisions were conducted by research students who taught undergraduates in groups. In Fenland there was an examination at the end of each year, whereas at Oxford the absence of an examination at the end of the second year (and for some undergraduates no examination during the first) gave time for the development of a wider outlook by undergraduates. Above all, there was no equivalent at Cambridge or anywhere else to the Oxford

Part 2. It was the unique combination of tutorials and Part 2 that made Oxford graduates in chemistry distinctive: they had been *required* to read, think, write, argue, and do experimental research.

After the First World War the flood of undergraduates combined with inflation led the University to approach the government for an emergency grant. As a quid pro quo the Asquith Commission was appointed and in summer 1920 it took evidence from Oxford's scientists, including Frederick Soddy, who was appointed in April 1919 as Dr Lee's professor, and responsible for inorganic and physical chemistry. An Oxonian who was best known for clarifying the concept of isotopes and for his work with Rutherford on atomic disintegration and radioactivity, Soddy quickly publicised his ambitions for a big school of chemistry based on a new University laboratory for inorganic and physical chemistry costing £100,000, a step on which all heads of science were then agreed. He attacked the five college laboratories as dissipating effort and resources. By late 1919 he had learned that there was to be no new building for him and submitted to the Registrar his first complaint that the University had broken faith with him. Next year in his evidence to the Asquith Commission he called again for a big University laboratory for physical chemistry and launched a vehement attack on the colleges who, he claimed, appointed teachers to the detriment of the University that required people absorbed in experimental research. His call for central University provision for physical chemistry was supported by the Natural Sciences Board, which also acknowledged that such a proposal made the future of the college laboratories vexed, and by Perkin, who regarded the college laboratories as wasteful and unsatisfactory. The only dissidents from this view were representatives of Queen's, Jesus, and Magdalen, who argued that the college laboratories enabled the tutorial system in science to be retained and remedied the deficiencies of the University laboratories. There is no record of Balliol, Trinity and Christ Church deviating publicly in 1920 from the line on the college laboratories taken by Soddy, Perkin, the Natural Sciences Board, and eventually by the Asquith Commission, which suggested that the college laboratories could be used as temporary homes for developing new subjects or exclusively for research.[28]

By autumn 1923 it had become abundantly clear that Soddy, still a young man in his mid-forties, would not acquire a University laboratory for physical chemistry, that the college laboratories would continue to be thorns in his flesh, and that he and most of the college tutors were at loggerheads. In one respect Soddy was unfortunate. When he arrived in 1919, the wily Perkin had already launched a campaign to secure the extension to his Dyson Perrins Laboratory. It was completed in 1922 at an unexpectedly high cost to the University of £39,000. Some members of the University thought that Perkin was a financial blackmailer and suspected that all chemistry professors were racketeers. Once Perkin had acquired his palace of organic chemistry, it was impossible for Soddy to make comparable claims for physical chemistry: having doubled its borrowing, the University was retrenching financially and chemistry had already been generously treated. Furthermore, there was little possibility of Soddy securing external endowment of a new University physical chemistry laboratory as long as the Balliol-Trinity and Jesus laboratories existed. In any case, unlike Perkin, he had taken a stand by 1920 against the dependence of research on commercial, industrial, governmental, and philanthropic support.[29]

Soddy was statutorily responsible for physical and inorganic chemistry throughout the University; but, given the existence of the college laboratories and of independent college fellows, he was not master in his own house. Two of the leading chemistry fellows had little time for Soddy perhaps because they had coveted the Dr Lee's chair. Sidgwick ostensibly disliked and denigrated Soddy, but Hartley was more subtle. He saw to it that Soddy had only one success in placing a protégé in a college fellowship when A. S. Russell, a non-Oxonian who had studied radioactivity under Soddy at Glasgow, was elected Dr Lee's reader in chemistry at Christ Church in 1919 and fellow the next year.[30] In 1921 Hinshelwood, a pupil of Hartley, was elected chemistry fellow at Trinity and became Hartley's partner in directing the Balliol-Trinity laboratory. In the very year in which Soddy became the first Oxford scientist to win a Nobel prize, they immediately arranged for Hinshelwood to have research accommodation in converted lavatories in Trinity, an act which no doubt infuriated Soddy who complained that he had no space and no equipment to teach physical chemistry. By 1922 the Balliol-Trinity Laboratory was a large enterprise: it had five staff, including four college fellows, all of whom were Balliol men (Hartley, Hinshelwood, Raikes, Bowen), twelve rooms, 55 laboratory places, and an annual expenditure of just over £2,000, which made it easily the most costly of the college laboratories (Table 5.1). It received informally from Soddy's departmental grant a subsidy equal to £2 per term for every non-Balliol-Trinity student. All told in 1922 Soddy was supposed to pay to the college laboratories £1,000 that was earmarked in his departmental grant from the University. Particularly furious at his lack of control over the Balliol-Trinity Laboratory, in summer 1922 he ceased to pay them. By the next summer the University decided to pay £600 directly to the Balliol-Trinity Laboratory in 1923-4 for teaching physical chemistry on behalf of the

Table 5.1 College chemistry laboratories: 1922-3.

	Balliol-Trinity	Christ Church	Jesus	Magdalen	Queen's	Total
Undergraduates	94	14	41	23	16	188
Researchers of all kinds	6	2	11	1	9	29
Staff	5	1	2	1	1	9
Annual cost (£)	2,170	560	940	630	650	4,950
Number of rooms	12	4	7	3	3	29
Number of ordinary places	44	23	32	16	13	127
Number of research places	11	5	14	3	10	43

Source: Document of 20 April 1923, OUA, MR/6/3/34.

University. In autumn 1924 the University went further. It began to pay the laboratory fees of all students in the Balliol-Trinity, Jesus, and Queen's laboratories, thus in principle increasing the subsidy to the Balliol-Trinity Laboratory, but Hartley agreed informally not to accept more than £600 per annum.[31]

Soddy felt let down by the University that had not fulfilled its promise to provide him with a new building, had deprived him of his statutory responsibility for physical chemistry leaving him with inorganic chemistry, and in subsidising the college laboratories out of the University grant to his department had in his view legalised robbery. No wonder that by the mid-1920s he had become intransigent, unyielding, obstinate, and aggrieved, characteristics he had not displayed strongly at Aberdeen. Contrary to the views implied or expressed by Hartley, who was devoted to the Balliol-Trinity Laboratory, Soddy was perhaps as much sinned against as sinning.[32] He rightly saw chemistry as anomalous compared with other sciences at Oxford. He continued to regard the college laboratories as an unfortunate legacy from the past and the tutorial system as outmoded. With the insight of a Namier he characterised Oxford chemistry as 'a scattered collection of teachers and laboratories interconnected respectively by a tangled network of personal relationships and private ownerships'. His own experience confirmed for him his 1918 view that ancient universities were paralysed by the past and as resistant as marble.[33]

Not surprisingly Soddy wished to retain control of his shrunken empire especially through appointments but even on that matter found himself at odds with the college tutors. In late 1923 when the Hebdomadal Council launched an enquiry into the teaching of chemistry, Soddy thought that two of his demonstrators, Applebey and Lambert, both college fellows, should either not cooperate with it or resign their demonstratorships. Applebey and Lambert refused to resign and, when Soddy threatened to sack them, they so resented this threat to their positions and income that they secured the support of the Vice-Chancellor who quickly poured emollient on troubled waters after Hartley had told him that Soddy was very unpopular because of his attempt to coerce his demonstrators.[34] In 1927 Lambert was involved in another conflict that this time reached the Privy Council. In 1920 the Natural Sciences Board had acted statutorily in appointing Lambert as Aldrichian demonstrator for seven years in Soddy's department to teach physical and inorganic chemistry, without consulting Soddy. In 1927 the Physical Sciences Board renewed Lambert's appointment, again without consulting Soddy who protested to the Registrar and Vice-Chancellor. Archibald Garrod, the Regius professor of medicine, supported Soddy and told the Vice-Chancellor that the exclusion of Soddy from the Board reflected 'the old struggle between the [college] tutors and the professors.' Refusing to compromise or withdraw, Soddy petitioned the Privy Council against Lambert's appointment, the University formally opposed him, and in June 1928 the Privy Council dismissed his petition.[35]

Secure in the Dyson Perrins Laboratory, Perkin tolerated the continuing existence of the college laboratories, only one of which (Queen's) covered organic chemistry and did so in a style different from his and unthreatening.

Perkin agreed with Soddy about centralising teaching and research in physical chemistry but left him to fight a solitary campaign. The college chemistry fellows, who jealously guarded their statutory independence, resented Soddy's accusations that they could not cover in tutorials the whole of chemistry and that they were feeble researchers. Using their legitimate power in the University's boards, Council, and Congregation, they were able to repel Soddy's attempts at subjecting them and their colleges and to retain the college laboratories. Once Soddy had made his aims clear, they quickly jettisoned their previous view in favour of a University physical chemistry laboratory because they feared Soddy would rule it as a dictator. Only when his retirement was imminent in 1936 did they begin to favour in public a proposal they had liked in 1920 but then quickly dropped.

Two of the college laboratories, Magdalen and Queen's, were closed in 1923 and 1934, respectively, on the initiative of the two colleges. Magdalen had the smallest number of researchers and of bench spaces of the five college laboratories in the early 1920s. Given the recommendation of the Asquith Commission in 1922 against college laboratories, given the extra bench spaces made available by the completion of the Dyson Perrins Laboratory that year, given the opposition of Perkin (a fellow of Magdalen) to college laboratories, and given the dire situation of botany whose professor was also attached to Magdalen, it made sense for the college to do a deal with the University in 1923 to promote botany by converting the chemistry laboratory to botanical purposes.[36]

At Queen's there was no such discussion about how the college could serve the University better: it was simply that Chattaway, the laboratory director, intimated to the college in 1932 that he wished to retire in 1934 aged 74. Chattaway had fulfilled his impeccable pedigree. After graduating at Oxford he had taken his PhD under Bamberger and Baeyer at Munich and subsequently studied with George Bredig at Heidelberg. An active publisher Chattaway specialised in the preparation and purification of organic substances. He did not lead a research school; but he had research collaborators and produced a small cohort of pupils who were indebted to him for experimental ingenuity. He was an exacting teacher who insisted that practical organic chemistry was a clean subject, which could be conducted in a drawing-room in evening dress. For many years he generated a good spirit in his laboratory that attracted undergraduates mainly from Queen's, Keble, and St Edmund Hall. In his final year he was assisted in the laboratory by a notable trio of Queen's men, Parkes, Irving, and Eric James. Chattaway and Queen's gave the University two years notice of his retirement and of the college's desire to use the laboratory, which then had 12 undergraduate and 8 postgraduate bench places, for other purposes.[37] As with the closure of the Magdalen Laboratory, the demise of the Queen's Laboratory involved a college decision in cooperation with the University. That was also the case with Christ Church and Balliol-Trinity whose laboratories were closed in 1941 when the new University Physical Chemistry Laboratory was opened, but not with Jesus whose laboratory was closed in autumn 1947.

At Christ Church from 1920 Russell taught inorganic chemistry to engineering students for a time and continued his own research on radioactivity, especially on the chemistry of protactinium, until the third storey of the laboratory was removed in 1930. The Lee's Laboratory was small compared with that at Jesus that, with seven rooms and almost fifty bench spaces, cost the college almost £1,000 pa. The University paid nothing to Jesus for teaching physical chemistry on its behalf, but it did give Chapman £200 a year as director of the laboratory. Jesus took its chemical responsibilities so seriously that in 1919 it elected H. J. George as research fellow and lecturer in chemistry and subsequently promoted him to a tutorial fellowship. Jesus's teaching and Chapman's research were devoted to physical chemistry, but with an intellectual range less than that of the Balliol-Trinity Laboratory. Towards the end of his long career Chapman's eccentricities made him a figure of fun but in the mid-1920s he and his wife were still pursuing fruitful research in their favourite field of photochemical reactions. In 1926 they pioneered the use of a rotating slotted disc to produce alternating short periods of light and dark, a technique that was subsequently widely used. In spring 1939 George died and Jesus once again showed its commitment to chemistry by quickly appointing a replacement, L. A. Woodward, an Oxonian who had found his research field by studying Raman spectra with Debye at Leipzig. When Chapman retired in 1944 aged 75, Woodward ran the laboratory until it closed in 1947. It was kept alive during the Second World War because it played its part in the British contribution to the atomic bomb project: it was used for research in the separation of isotopes by gaseous diffusion. It survived after the war because it was one of Jesus's characteristic and attractive features that had ensured that one-seventh of all Jesus undergraduates read chemistry. It was closed only after much controversy in Jesus about its fate.[38]

The Balliol-Trinity Laboratory gave the lie to Soddy's accusation that college fellows were not interested in research at the laboratory bench. The accommodation there was neither palatial nor purpose-built, but rich in improvisation. The laboratory was bizarre in that in the 1920s and 1930s it occupied a changing suite of converted cellars, washrooms and lavatories in the two colleges, Figure 5.2. Hinshelwood was the main beneficiary of two conversions, occupying from 1921 to 1929 a former lavatory whose fittings were still visible and from 1929 to 1941 a former bath-house. In these externally squalid buildings, one of which was amazingly described by a German visitor as an Institute, Hinshelwood did much of the research that led to his Nobel Prize. He was content with his accommodation. As a wag commented, "That is the worst of being brought up in a cellar; a converted bathroom is then positive luxury".[39] Balliol and Trinity provided research accommodation not only for their own fellows but for two fellows of other colleges (Bowen, University; Raikes and his successor Wolfenden, Exeter), all Balliol men who helped with teaching. The financing of the laboratory was complicated but there is little doubt that it was subsidised by the two colleges who paid for conversion and running expenses and by the demonstrators. In return for teaching on behalf of the University all honours undergraduates who wished to do practical physical

Figure 5.2 The Balliol/Trinity laboratories, see text. (Reproduced with permission from the President and Fellows of Trinity College, Oxford.)

chemistry, the laboratory received from the early 1920s £600 pa, of which £350 was given to Bowen and £150 to Raikes and then Wolfenden. The University paid Hinshelwood £200 pa as head of the laboratory in the 1930s, but Bell conducted practical classes for nothing.[40] All five of them received much less than if they had been college fellows who were demonstrators in a normal Oxford laboratory such as the Dyson Perrins: they were penalised financially because they were college fellows who taught in a college laboratory peopled predominantly by non-Balliol-Trinity undergraduates.

Space was so tight in the Balliol-Trinity Laboratory that undergraduates and researchers often worked on adjacent bench spaces, which helped to generate a friendly family atmosphere. Facilities were slender. There was a lab-boy, the invaluable James Warrall, but no technical assistance, no secretary, no workshop, and no glass-blowers. Everything possible was home-made in the laboratory. The apparatus was simple and light, being mainly glassware plus electrical equipment such as galvanometers. There was no large, heavy, commercially made apparatus. Bowen was perhaps the most inventive researcher. In his research on photochemistry his first source of light was an old street light from Oxford City lighting department; and in that on fluorescence he used old biscuit tins painted grey and a set of false teeth which Hartley scoffed at but Bowen turned to good use: they did not fluoresce in ultra-violet light whereas the natural ones did. Hinshelwood used immersed electric light bulbs to heat constant-temperature baths. Even undergraduates often designed and assembled apparatus from scratch, which gave knowledge of how it worked as well as a sense of adventure. Such skills were highly prized by Hartley who at a tutors' meeting in Balliol praised one of his pupils for his brilliance because he was the laboratory's best glass-blower.[41] It could be argued that this do-it-yourself approach, which was partly a result of shortage of money, was first-class training in research and developed initiative.

Before the war Nagel who was a non-researcher ran the Balliol-Trinity Laboratory, leaving research to Hartley. But from 1921, when Hinshelwood succeeded Nagel at Trinity and joined forces with Hartley, the new director of the laboratory, research received greater emphasis. Though Hartley had studied in Munich under Willstätter and Groth, his greatest debt was to T. W. Richards of Harvard who became his role model. Richards, with whom Hartley worked for an inspiring fortnight in 1902, was not a profound theorist but an experimentalist who focused his research on the exact determination of chemical constants and the testing of accepted or provisional generalisations. Obsessed with methods of purification, Richards gave to chemistry new standards of precision measurement and as a meticulous director of research encouraged a wide-range of investigations in his laboratory from which issued a constant stream of men trained in his methods and imbued with his ideals. In the Balliol-Trinity Laboratory Hartley tried to follow Richards in all these respects. In the 1920s, for example, his main research was on precision measurements concerning the electrochemistry of non-aqueous solutions. Using standardised procedures in accordance with Richards's approach, by 1930 Hartley and his researchers had tested but not contributed directly to the Debye–Hückel theory of strong electrolytes. Hartley's emphasis on procedures and precision was easily adapted to the Part 2 of the chemistry degree. As the Balliol-Trinity Laboratory techniques were highly standardised, they could be learned and deliver useful results in less than a year. At the same time problems that required more technical development for their solution could be studied by postgraduates and, with a skilful supervisor such as Hinshelwood, by postdoctoral researchers.

Hartley was a busy and efficient man. Through his summer schools in physical chemistry, then neglected in schools, which he ran in the 1920s for school teachers and through his activities for the Science Masters' Association, he directed able students to Oxford and Balliol and away from Cambridge. In the 1920s the Balliol-Trinity Laboratory was very much a Hartley enterprise, the four senior demonstrators being Hartley and three of his former pupils (Hinshelwood; Bowen; Raikes followed by Wolfenden). At Balliol he was succeeded as chemistry fellow by two more pupils, Oliver Gatty and Bell. Hartley left Balliol in 1930 because the war had given him a taste for high command in industrial chemistry, a taste that could not be exercised at Oxford. Instead he pursued his interests in applied science through consultancies and especially a directorship of the Gas, Light, and Coke Company. He was therefore persuaded by Lord Stamp to join the London, Midland, and Scottish railway, then in the throes of internal struggles, to organise its research. As a man of action who after the First World War appeared in Oxford still wearing his army uniform and styled himself General Hartley, he wanted a large and new field of operation. Having acquired an FRS in 1926 and a knighthood in 1928, Hartley probably realised that he could go no further as an academic chemist.[42]

Hartley was adept after the first war in placing his Balliol pupils as fellows in other colleges, his successes being at Exeter with Raikes (1919), at Trinity with Hinshelwood (1921), at University with Bowen (1921), and again at Exeter with Wolfenden (1927). This Hartley mafia was consolidated by Hinshelwood who placed his own Trinity pupils as fellows in other colleges besides ensuring that one of them, J. D. Lambert, replaced him at Trinity in 1938: in 1930 Thompson succeeded Applebey at St John's and in 1939 L. A. K. Staveley followed A. F. Walden at New College. In both cases a physical chemist replaced a general or inorganic chemist. Thus, by 1939 members of the Balliol-Trinity axis, focused on physical chemistry, occupied not just the Dr Lee's chair of chemistry but also six college fellowships. Only Sidgwick at Lincoln rivalled Hartley and Hinshelwood in placing his protégés in college fellowships with T. W. J. Taylor at Brasenose (1920), Hammick at Oriel (1921), Sutton at Magdalen (1936), and Woodward at Jesus (1939).[43]

Like Richards, Hartley did not attempt to pull the whole of his laboratory behind him to work on his pet research field. Instead he encouraged work on kinetics by Hinshelwood, on photochemistry by Bowen, and on acid-base catalysis by Bell. In kinetics Hinshelwood was continuing the Oxford tradition begun by A. G. V. Harcourt. Apart from military service in the First World War, his whole career was spent at Oxford as Balliol undergraduate, Balliol research fellow 1920–1, fellow of Trinity 1921–37, and Dr Lee's professor of chemistry 1937–64. Much of the work on kinetics for which he received a Nobel Prize was done between the wars. This bald statement ignores the key to Hinshelwood's success, which was his fertility in developing six sub-fields in the space of sixteen years. In the 1920s he launched research on surface reactions (1921), unimolecular reactions (1924), and gaseous explosions (1927) which he explained with his important notion of branching chains. In 1930 he began work on reactions in solutions and on the mechanism of composite reactions. Just before the Second World War (1937) he embarked on the difficult task of examining the kinetics of bacterial growth. In much of this research he was assisted by Oxford undergraduates who worked with him for Part 2 on manageable topics that fitted into his large purposes. Of his 48 research collaborators between the wars, 22 were Trinity undergraduates, with 2 from Balliol and 10 from other colleges. Hinshelwood was able to draw publishable work even from two undergraduates who obtained thirds in finals. Thus, he exploited a unique local feature (Part 2) in an unusual site (a college laboratory) to produce work of international calibre. He attracted to this locale six Rhodes scholars, including two with PhDs, who took a DPhil or BSc under him. On a couple of specific topics, he enjoyed the services of post-doctoral workers, E. A. Moelwyn-Hughes and A. T. Williamson, who worked on solution reactions and gaseous explosions, respectively. He took on very few Continental collaborators, the chief one being Klaus Clusius. Hinshelwood received many applications from foreigners to work with him but his limited accommodation gave him the perfect excuse for rejecting most of them. In any case Hinshelwood was a chemist who believed that the individual was the source of the greatest scientific innovations, so he preferred working with a small group of congenial collaborators: large-scale bureaucratically organised research was anathema to him. Temperamentally he was reserved, reticent, and private, a lone wolf who did not communicate freely and widely with his scientific colleagues at Oxford. He was a reluctant candidate for the Dr Lee's chair, only accepting it in January 1937 to prevent Norrish of Cambridge from moving to Oxford. In the 1920s and 1930s he preferred to make his mark in Oxford by his research publications rather than through wide personal contacts. A bachelor, he was comfortable in his Trinity fellowship and prospered as a connoisseur in his college. He was an accomplished linguist and a painter who appreciated Chinese ceramics. In the common-room he amused his colleagues with nursery rhymes and sonnets about college affairs and in his college laboratory displayed effortless mental superiority by reading Dante in the original.[44]

Hinshelwood also started the tradition of Oxford's college fellows, who were physical chemists, writing synthetic works that summarised their research and, as a delegate of Oxford University Press from 1934, he helped it to flourish. He showed the way in 1926 with his first book, which discussed the kinetics of chemical change in gaseous systems. It avoided encyclopaedic compendiousness by focusing on the clear exposition of general principles illustrated by selected examples of different types of reaction. It also provided a rationale for hitherto incomprehensible features. He was followed by Thompson on spectroscopy (1938), Bell on acid-base catalysis (1941), and Bowen on photochemistry (1942). None of them, however, had such a wide range as Hinshelwood, whose penchant for synthesis reached its culmination in his book on *The Structure of Physical Chemistry* in which he made strong claims about physical chemistry as a liberal education, and about science as a humanistic enterprise, in ways that were congenial to his Oxford milieu. As a firm believer in 'a liberal occupation with wide studies', Hinshelwood emphasised the structure and continuity of the whole of physical chemistry and examined its various parts, not as a series of specialised topics, but in their relation to one another. That task in his view involved artistic judgment. In a parallel way Hinshelwood argued that, because science was basically the attempt by the human mind to order facts into satisfying patterns, the imposition of design on nature was an act of artistic creation. For Hinshelwood this view of science as 'a construction of the human mind' was totally compatible with his own experimentalism in the Balliol-Trinity Laboratory because some parts of the construction were closely related to 'things of direct experience and observation'.[45]

For almost two thirds of the twentieth century Balliol enjoyed the services of just two chemistry fellows, Hartley and his pupil Bell. Like Hinshelwood Bell was a beneficiary of Hartley's flexible attitude to the research fields adopted and pursued by his Balliol pupils. For his Part 2 Bell worked on interfaces, an interest that led Hartley to suggest that he should go to study at Copenhagen under J. N. Brønsted whose setup had been praised by W. F. K. Wynne-Jones, a previous Hartley pupil who had spent a year there. While an undergraduate

Bell gave a paper at a Faraday Society meeting held in 1927 in Oxford, met Brønsted, and then next year went for four years to Copenhagen where Brønsted fostered his interest in acid–base catalysis and Bohr's colloquia held in the adjoining Institute for Theoretical Physics launched his interest in quantum theory and its application to chemical kinetics. Bell was funded by a University studentship and by the Goldsmiths' Company with which Hartley had close contacts. Bell's return to Balliol in 1932 as Bedford lecturer in physical chemistry was arranged by Hartley. On Gatty's resignation as chemistry fellow in 1933, Bell was on hand to replace him.[46]

Three of Hinshelwood's pupils who became college fellows (Thompson, J. D. Lambert, and Staveley) followed Bell in finding their research feet through postgraduate experience abroad or outside Oxford. Thompson's initial research under Hinshelwood on kinetics was so promising that, before he graduated in summer 1929, St John's elected him to a research fellowship for 1929–30 to enable him to study in Berlin under Fritz Haber and to be a paying guest with Max Planck, to gain a PhD, and to return in 1930 as tutorial fellow. Initially Thompson continued his work on reaction kinetics but soon changed to visible and ultra-violet spectroscopy. This move was encouraged by Hinshelwood and supported by a Royal Society grant but was opposed by Soddy in whose old chemistry department Thompson worked. To learn the new techniques Thompson drove twice a week to Imperial College, London, to work with Alfred Fowler. By the late 1930s Thompson's main interest was infra-red spectroscopy; with the aid of a Leverhulme Travelling Fellowship, he spent a year at the California Institute of Technology learning the latest techniques from R. M. Badger. In finding spectroscopy as his enduring research field and in achieving his FRS in 1946, Thompson was greatly indebted to experience in London and Pasadena.[47] J. D. Lambert, who succeeded Hinshelwood as chemistry fellow at Trinity in 1938, graduated in 1934, worked for two years as an assistant to Hinshelwood paid for by a Royal Society grant, and with Hinshelwood's help then spent 1936–7 on a University studentship working at Göttingen under Arnold Eucken who introduced him to the then new field of molecular energy transfer in gases studied by using ultrasonic waves, which became his enduring research interest. On his return he was made lecturer in chemistry at Trinity for a year and then fellow.[48] Another Hinshelwood pupil, Staveley, who graduated in 1936, worked for a year with his tutor who encouraged him to begin a line of research not then pursued in Oxford. Financed by a grant from the Goldsmith's Company, Staveley spent 1937–8 at Munich studying under Klaus Clusius, who had spent a year in Oxford with Hinshelwood in the late 1920s. From Clusius Staveley learned the techniques of experimental chemical thermodynamics at low temperatures, which remained his life-long research field. On his return to Oxford Staveley was made lecturer in chemistry at New College for a year and then in 1939 tutorial fellow.[49] The cases of Lambert and Staveley are instructive. Both found it possible and advantageous to visit Nazi Germany to gain essential techniques in an aspect of physical chemistry then ignored in Oxford, and at a time when Jewish physicists and mathematicians were leaving the Third Reich in droves.

The Hartley–Hinshelwood tradition of encouraging prospective college fellows to study abroad was also adopted by Sidgwick who used his friendship with Petrus (Peter) Debye to send to Leipzig in autumn 1928 two of his Lincoln pupils, Woodward and Sutton. Woodward went to Leipzig on a DSIR senior research grant to research under Debye who introduced him to the then new field of Raman spectroscopy. Having gained his doctorate in 1931, Woodward took three jobs outside Oxford before returning to it in 1939 as chemistry fellow at Jesus where he pursued Raman spectroscopy as his main interest.[50] A similar interest in structure led Sutton in 1928–9 to spend six months in Leipzig on a DSIR studentship learning from Debye the techniques of measuring dipole moments that revealed the detailed structure of molecules. On his return to Oxford Sutton set up his apparatus on a wooden cabinet on wheels, christened by Hammick 'the commode', and ran Sidgwick's research group when he was in the USA in 1931. With the backing of Sidgwick and Robinson, Sutton was elected a fellow by examination in organic chemistry by Magdalen in 1932. Again supported by Sidgwick and Robinson, Sutton spent 1933–4 on a Rockefeller fellowship working in Linus Pauling's group at the California Institute of Technology where he learned the then novel technique of electron diffraction from Lawrence Brockway. On Sutton's return he pursued work on molecular structure, using the two techniques of dipole moments and electron diffraction, both of which he had learned abroad. In 1936 he was promoted to chemistry fellow at Magdalen.[51] The careers of Bell, Thompson, Lambert, Staveley, Woodward, and Sutton, make it abundantly clear that it is wrong to assert, on the basis of the admittedly spectacular case of Hinshelwood, that physical chemistry at Oxford was entirely indigenous. That was not so. Most Oxford graduates who became fellows of colleges in the 1930s learned about or discovered their enduring field of research and the associated techniques while undertaking postgraduate study under acknowledged experts in London, in Denmark, in Germany even during the Third Reich, and in the USA at the California Institute of Technology. Thus, physical chemistry at Oxford was constantly diversified by topics and techniques imported mainly from abroad.

It was the success of physical chemistry that led to the redefinition of the Dr Lee's chair in 1936–7, once Soddy had intimated his resignation effective at the end of 1936. That redefinition reflected in part the power of the Balliol-Trinity and Lincoln axes, controlled by Hinshelwood and Sidgwick who put their men, mainly physical chemists, into college fellowships. But it was more than just a mafia operation at a time when fellowships were rarely advertised. In the Oxford context physical chemistry enjoyed perceived advantages over inorganic chemistry, advantages that also kept at bay challenges from organic chemistry. Physical chemistry had wide scope: it often dealt with general processes involving many substances, discovered general empirical laws, and created general theories. In contrast, inorganic and organic chemistry seemed limited: they were piecemeal, descriptive, and often merely specific; they could be dismissed as cookery and smells. Consequently, physical chemistry was more amenable to comprehensive generalisation and intellectual synthesis. Oxford's physical chemists wrote books as well as papers, whereas their colleagues in inorganic and organic chemistry merely published papers. The physical chemists were more appropriate agents of the Oxonian aims of a liberal education that would produce a national elite. In terms of college teaching it was claimed with some justice that physical chemists could teach inorganic and organic chemistry but that inorganic and organic chemists could not teach physical chemistry. As long as comprehensive teaching was seen as a primary function of college fellows, that was a significant difference. Physical chemistry was often mathematical in its language and deductive in its arguments. As such it was a better mind trainer as part of a liberal education than inorganic and organic chemistry, which were often merely factual and required only a good memory. In the laboratory too, physical chemistry could claim an advantage. At Oxford between the wars there was a strong emphasis on precision measurement that the organic chemists eschewed and the inorganic chemists gave less attention to. Through its concerns with instrumentation, measurement, and purification of materials, physical chemistry trained practical chemists effectively. In all these ways physical chemistry was, or was represented as, the most appropriate and flexible of the three branches of chemistry for producing liberally educated chemists who would becomes leaders in universities, schools, and industry.

As soon as Soddy had given notice of his impending resignation of the Dr Lee's chair, the University took the advice of the Board of Physical Sciences, the sub-faculty of chemistry, and an advisory committee of Sidgwick, Hinshelwood, and Chapman. It was quickly agreed that inorganic and physical chemistry could be combined in a new building and that in the new appointment the importance of physical chemistry should be emphasised by appointing a physical chemist who would be responsible for both physical and inorganic chemistry. Hinshelwood was clearly the man the electors had in mind and he was elected in January 1937. The only dissident was Soddy who objected publicly to the restriction of the chair: he thought inorganic chemistry was the basis of chemistry in all its forms, including the industrial, a point strenuously rebutted by Sidgwick. Soddy also stressed that appointing a physical chemist to the chair would not solve the problems of control and of accommodation in physical chemistry: the new Dr Lee's professor would have no University laboratory to run and could only secure control by encroaching on the Balliol-Trinity and Jesus laboratories.[52]

The University's response was two-fold. Firstly, Hinshelwood was elected Dr Lee's professor under *ad hoc* arrangements. Though technically a professorial fellow of Exeter, he remained the senior science fellow of Balliol and Trinity, had the same authority in their laboratory as before, and was to retain his rooms in Trinity as long as he was in charge of it. Secondly, as soon as Soddy was out of the way, the proposal for a new University laboratory for inorganic and physical chemistry was rejected by the physical chemists who wanted a new building devoted entirely to physical chemistry. In a remarkable *volte face*, the college physical chemistry laboratories were suddenly seen as makeshift, poorly endowed, and limited in equipment and accommodation. When the University Appeal was launched in February 1937 it stressed that a new laboratory for physical, but not inorganic, chemistry was a top priority.

By late 1937 Lord Nuffield had given £100,000 to pay for it and ICI had promised £10,000 for maintenance. In spite of the difficulties caused by the run-up to the war and the war itself, a University physical chemistry laboratory, for which Soddy had campaigned so fruitlessly, was ready for occupation in the darkest year of the war, 1941, and the Balliol-Trinity Laboratory was closed but not forgotten. Its contribution to physical chemistry at Oxford was commemorated by the arms of the two colleges carved on the front of the new building.[53]

5.4 THE MANCUNIAN INHERITANCE

In the 1920s Perkin continued to make his presence felt at Oxford. He maintained his reputation as the leading British organic chemist of his generation. He showed by personal example the overriding importance of research. As an effective research supervisor he attracted researchers and demonstrators from outside Oxford to work in the Dyson Perrins Laboratory where he was revered as Pa Perkin. He maintained his interest in industrial research: British Dyes paid for an Oxford colony until 1925; and from 1923 to 1925 he was advisor to the research staff of the British Dyestuffs Corporation at Blackley, Manchester, which he visited almost weekly.[54]

There were, however, two disappointments for Perkin. The first was that he enjoyed only one success in securing the election of a Perkinian organic chemist to a college fellowship. Though Queen's predictably rewarded the loyal Chattaway with a fellowship in 1919, Magdalen elected Hope to a tutorial fellowship the same year, thus breaking the monopoly of physical and inorganic Oxonian chemists. Perkin used his own position as a professorial fellow of Magdalen in favour of Hope, one of his Manchester graduates and a member of the British Dyes team. Subsequently, Perkin had no success in placing his protégés in fellowships, some of which went to Oxonian physical organic chemists who worked in the Dyson Perrins Laboratory and as college fellows occupied independent fiefdoms even in the laboratory for which Perkin was responsible overall. Perkin did not like physical chemistry but tolerated it if applied to organic compounds. Not surprisingly, whimsical physical chemists tried to disguise their work by making it smell like classical organic chemistry: Hammick, a demonstrator in the Dyson Perrins Laboratory, used to let 'the Old Man' have a nauseous whiff of pyridine from his bench in order to deceive Perkin into thinking that proper organic research was being done. It is significant that Sidgwick, a long-serving demonstrator, relinquished this post in 1923; he regarded Perkin's exclusive concern with the analysis and synthesis of natural products as narrow and passé.

Another disappointment for Perkin was that the introduction of Part 2 increased the grip of the college tutors on their more ambitious charges whom they usually steered into physical chemical topics for their research year. The scheme, on which Perkin laid such store, thus backfired: it did not produce a large crop of graduate organic chemists agog to take DPhils in organic chemistry under him. After the war Perkin found only two of his organic demonstrators and researchers from among recent Oxford graduates, namely, Sydney Plant and Harry Ing. Plant had graduated under war-time conditions in 1918 from St John's and Magdalen. In 1919 he became a demonstrator while working for his DPhil under Hope (also appointed a demonstrator that year) and Perkin; after Hope was taken ill in 1925, Plant became the general factotum in the Dyson Perrins Laboratory though never a college fellow. In 1921 Ing began a five-year stint of demonstrating while an undergraduate. He took his DPhil under Perkin, left to be a research fellow at Manchester under Robinson, and eventually returned to Oxford as reader in chemical pharmacology.[55]

The majority of demonstrators in organic chemistry came from outside Oxford and often had a connection with Robinson or Manchester, John Gulland, R. D. Haworth, and Wilson Baker being cases in point. Gulland, a demonstrator in the Dyson Perrins Laboratory from 1924 to 1931, was an Edinburgh graduate who did research on alkaloids under Robinson at St Andrews and then at Manchester. Haworth studied with Lapworth, Perkin's brother-in-law, at Manchester, where he took his BSc in 1919 and his PhD in 1922. Baker, a Manchester BSc (1921) and PhD (1924), had collaborated with Lapworth and Robinson there before coming to Oxford as a demonstrator in 1927, the result of Robinson's having a word with Perkin over dinner. The arrival of Haworth in 1921 as an 1851 Exhibition Research Scholar permitted Perkin to initiate at last, in 1922, a loose and intermittent lieutenant system in which some of his colleagues gave detailed supervision of research. Perkin suggested the main, though flexible, lines often as small parts of a large project concerned with synthesising a substance such as morphine. Of his four subalterns, Hope, Clemo, Plant, and Haworth, only Plant was entirely an Oxford product. Clemo was never a demonstrator but part of the British Dyes team, while of Haworth's five years at Oxford four were supported by research studentships. Haworth and Clemo found it expedient to leave Oxford in the mid-1920s, being followed by Gulland in 1931; as distinguished professors of organic chemistry at Sheffield, Newcastle, and Nottingham, respectively, they spread the Perkin gospel. Of the imported demonstrators only Hope and Baker secured college fellowships, and Baker did not replace Chattaway as fellow in chemistry at Queen's until 1936, after Perkin's death.[56]

The college structure, the dominance of the physical chemical college fellows, and their control of their pupils' research through Part 2 ensured that in the 1920s relatively few Oxford graduates embarked upon research for a DPhil in organic chemistry. There was no equivalent in Perkin's Oxford period to the way in which he had nurtured the research careers of promising young undergraduates at Manchester, such as Robinson and W. N. Haworth, both of whom were to win Nobel prizes. Consequently, as with his British Dyes team, his demonstrators, and his research lieutenants, Perkin recruited many of his graduate students from outside Oxford. Among the subsequently well-known migrants were Thomas Stevens from Glasgow, who eventually joined R. D. Haworth at Sheffield; William Davies from Manchester, who as professor of organic chemistry developed laboratory research at Melbourne; Osman Achmatowicz from Poland, who became a professor at Łodz Polytechnic and eventually an official in the Polish ministry of higher education; Louis Fieser and Joseph Koepfli, from the United States, who became professors at Harvard and Caltech; and V. M. Trikojus, who became professor of biochemistry at Melbourne. Other graduate students came from as far as Sweden, India, and Japan.[57]

Perkin was a persuasive and efficient administrator who ruled his laboratory without putting pen to paper. In his dealings with the University he showed unscrupulous determination mediated by worldly wisdom. Perkin accepted with some grace the peculiarities of the Oxford system, which gave a professor charge of a university laboratory but which in part staffed it with college fellows who were statutorily independent of him and had as great a say as he had in the subfaculty of chemistry and in the Natural Sciences Faculty Board. Accepting the clear limitations of these arrangements, Perkin persuaded the University and Dyson Perrins to provide for him a laboratory that was the first major step towards the creation of the Science Area at Oxford. One secret of Perkin's persuasiveness was noted by one of his Magdalen contacts: 'Perkin often gave the impression of being only imperfectly acclimatised and of maintaining a good natured suspicion of those who professed to be bound by statutes or regulations . . . He could not have fitted better into an Oxford college with its widely different associations if he had been a member of such a body for the whole of his life'.[58] Perkin's adaptability was based partly on his character and partly on his wide interests in music, horticulture, hospitality, and travel, all of which mollified the suspicion that his Germanic emphasis on research engendered in some quarters. At Manchester he was such an accomplished pianist that he played duets with the violinist Adolf Brodsky, the leader of the Hallé Orchestra 1895–6 and principal of the Royal Manchester College of Music 1895–1929, who gave the first performance of Tschaikovsky's violin concerto. At his Oxford home Perkin soon removed a partition wall in order to create a long room for chamber music performances. To the end of his life he practised at his piano every day before breakfast. He believed that like organic chemistry a Beethoven sonata needed to be worked at. For diversion on a train journey he used to read the score of a string quartet. In his horticultural work he also attained as a devotee a high professional standard, specialising in flowering plants, of which many were donated to the University Parks. He kept a good cellar and was extremely hospitable, especially at Magdalen, where he gave many lunch and dinner parties. In the long vacation he and his wife regularly visited the Swiss and Italian lakes. These wide interests made him the ideal man to introduce Germanic research practices into Oxford.

Perkin was succeeded in 1930 as Waynflete professor by his favourite chemical son, Robinson, who was proud to be a leading member of what he called the Perkin family of organic chemists.[59] As an undergraduate at Manchester he was inspired by Perkin's lectures to become an organic chemist in the Perkin mould. Having spent no fewer than seven postgraduate years in Perkin's department at Manchester, where they began their long collaboration, Robinson embarked on an academic odyssey that took him to chairs at Sydney, Liverpool, St Andrews, Manchester, and University College, London, before he settled in Oxford in 1930. In 1947 Robinson was awarded the Nobel Prize for his work on plant products, especially alkaloids, a topic that he first attacked with Perkin in Manchester. Robinson's debt to Perkin may be gauged from his publications: of his first thirty papers twenty-four were written with Perkin; all told they published sixty-four joint papers. Not surprisingly as Waynflete professor Robinson was expected to continue Perkin's modes of work, see reference 82.

In several ways Robinson continued the Perkin regime, but he was more intense, more mercurial, more innovative, and less equable than his chemical father. Like Perkin's his research was focused on elucidating the structure of naturally occurring substances, especially alkaloids and colouring matters, using the established procedures of degradation and synthesis. The former involved breaking down a substance into fragments that could be identified; synthesis involved making the substance from chemical compounds of known composition and structure by a series of controlled reactions whose course was indisputable. Like Perkin he did not employ new physical methods of investigating structures. He maintained Perkin's emphasis on research as the prime academic function, but spent less time at the laboratory bench because he had more research students (about 30 at any one time) than Perkin. In the 1930s his wife, an organic chemist whom he had met at Manchester, spent more time at the bench than her husband who managed only short spells of practical work himself.[60] He confined himself to exploratory experiments in test-tubes, leaving the follow-up to a collaborator. As a more innovative researcher than Perkin, he ignored the University terms contumaciously. Whereas Perkin's forte was the rapid exploitation of new reactions and reagents discovered by others, such as Arthur Michael (1887) and Victor Grignard (1900), Robinson's was the devising of methods of synthesising such compounds as steroids. One favourite technique involved the synthesis of a substance containing four benzene rings and then modifying it. Another was to obtain a three-ring structure containing the chemical groups that would allow a fourth ring to be added.

In synthetic work Robinson was not as persistent as Perkin. On occasion he abandoned synthetic routes (of which he was often a brilliant deviser with his chess-player's ability to see several moves ahead) which were later shown by others to be practicable. His restless volatility led to his devising a large number of synthetic reactions and procedures but completing a smaller number of actual syntheses. Again in contrast with Perkin, Robinson worked in theoretical chemistry and was responsible in 1926 for an electronic theory of organic chemistry. But he had not the patience to consolidate and popularise it. His first general account of it appeared in 1932 in an accessible form and then he dropped the topic for seventeen years, being far more interested in the steroids research began in 1932. Meanwhile, from 1926 Ingold developed a rival treatment, became wholly absorbed in the study of reaction mechanisms in organic chemistry, kept his ideas continuously on display, and explained the application of his ideas to numerous specific reactions. Robinson's mercuriality of temperament allowed Ingold to be widely seen as the originator and doyen of the electronic theory of organic chemistry and led to poisonous relations between the two men because Robinson, who was legendary for his impatience and irascibility with opposition, accused Ingold of plagiarism. There was also a less obvious source for Robinson's unconcealed hostility to Ingold: whereas Robinson was faithful to Perkin's dismissal of physical methods, Ingold was the leading early exponent in Britain of the new chemical physics.[61]

Robinson inherited from Perkin the capacious and well-endowed Dyson Perrins Laboratory. In the 1930s it received a major extension only once, in 1939–40, at a cost of £29,000 which was raised by Robinson from the Rockefeller Foundation (£23,000) and ICI (£6,000).[62] Robinson had not pushed his case in the University Appeal of 1937 because he thought the prime need was for a new University physical chemistry laboratory; and in 1938 he deliberately went outside the University for money, trusting successfully to Rockefeller's interest in the biological implications of organic chemistry (especially the synthesis of proteins) and to his strong connections with ICI. The need for expanded accommodation had been generated by Robinson's reputation, his productivity, the presence by the end of the decade of sixty research students in the laboratory, and the closure of the Queen's Laboratory in 1934. Robinson's high stature attracted post-doctoral researchers of the calibre of Alexander Todd, a Nobel Prize winner in 1957, and Norman Rydon, whose work with Robinson on oestrogens led to the future development of the 'pill'. From Australia he drew J. W. Cornforth, Nobel Prize winner in 1975, and Arthur Birch, deviser of the Birch reduction that was rapidly used by pharmaceutical companies to produce steroids. Like Perkin, Robinson secured few of his researchers from Oxford graduates in chemistry: they came mainly from elsewhere in Britain and from abroad. It was Robinson who was responsible for the international composition of the researchers in the laboratory: in Todd's short Oxford period he was struck by the presence of workers from Australia, New Zealand, Sweden, Japan, and Switzerland. As a supervisor Robinson, known to his worshippers as the Waynflete Wonder, was inspiring but not frequent in making contact with his students from some of whom he received the sobriquet of visiting professor. Robinson's own productivity was extraordinary. In the years 1931–8 inclusive he and his collaborators published no fewer than 226 papers, *i.e.* a paper per fortnight maintained for eight years. No wonder that by force of example he inspired workers in the Dyson Perrins to produce a maximum in the 1930s of 84 papers in 1934–5 and in the late 1930s he employed Springall, formerly one of his DPhil students, as a scientific gentleman's gentleman to help in publishing papers.[63]

In the 1930s Robinson secured only two new University demonstrators, King and J. C. Smith. Both came from outside Oxford, from London and Manchester, respectively, where they were known to Robinson, to join the existing quintet of Hammick, Plant, Taylor, Baker, and the ill Hope. Again like Perkin, Robinson was reluctant to use his demonstrators as research lieutenants. He was often possessive and dictatorial with his research students, though when flooded with them he would on occasion use a trusted demonstrator: in the mid-1930s King helped Robinson with the supervision of researchers working with Rockefeller money. Again like Perkin he was unsuccessful in placing his protégés into college fellowships in the 1930s, the solitary exception being Baker who was elected at Queen's in 1936 to replace another organic chemist, Chattaway. One result was that King, who served as departmental and University demonstrator for 17 years but was never a college fellow, left in chagrin for a chair at Nottingham.[64] Robinson resented a feature that Perkin accepted with some grace, *i.e.* that he had no control of college fellows who were physical chemists doing research in his laboratory: they occupied independent fiefdoms. Hammick, egged on by Sidgwick, was the leader of the independent college fellows, the others being Taylor and Sutton. Robinson resented the way in which Hammick, a tutor at Oriel, Corpus Christi, and Wadham, sometimes had eight Part 2 students working in the Dyson Perrins Laboratory, while his own demonstrators, who were not college fellows, had none or very few. In retaliation, Hammick told Robinson that he and the colleges would decide with whom his pupils worked and he won. Furthermore, the college fellows in the laboratory, like everyone else except undergraduates, were not controlled by Robinson through rationing of common apparatus and chemicals. Until after the Second World War there was no system for requisitioning and issuing apparatus and chemicals. Like Perkin, Robinson was proud of his paperless administration based on trust; but he relied heavily on the steward of the Dyson Perrins Laboratory, Fred Hall. Robinson was too impatient to try to force his views on the college fellows through staff meetings: he held only three in twenty five years.

Robinson also continued Perkin's enthusiasm for industrial research and for commercial consultancy. Robinson had worked as an industrial chemist for a year when he was director of research at British Dyestuffs Corporation, Huddersfield, 1919–20. This experience gave him an understanding of industrial research, useful contacts, and a comprehensive knowledge of dyestuff chemistry. After British Dyestuffs had been absorbed by ICI, Robinson did research for ICI and gave advice in exchange for money, information, and for chemicals, both stock and those specially made for him. He was a powerful figure from 1929 in ICI's Dyestuffs Group Research Committee, for which research was done in Oxford, and in the creation of the pharmaceutical division of ICI via the medical section of ICI's Dyestuffs Group at Blackley. At Oxford Robinson continued to take out patents, registering nine, sometimes in collaboration with ICI.[65]

Robinson helped several German organic chemists who, as non-Aryans, faced persecution, dismissal, or expulsion by the Nazis from spring 1933. He quickly gave free accommodation to two organic chemists, Fritz Arndt from Breslau and Arnold Weissberger from Leipzig, both of whom had been dismissed from their posts. They were paid £600 pa and £400 pa, respectively by ICI for two years and they received expenses of £268 from the Rockefeller Foundation. In 1934, Arndt moved to a chair at Istanbul where he stayed until he retired. Weissberger had greater difficulty in finding a permanent home. Though he published articles and an important book on organic solvents in 1935 and supervised research in the Dyson Perrins Laboratory, his ICI grant was not renewed in autumn that year so in 1936 he left academia reluctantly for a post as an industrial researcher in synthetic organic chemistry with Kodak in Rochester, USA. Herbert Appel, another organic chemist from Leipzig, was not officially dismissed though classed as a non-Aryan but he sought sanctuary first in Switzerland and then for a year from October 1934 working with Robinson on anthocyanins before moving on to Birmingham for a year. Robinson also helped Gerhard Weiler, a specialist in small-scale organic chemical analysis, a technique then new to the Dyson Perrins Laboratory. He arrived from Berlin in March 1934 with his apparatus, was installed in the Dyson Perrins Laboratory, and quickly established his micro-analytical enterprise as an important resource for many scientists in Oxford and in industry. By the late 1930s he was in commercial partnership with Fritz Strauss selling micro-analytical services from private premises in Oxford. He was naturalised in 1946 and ended his days in Oxford. In late 1938 Robinson tried to help a prospective refugee, Richard Willstätter, an aging organic chemist and Nobel prizewinner, who had resigned from his chair at Munich in 1924 in protest against growing anti-semitism. Robinson deeply admired Willstätter's work on natural colouring matters and realised that some pretext was needed to persuade the Nazis to allow Willstätter to leave Germany. Robinson officially invited him to Oxford to lecture in 1939, telling the Registrar that Oxford should grab him before the Americans fell over each other to do so.[66] In the event Willstätter was offered sanctuary and unique facilities by the Swiss government in January 1939 and migrated in March to Locarno where he died during the war.

    The legacies of the Perkin–Robinson approach to organic chemistry were a continuing insistence on Baeyer-like degradative and synthetic methods and a marked suspicion of both physical methods and theoretical chemistry. Perkin and Robinson were good bench chemists, proud of their ability to induce a reaction to 'go' and gums to crystallise. They relished their ability to *make* many substances, both naturally occurring or not, and were proud that the latter were useful in dyestuffs, medicine, agriculture, and petrochemicals. In contrast, physical chemists, non-makers and often non-applied in their research, seemed merely effete. In the 1930s Robinson was cautious about using infra-red and ultra-violet spectroscopy, X-ray crystallography, and measurements of dipole moments in his own research. Having withdrawn in 1932 from work on reaction mechanisms in organic chemistry, Robinson was suspicious of Ingold's work on this subject, which was characterised in Oxford as an Ingoldsby legend.

    Robinson's approach to the problems of molecular structure was supplemented by that of Sutton, who worked in the Dyson Perrins Laboratory, and Hodgkin who researched in the mineralogy department. Though both were helped by Robinson at crucial stages of their careers, he and his collaborators did not adopt Sutton's techniques of dipole moments and electron diffraction or Hodgkin's use of X-ray crystallography. In his year at the California Institute of Technology Sutton was impressed by Pauling's team-research that he regarded as professional compared with Oxford's amateur approach, which was satisfactory for doing research but not for research training. Pauling employed mathematical physicists and experimenters in several fields, and acted as a link figure. For Sutton, Pauling's great strength was his rare combination of chemical instinct and mathematical and physical techniques. Sutton wanted to found at Oxford a group devoted to molecular structure using physical methods à la Pauling but found Robinson unresponsive: Robinson proclaimed that he was an organic chemist, that it was up to Sutton to mobilise other physical chemists in promoting his scheme, and that the long-term solution for Sutton was a new University physical chemistry laboratory.[67] In Hodgkin's case it appears she was content with her location and made no attempt to transfer to the Dyson Perrins Laboratory. During the war, however, she joined the team working at Oxford under Robinson on the structure of penicillin. By 1945 she had beaten Robinson in determining its structure and shown that Robinson's preferred structure was inadequate. Similarly by 1956 she had beaten Todd in elucidating the structure of vitamin B12. Thus, the classic degradative and synthetic methods employed by Robinson and his pupil Todd began to be outflanked in Oxford by the new physical method of X-ray crystallography. Elsewhere the Perkin–Robinson style of research was lethally challenged in the 1950s at Harvard by R. B. Woodward who combined their methods with molecular orbital theory and new physical resources (such as infra-red and nuclear magnetic resonance spectroscopy) to synthesise vitamin B12, chlorophyll, strychnine, cortisone, and cholesterol.

5.5 THE DR LEE'S CHAIR AND OLD CHEMISTRY

Inorganic chemistry, the responsibility of Soddy and then Hinshelwood, was pursued in the 'Glastonbury Kitchen' building attached to the Museum, Figure 5.3. It was physically and mentally separate from physical chemistry, which was located in two college laboratories, and from organic and physical organic chemistry, which were in the Dyson Perrins Laboratory. In the 1920s the inorganic chemistry laboratory was informally known as the Old Chemistry Department. In Sidgwick's view this odd name was exactly right because that was what Soddy lectured on – old chemistry! Between the wars the Old Chemistry Department at Oxford did not make a big national or international impact, the exceptions being Hume-Rothery in metallurgy and Harold Thompson in infra-red spectroscopy. The most curious feature, at least *prima facie*, was that Soddy, appointed to the Dr Lee's chair at Oxford in 1919 and in 1921 the first Oxford professor to be awarded a Nobel prize, did not set up and lead the expected school of radiochemistry there.

The reasons for Soddy's failure to do so are not as simple as some critics have made out. It is well known that Soddy did not have a wide circle of friends among fellow scientists at Oxford, that he was averse to compromise, and that he was at loggerheads with the colleges, the college tutors in chemistry, and the University. In the 1920s he was at odds with two of his senior demonstrators, Lambert and Applebey, who thought him perverse and tactless. These general features of Soddy's regime do not, however, explain why his early attempt to found a school of radiochemistry and his own published research in that subject

Figure 5.3 The Old Chemistry Department in 1937, showing the original laboratory of 1860 (The Abbot's Kitchen), see Figure 4.1; the extension of 1879 behind and to the left of it and the Radcliffe Science Library to the right. (Reproduced from Oxfords Special Number, February, 1937).

had foundered by the mid-1920s. Part of the explanation lies in Soddy being a solo worker without a research assistant, collaborator, or lieutenant. In autumn 1919 when A. S. Russell, who had done important radiochemical research with Soddy in Glasgow, was appointed Dr Lee's reader in chemistry with Soddy's warm support, it seemed that the core staff of a school had been assembled. Yet Russell never became a demonstrator or informal collaborator in Soddy's department, lectured there only once on radiochemistry, pursued his own research independently it seems in his Christ Church laboratory, was swamped by teaching (55 hours a week at worst), and by the 1930s had turned away from the little research in radioactivity he had managed to do to the study of intermetallic compounds in mercury. Soddy and Russell remained friendly but as researchers they went their separate ways.

It had also gone unnoticed that Soddy's attempts to secure radioactive materials for research, from the Imperial and Foreign Corporation of London in 1921, were doused with cold water by Lewis Farnell, the Vice-Chancellor, so a rare opportunity was lost. In autumn 1921 the Corporation, to which Soddy was scientific advisor, projected a laboratory in England to house radium. Soddy wanted this laboratory to be located in Oxford and, given his unsuitable accommodation, in a separate building for which the Corporation would partly pay. Soddy was agog at the unique opportunity of working with the two grammes of radium, worth £70,000, which he had brought from Czechoslovakia to London where it was lodged in the Czech Embassy. In his two years at Oxford he had had no more than 30 milligrammes of radium bromide to work with. He looked forward to having a new radiochemical laboratory as a University institute, with facilities and staff paid for by the Corporation, and to training people there. By October 1921 a draft agreement between the University and the Corporation had been drawn up by Soddy. It permitted Soddy to use the Corporation's equipment and materials for its scientific and technical work for which he would be paid, even though he would be working in University premises. As a *quid pro quo* the University insisted without prior warning that any scientific research done by Soddy using the Corporation's equipment and materials should be published and should not become private knowledge owned by the Corporation. On hearing of this insistence, prompted by Farnell, that there be no bar on publication, the Corporation dropped the Oxford scheme, leaving Soddy furious. Not for the first time the University, it seemed to him, had let him down. More than that, the University's insistence on there being no restraint on publication had scuppered his ambitious scheme, yet in the Dyson Perrins Laboratory no similar insistence applied to commercial firms doing research there or to Perkin's own industrial research; there seemed to be one University law for Perkin and another for Soddy. And, of course, Soddy was frustrated at the loss of a golden opportunity to investigate the unworked residues from the mines of St Joachimsthal: in his view they constituted an El Dorado for research.[68]

Soddy's chagrin with the University for denying him the patronage of what he called the Czech Radium Corporation and his institutional separation from Russell did not facilitate Soddy's ambitions for research on the purification, extraction, and analysis of radioactive elements and their compounds. His hostility to the college tutors ensured that in the 1920s he attracted only three Oxford graduates, two of them women, to work with him. But he did draw from Japan Satoyasu Imori who became the father of Japanese radiochemistry, a couple of Rhodes scholars, Miss Hitchins from Glasgow who became his private assistant, Paolo Misciatelli as a Ramsay research fellow to do a DPhil, and J. K. Marsh from Belfast as a post-doctoral researcher.[69] In 1927 Hitchins left for Africa, Misciatelli returned to Italy, and Marsh went back to Ulster. Without these collaborators Soddy's publishing career as a radiochemist finally ended, though it had been petering out before then. In compensation, Soddy spent a lot of time in his laboratory workshop, devising many mechanical gadgets that he patented, the best known being his continuously variable dividing engine that ruled scales for spectrographs. This talent for design was generously put at the service of the University in the late 1920s when he reconstructed the Old Chemistry Department. Acting as his own architect, designer, draughtsman, and engineer, Soddy gave particular attention to bench reconstruction, lighting, drainage, and ventilation. Whenever possible he used his own workshop to make artefacts he had designed, he supervised the work done, and he obtained discounts on materials. He avoided the use of expensive and bungling contractors such as those who had cost the University so much when they built the extension to the Dyson Perrins Laboratory in the early 1920s. Showing no sign of chagrin at having to wait ten years for a University grant to modernise his department, Soddy did so in 1929–30 at a cost of just under £7,000, his ingenuity saving the University thousands of pounds. Showing no loss of spirit, he continued to press the University, though without success, for extensions to his empire until just before his resignation in 1936 after the death of his wife.[70] Irritatingly fertile in producing, amending, and adding to his designs, Soddy was literally a department builder, an achievement that his detractors wrongly ignore.

He was not, however, a discipline-builder in physical or inorganic chemistry. This failure was the result of the difficulties he met at Oxford. It was also caused by Soddy's adoption of the roles of sage, prophet, and critic, which resulted in many publications on social and economic problems in their relation to scientific progress. It has often been said that Soddy went off the rails in 1919, foolishly risking the status of an amateur crank in economics and stupidly forsaking chemistry in which he had special gifts. It makes more sense, however, to see him as one who was so devastated by the waste, hatred, and futility of the First World War that he became a seer and propagandist, albeit a solitary one, concerning the social problems of science. He wrestled in particular with the question of how it was that the prospective benefits of science were often vitiated. He concluded that the main villains were the banks, so from the early 1920s he devoted much of his energy and time to writing and speaking about monetary reform. He aligned himself with such fringe movements as Social Credit and New Europe. As a chemist who pondered about the relations between science and life, he had concluded by 1919 that the chief problems facing humankind were not physical but moral. Accordingly, he also concerned himself with the social responsibility of scientists. He was particularly prominent in 1935. As an unconventional socialist and active member of the National Union of Scientific Workers in the 1920s, he associated himself with younger left-wing scientists, such as Bernal, Blackett, and J. G. Crowther, in deploring the misapplication and frustrations to which science was subject. He also led a well-publicised attack on the Royal Society of London that he depicted as a private bureau run by divine right by powerful officers and a self-selecting Council. He was supported by ten of his fellow scientists at Oxford, including Robinson and Chapman, but not by either Sidgwick or Hinshelwood. This was perhaps the only occasion on which he was not isolated while at Oxford. Usually his chemical colleagues regarded his behaviour as abrasive, awkward, and inconsistent, and they deplored his assumption of the role of seer as a sterile aberration. Soddy, however, saw himself as 'the pioneer and bearer of a new evangel' and not as a crank or imposter who had strayed from the path of pure science.[71]

The Old Chemistry Department under Soddy was different from the Dyson Perrins under Perkin and Robinson in that the professor and senior demonstrators were not leaders of research groups or innovative solo researchers. These roles were filled in the 1930s by H. W. Thompson, fellow of St John's and departmental demonstrator, and by William Hume-Rothery, a metallurgist who was a guest in the department. Soddy's enduring senior demonstrators were Bertram Lambert, Ernald Hartley, and Walden, while Applebey left in 1928 to be succeeded by F. M. Brewer. All were Oxonians and hoped that local esteem could still be gained by routes other than those of research, research-leadership, and publication. After bitter conflict with Soddy in the 1920s Lambert assumed the role of mediator and peace-maker between Soddy and his staff. He continued his war research on the absorption of gases by solids but was a slow publisher. After Soddy's retirement he received his reward when he was made administrative head of the Old Chemistry Department. Applebey also sparred with Soddy in the 1920s and was lured in 1928 to be research manager at ICI's Billingham plant. Walden never undertook original work, preferring to be renowned locally as 'Teacher'. Ernald Hartley and Brewer, however, had both begun promisingly as researchers but as demonstrators did not maintain that role. Hartley was indeed famous for the research he had done for thirteen years before the first war with Lord Berkeley who had invited him to work in his country-house laboratory at Foxcombe Hall, near Oxford, on the osmotic pressure of solutions. After war service with Lambert devising gas respirators, he led a quiet life as a demonstrator, publishing little and enjoying country-life at his Frilford House home, eight miles from Oxford. Apparently the time he spent travelling limited the hours he could spend in the laboratory. The fruits of his University labours, in contrast with those done with Lord Berkeley, were the men he helped rather than the papers published. Brewer, the only one of the five main demonstrators to graduate after the first war, was a pupil of Sidgwick who used his American contacts to send Brewer for two years to work at Cornell University as a Commonwealth Fund fellow under L. M. Dennis, who gave him an interest in the chemistry of germanium. On his return to England he trod water for a year at Reading University before becoming a key man in the teaching and organisation of inorganic chemistry at Oxford from 1928. Brewer was not a great researcher and did little work at the bench but he gained the rewards of a readership in 1955 for his administration and of being Mayor of Oxford 1959–60 as the culmination of his civic labours.[72]

Though all these demonstrators had Part 2 undergraduates to supervise, and Lambert, Walden, and Applebey had the additional advantage of being college fellows, none of them used Part 2 of the chemistry degree as Hinshelwood did as a source of research pupils and of publications. The first demonstrator in Soddy's reign to do this was Thompson. A college fellow by the age of 22, the blunt and ambitious Thompson used the Part 2 system to swell his own publications and to push his St John's pupils into print: all four finalists from St John's in 1933 co-published with him. Not surprisingly he taught subsequently distinguished pupils who in the 1930s included Christopher Kearton, Fred Dainton, and Jack Linnett, his favourite research collaborator in the 1930s. Linnett was a St John's graduate tutored by Thompson who supervised his research for a DPhil 1935–7. Then Linnett was awarded a Henry Fellowship that enabled him to work for a year on infra-red and Raman spectroscopy at Harvard University under George Kistiakowsky and Bright Wilson, from whom he also learned much about the application of quantum mechanics to chemistry. On his return to Oxford in autumn 1938 Linnett's career prospered as a research fellow, departmental and University demonstrator, college fellow, and University reader until in 1965 he left for the chair of physical chemistry at Cambridge.[73]

The pioneering work in the science of metallurgy undertaken by the totally deaf William Hume-Rothery was kept alive at Oxford until 1938 by his loyalty to the University, his own pocket, and by the prescience of various external funding bodies. Only in 1938, after he had been elected FRS in 1937, was he made a University lecturer in metallurgical chemistry and a fellow of Magdalen by special election. The establishment of a department of metallurgy and an undergraduate degree in the mid-to-late 1950s owed much to Hume-Rothery's efforts, abetted by external pressure. With some justice he became the University's first professor of metallurgy in 1958 and a professorial fellow of St Edmund Hall. His long slog to recognition began in 1922 when, having gained a first in chemistry at Oxford, he was advised to take up metallography by Soddy who was then keen to develop metallurgy at Oxford. Soddy arranged for Hume-Rothery to do a PhD under Sir Harold Carpenter at the Royal School of Mines in London, after which it was intended by Soddy that Hume-Rothery should return to the Old Chemistry Department to promote the subject. While Hume-Rothery was in London, he developed an interest in inter-metallic compounds that did not conform to the rules of valency then current in inorganic chemistry and he began to speculate about the role of free electrons in the solid state. On his return to Oxford in 1925, supported for four years by a Senior Demyship (*i.e.* graduate scholarship) at Magdalen, his undergraduate college, he discovered that Soddy, piqued by his disputes about the Balliol-Trinity Laboratory, would not accommodate him. Desperate to settle in Oxford, Hume-Rothery secured from Perkin a temporary home for almost four years in the Dyson Perrins Laboratory topped up by occasional visits to the laboratory of Engineering Science. Then in autumn 1929 he moved to the Old Chemistry Department, where he was to stay as a guest for many years, through the influence of Lambert, who had been his Part 2 supervisor. External funding enabled Hume-Rothery to maintain his research once his Magdalen post had ended. For three years he held a research fellowship awarded by the Armourers' and Braziers' Company. In this period he published his book on the metallic state in which he used the lattice theory of Frederick Lindemann, professor of experimental philosophy, a supporter and confidante, as a point of departure from which he began to work out empirical rules of alloy formation. As he thought that their meaning would have to be left 'to some Cambridge wave mechanic', Hume-Rothery was well aware that he was so far the Kepler but not the Newton of physical metallurgy. As such he was critical of the insular approaches usually taken by metallurgists and physicists to the features of metals and alloys. The former merely amassed commercially useful data, whereas the latter had investigated underlying principles but their work was vitiated by lack of chemical knowledge, by the use of unsuitable specimens, and by ignorance of practical metallurgy.

Hume-Rothery's research at Oxford would have ended in 1932 but for the lucky accident of the establishment that year of the Warren Research Fund of the Royal Society of London for research in engineering, chemistry, physics, and metallurgy. Against hot competition, Hume-Rothery was elected in June 1932 to a Warren fellowship, tenable for four years, renewable on application by the fellow for a further three years, at a salary of £700 pa. Hume-Rothery probably owed his election to two Oxonians, Sidgwick and Egerton, who were members of the Warren Fund Committee. Sidgwick was interested in Hume-Rothery's early work and had introduced him to Lindemann's lattice theory, while A. C. G. Egerton, then reader in thermodynamics at Oxford, was interested in chemical technology and as one who had been thrown out of the Old Chemistry Department by Soddy in 1921, he had special sympathy with Hume-Rothery's situation there. The Warren fellowship gave Hume-Rothery security, status, and salary not just to 1939 but with one short break to 1955, even though officially it was a limited-term appointment. Supported by considerable grants from the Royal Society for apparatus and materials, which were topped up by Hume-Rothery from his own pocket, on occasion rising to £70 pa, he built up from 1933 a small research group, recruiting mainly Oxford chemists via Part 2 and the occasional Oxonian physicist to work with him, sometimes for a DPhil. In his X-ray work on the determination of lattice and atomic constants, he was greatly helped by H. M. Powell from the nearby department of mineralogy. His growing reputation was confirmed with his second book that dealt with the structure of metals and alloys. With its accounts of atomic structure and the theory of the metallic state, this work which was inspired and published by the Institute of Metals, gave the first general and accessible account of the principles of structural metallurgy for industrial metallurgists as opposed to physicists. After Henry Tizard had proclaimed as part of the University Appeal of 1937 that special encouragement should be given to metallurgy and after Hume-Rothery had been elected FRS that year, his position at Oxford was regularised in 1938. He even secured a departmental demonstratorship in 1937 for his favourite pupil and collaborator, G. V. Raynor. Though his accommodation in the Old Chemistry Department was not palatial, being restricted before the war to one room and a covered yard, Hume-Rothery saw the immediate future of metallurgy as best pursued there. His career to then supplies an extreme example of the importance of external funding. It enabled him in the 1930s to become a leading figure in British scientific metallurgy. Hume-Rothery's career also shows that the lack of formal recognition at Oxford until 1938 was in part compensated by the informal help he received at crucial points in his career from Oxonians such as Soddy, Perkin, Lambert, Lindemann, Sidgwick, Egerton, Powell, and Tizard.[74]

5.6 THE CHEMICAL SYNTHESISER

Oxford chemistry was tripartite in terms of sites. Physical chemistry prospered in two college laboratories, organic chemistry burgeoned in the Dyson Perrins

Laboratory, and inorganic chemistry was housed in the Old Chemistry Department. If there was a unifying force it was provided by Sidgwick. His synoptic vision, which embraced the main three branches of chemistry, was revealed in his publications, his contributions to colloquia and seminars, and his dominance of both the Alembic Club and the Dyson Perrins tea club. It was rightly said after his death that he was a strict but not severe father of the whole Oxford school of chemistry.

Sidgwick was a late developer. Having secured firsts in chemistry and Greats at Oxford, he spent three years in Germany learning physical chemistry from Oswald at Leipzig and organic chemistry from von Pechmann at Tübingen. On his return he was elected in 1901 tutorial fellow at Lincoln, a post he held to 1948. The security conferred by his fellowship enabled Sidgwick to find his chemical feet slowly and to take his time when researching and writing synoptic works. Initially there was little direction to his research. He was not an ardent experimenter at the bench. But in 1910, with the publication of his book on the organic chemistry of nitrogen, he discovered at the age of 37 that his special gift was his ability to bring to chemistry 'not so much the burning and single-minded zeal of the discoverer as the panoramic learning of the scholar'. Sidgwick's first work of arm-chair or literary chemistry brought physical, inorganic and organic chemistry together and it disproved, by precept and example, the view held by organic chemists such as Perkin that physical chemistry was all very well but it did not apply to organic substances. The book, which grew out of an Oxford lecture course, established his international reputation.

It was extended in 1927 by his famous work on the electronic theory of valency that eventually sold over 10,000 copies. The immediate stimulus for it was the visit in 1923 of G. N. Lewis, of the University of California, to Oxford where he stayed with Sidgwick. From their discussions and the publication of Lewis's *Valence and the Structure of Atoms and Molecules* later that year, Sidgwick developed an electronic theory of valency that applied to the whole of chemistry. The most original part of the work dealt with inorganic coordination compounds: Sidgwick showed that the so-called coordinate link was the same as the covalency of carbon, which he explained electronically, thus once again bridging the gap between inorganic, organic, and physical chemistry. As his book presented a much more comprehensive analysis of the uses of the electron in chemistry than had Lewis, it transformed inorganic chemistry from a jumble of unrelated facts into a subject with intelligible principles because it showed how the number of electrons outside the nucleus of an atom of a given element controlled its chemistry. Its magisterial scope inspired many readers, including schoolboys such as Dainton, M. J. S. Dewar, and A. G. Ogston who were drawn to study at Oxford by it.

In 1933 yet another work of synthesis appeared with the publication of his book on the covalent link in chemistry. Like the 1927 book this one took the whole of chemistry for its province. It grew out of his invitation from L. M. Dennis to visit the USA in 1931 as George Fisher Baker lecturer in chemistry at Cornell University for which he was paid $5,000 salary. About a third of the book was concerned with dipole moments, the importance of which had been communicated to him in 1928 by Debye while his guest at Oxford. In 1937 the second edition of his *Organic Chemistry of Nitrogen* appeared. It was revised and re-written by T. W. J. Taylor, fellow of Brasenose and a former pupil of Sidgwick, and Wilson Baker, fellow of Queen's, to whom Sidgwick had entrusted the task having written four draft chapters himself. As eleven colleagues, past and present, collaborated as drafters and as commentators on drafts, the book was very much an Oxford joint-stock effort and a tribute to Sidgwick's popularity and wide view of chemistry. Shortly before the Second World War he embarked on another great synoptic work, on no less than the chemical elements and their compounds, which finally appeared in 1950 when he was seventy seven. As a college fellow Sidgwick could afford to bide his time: he published little of major importance until early middle age. As a college fellow in chemistry and the solitary science tutor in a college that had no mathematics fellow, Sidgwick was forced to take a broad view of his subject which helped him to compose four major synoptic books.

In the 1930s he was active on the national stage as chairman of the chemistry research board of the DSIR (1932–5), president of the Faraday Society (1932–4), a vice-president of the Royal Society (1931–3), and president of the Chemical Society (1935–7) where he dispatched business with bewildering celerity. He had previously been for seven years the first specially appointed chairman of the Publication Committee set up by the Chemical Society to referee papers submitted to the Society's journal. He was also visible in Oxford. He was the leading representative of science in the delegacy of the University Press. Famed for his sallies against fallacy and pretension, in his dealings with colleagues he was always illuminating, usually pungent, and sometimes disconcerting. Secure in his college fellowship, he was free from professional rivalry and personal vanity. He enjoyed biting people verbally, referring once to Wadham men as gutta percha from the neck upwards. As a bachelor don he was a renowned host to visiting scientists and persuaded his college to set aside a permanent guest room for his use. Through his hospitality he helped to make Oxford chemistry known and respected. For instance, when J. B. Conant of Harvard University visited Oxford, Sidgwick gave a dinner party in Lincoln for him, Lindemann, Einstein, and Sutton. Sidgwick exploited his powers of badinage in promoting chemistry socially at Oxford. He was proud of Oxford's chemistry, averring that if someone in Cambridge lit a Bunsen burner it was national news while if someone in Oxford isolated a new element it would be ignored by the press. He was therefore the ever-present life and soul of the Alembic Club, the University's chemical society, between the wars. He kept it going during the First World War and afterwards made it a very useful institution for both junior and senior members. It met once a week, with about two distinguished visitors a term, the remaining speakers being Oxonians from all the laboratories. There was a strong emphasis on research in progress and on research tactics and strategy. When a speaker was ill, Sidgwick was happy to step into the breach, sometimes mischievously so. In 1936 when Rideal was unwell Sidgwick spoke on the resonance theory of organic chemistry and glossed the ideas about the structure of benzene of Ingold, Robinson's bête-noire. As a discussant he was at home with all branches of chemistry. While president of the Chemical Society Sidgwick made some of the Club's meetings joint affairs with the Chemical Society and the Royal Institute of Chemistry. The Club often met in the lecture-room at Jesus and would then migrate after the formal proceedings to Sidgwick's rooms in Lincoln, a stone-throw's distance, where discussion would continue informally.

Sidgwick was also prominent in the joint chemistry–physics colloquia where he revealed his instinct for applying mathematics and physics fruitfully to chemical problems. He loved jousting with the destructive Lindemann and relished correcting the errors of specialists. When he tapped on his bald head with talon-like hands, the audience was agog for the chemical vulture to pounce. Though 'Sidger' was acerbic, he was sociable: whenever possible he presided over the tea club of the Dyson Perrins Laboratory, demolishing the sloppy English of those present, delivering impromptu expositions of chemical topics of all kinds, and attacking cuttingly the electronic theory of organic chemistry held by Robinson, the head of the laboratory where Sidgwick poured the tea. It was this sort of collegiality, allied to his synoptic view of chemistry, which made Sidgwick the dominant figure in Oxford chemistry as a whole between the wars, and not Hinshelwood, Robinson, or Soddy. His unique contribution was recognised by the University in 1935: he was elevated to a personal chair, a rare type of promotion then.[75]

5.7 X-RAY CRYSTALLOGRAPHY

Between the wars the department of mineralogy and crystallography led a precarious existence. It was not a final honours school, it recruited few undergraduates, and at best had only two staff (in the form of University demonstrators) in addition to the professor, Bowman, a non-researcher. Though marginal in the University, it provided in the 1930s a home for Powell and Hodgkin, two X-ray crystallographers who researched on the internal structure and chemical composition of puzzling compounds.

Until about 1930 the research of Bowman's department was mainly focused on crystal indexes, which involved studying the external characteristics of crystals. This tradition of morphological crystallography was found wanting by Powell, who had graduated in chemistry in 1928 and had been appointed a departmental demonstrator in 1929. For the first six months of 1930 Powell went to Germany to learn X-ray techniques from Ernst Schiebold at the University of Leipzig. On his return he began research on alkyl thallium compounds and in autumn 1931 was joined for a year by his first research student, Dorothy Crowfoot, an Oxonian chemist whose subject for her Part 2 degree was an extension of Powell's research to dialkyl thallium halides. Under Powell, X-ray plant was installed in 1932–3 in a different room of the University Museum. It was in this X-ray laboratory that Powell and Crowfoot pursued their research. Powell studied the structure of complicated inclusion compounds, for which he coined the term clathrates. He collaborated generously with Hume-Rothery on the structure of alloys, with Sidgwick on stereochemical types and valency groups, and with promising pupils such as A. F. Wells. Such was Powell's reputation relative to that of Crowfoot in 1944 that he defeated her and other strong contenders in a contested election for the readership in chemical crystallography. In 1947 she turned the tables being elected FRS three years before he was. His FRS did not lure any college into electing him as a fellow: that happened later when he was 57 years old when Hertford responded to his elevation to the chair of chemical crystallography in 1964.[76]

In contrast, Crowfoot's career shows the vital importance for her of the college system. Her father, J. W. Crowfoot, was an Oxford classicist who pursued a distinguished career in North Africa and Palestine, as an educator and archaeologist, becoming Principal of Gordon College, Khartoum, director of education and archaeology in Sudan, and director of the British School of Archaeology in Rome. Her mother was an expert on ancient textiles. Crowfoot went to Oxford simply because her father had been there. She landed at Somerville because her mother chose it and her father had met and was very impressed by Margery Fry, its principal. She decided to read chemistry because it was well taught at her school and she had been particularly impressed by two books, one of W. H. Bragg's books on X-ray crystallography and T. R. Parsons's text-book of biochemistry. Before arriving in Oxford she was thinking about X-ray work on the structure of biochemically important molecules. As an undergraduate she was excited by the lectures on complicated organic chemical structures given by Robinson. Through Brewer, her final year tutor, she was told that Powell had begun research in X-ray crystallography in the mineralogy department so she did her Part 2 under him, graduating in 1932 with a first.

From 1932–4 she worked at Cambridge, with Bernal, a practised Lothario, learning about the application of X-ray crystallography to the study of the structure of big organic molecules such as steroids and proteins, a new and exciting field that became her speciality. She was enabled to serve her intellectual apprenticeship under Bernal because Somerville elected her for 1932–3 to a Harcourt research scholarship, available only to Somerville graduates every other year and worth £100 pa, and then in spring 1933 made her a research fellow in natural science tenable for two years from October 1933. Funded by this fellowship she returned to Oxford in autumn 1934 and chose to begin her own independent work as an unpaid and unpaying freelance guest-researcher in the department of mineralogy doing just what she wanted. She needed more apparatus and, realising that neither Bowman nor Powell was adept at securing funding, she turned to Robinson who rapidly persuaded ICI to give her £600 for two X-ray tubes, a transformer set, and two goniometers (used to measure the angles of crystals). Through Powell's contact with Lindemann, alternating current was laid on via a cable from the Clarendon Laboratory. In spring 1935 Crowfoot used the new facilities to begin work on the structure of the protein insulin, using crystals supplied by Robinson. In return she helped Robinson by identifying sterol derivatives for him. In 1935 Somerville formally elected her as tutorial fellow in natural science, which gave her further financial and intellectual independence. At the outbreak of the Second World War she was one of only two tutorial fellows in science in the five women's colleges, the other being Jean Orr-Ewing at Lady Margaret Hall. Somerville also provided the means by which she met her future husband, Thomas Hodgkin, an Oxford historian. They first met in March 1937 at the London home of Fry, retired principal of Somerville, who was providing a haven for her cousin, Thomas Hodgkin, a communist who had recently been deported from Palestine. They married in December 1937, the first of their three children being born in 1938; but Crowfoot managed to sustain her left-wing politics and to combine motherhood and research. It was in 1937 that she secured her first research student, D. P. Riley, in an unusual way. Even though she was a college fellow, as a woman she was denied membership of the Alembic Club. In 1936 she was asked by Riley on behalf of the junior section of the Club to talk about the research she had done at Cambridge. He was so impressed by her account that he asked her to supervise his research for his Part 2 for 1937–8. Then he became her first DPhil student, working on the crystal structure of proteins. By the outbreak of war she was beginning to withdraw from collaborative work with her mentor, Bernal, and had established an independent reputation: in 1939 W. H. Bragg and M. Perutz were happy to supply to her X-ray data about haemoglobin for Riley's calculations, and she was widely regarded as Britain's leading authority on the determination of molecular weights using X-rays. Her scientific life was focused on research: before the war she gave only one course of lectures and, unlike Powell, avoided a demonstratorship.[77]

5.8 CAREERS: THE LURE OF INDUSTRY

Though many college fellows in chemistry did not enthuse publicly about organised applied science, Oxford's chemistry graduates were prominent as managers, directors, and researchers in the larger chemical firms. Between the wars a good post in industry was as desirable as taking a DPhil, assuming an academic post elsewhere, or becoming a school teacher. Several graduates of the 1920s and early 1930s became chairmen of important industrial and pharmaceutical enterprises: Christopher Kearton of Courtaulds, Peter Allen of ICI, Sydney Barratt of Albright and Wilson, Michael Perrin of the Welcome Foundation, and Kenneth Hutchison of the South Eastern Gas Board and the leading transformer of the gas industry in the 1950s. Some became research directors for a whole company: witness Geoffrey Gaut at Plessey, Bryan Topley at Albright and Wilson, and at ICI J. D. Rose who followed yet another Oxonian, Wallace Akers. The case of Akers, like that of Ernest Walls, managing director of Lever Brothers, reveals that even before the first war Oxford chemistry graduates were not spurning a career in industry.[78]

Their suitability had been particularly appreciated from 1907 by Brunner Mond whose chief chemist and then research manager at its Winnington plant in Cheshire, Francis Freeth, assiduously and snobbishly collected promising men from Oxford and made Winnington into a notable research department in the British chemical industry between the wars. Brunner Mond was attractive to Oxford graduates who were aware that it was a prosperous firm that paid good salaries and offered alluring prospects of advancement. At Winnington there was the added inducement of membership of the Winnington Hall Club that in some ways was an extension of college life into the fertile Cheshire plains. The Club, which excluded non-graduates and commercial staff, provided mechanisms for developing chemical ideas and social relations. With its dining rooms, bars, croquet lawns, guest rooms, and cordial informality, the Club offered remarkable opportunities for vertical and horizontal communication in a pleasant building with a Tudor wing. The Club, to which members were elected on the basis of their scientific and social qualifications, helped to generate corporate pride and loyalty among the scientifically trained staff of Brunner Mond. Winnington and its Club were dominated by Oxford chemists, a feature that was continued when Winnington became in 1926 the Alkali Group of ICI, formed that year by the merger of Brunner Mond, Nobel Industries, British Dyestuffs Corporation, and the United Alkali Company. At Winnington there was a small contingent of Cambridge engineers but the neighbouring universities of Liverpool and Manchester were sparsely represented. In 1919, with the development of Brunner Mond's synthetic ammonia plant at Billingham, further opportunities arose for Oxford graduates. Though the demands of high-pressure technology there led to the employment of more engineers that at Winnington, Billingham had its own club for its graduate scientists and managers at Norton Hall, modelled on that at Winnington.

The formation of ICI brought no change in recruitment practices for Winnington and Billingham, except that the visits to the Oxford laboratories were usually made not by Freeth but by Kenneth Gordon from Billingham and by H. E. Cocksedge, an Oxford chemist who replaced Freeth in 1926 as research manger at Winnington. Sometimes they were joined by yet another Oxford chemist, W. H. Demuth, who was development director at Winnington. Such men recruited fellow Oxonians not just on a mafia basis: they were keen to grab young chemists who had gained a year's useful research experience during Part 2 and in some cases were already co-publishers of papers with their tutors. This was so with Peter Allen, tutored by Hinshelwood, and Kearton, tutored by Thompson; Allen went to Winnington, Kearton to Billingham.

ICI in general and its Winnington and Billingham plants in particular were attractive to Oxford chemistry graduates who perhaps knew that a few of their leading teachers, such as Hinshelwood, Perkin, and Robinson, were consultants for ICI. The move in 1928 and 1930 of two Oxford chemistry dons, Applebey and Hartley, to research posts with ICI (Billingham) and the LMS Railway, respectively, sent out a firm signal to undergraduate chemists that a career in industry was worthwhile. No wonder that between the wars seventeen Jesus men, mainly chemists, joined ICI or its predecessors. The Alkali Group based at Winnington was in a strong position within ICI. Its starting salaries were the highest in any group of the company and its promotion prospects best as a result of its being the most profitable group. Oxonians did well at Winnington. Allen reached £1,000 pa eleven years after going there. Lincoln Steel received

£1,400 pa before he was 30. William Lutyens became chairman of the Alkali Group when just 40, and was succeeded by Digby Lawson and Steel. Though the Alkali Group employed in the 1930s about a fifth of the number of chemists to be found in Dyestuffs, about a third of those in General Chemicals, and about half of those in Fertilisers (Billingham) and Explosives, it was disproportionately well represented and powerful in ICI directors appointed from 1926 to 1952. Of the 29 executive directors, nine came from Winnington and six from Billingham. Only one came from Dyestuffs, which employed about 40% of ICI's chemists. As Winnington discriminated so heavily in its recruitment in favour of Oxford and other groups of ICI cast their nets more widely, that meant that Oxford chemists were prominent as directors of ICI. There was a wave appointed in the early 1940s (Akers, Lutyens, Lawson, and Steel) followed by another in the early 1950s (Allen and Prichard). If Winnington gave opportunities to Oxonians for managerial advancement, it also looked after researchers. For example, C. W. Bunn was recruited to Winnington in 1927 by Cocksedge who sent him back to Oxford to study crystal morphology. An expert crystallographer, who joined Hodgkin in Oxford during the war in the successful attack on the structure of penicillin, Bunn stayed with ICI as a researcher for the whole of his career. Though Winnington was the most attractive destination for Oxford chemists, there was a colony of them, mainly from St John's, at Billingham in the 1930s: Akers was chairman, Gordon general manager, Walter d'Leny technical director, Applebey research manager, and Kearton a rising star recruited by Applebey. Indeed of about 60 Oxbridge graduates employed at Billingham, six were from St John's.[79]

It was at Winnington, the home of the Alkali Group of ICI, that polythene, ICI's major discovery before the Second World War, was taken from laboratory experiments to large-scale production. Several Oxonians played an important role. Firstly, in 1931 Robinson, who was a consultant to ICI's Dyestuffs group, suggested that several reactions be tried under very high pressures without catalysts. One reaction in his list was that between ethylene and benzaldehyde. The work he outlined was done at Winnington by R. O. Gibson and E. W. Fawcett who studied some 50 reactions. In 1933 they subjected ethylene and benzaldehyde to 2,000 atmospheres' pressure and produced a small amount of polythene as the accidental consequence of looking for the chemical result of a specific reaction. Subsequent experiments on polythene were not reproducible and explosions so common that Dyestuffs Group withdrew its support of the work late in 1933.

Meanwhile Perrin, an Oxford graduate in chemistry (1928), had been recruited to Winnington by Freeth who placed him as an ICI employee in the Amsterdam laboratory of Michels, an expert on high-pressure techniques. In 1932 while on holiday in England, Perrin and J. C. Swallow, deputy director of research at Winnington, wrote a report on the desirability of studying the mechanisms of reactions at 20,000 atmospheres. In late 1933 Perrin returned to Winnington where he did high-pressure work under Swallow. Influenced and advised by Hinshelwood, Perrin began to study the effect of high pressure on reactions in solution, initially with Gibson and Fawcett, as a general phenomenon of physical chemistry. The interest in high pressures led him in late 1935 to repeat the 1933 experiment but without benzaldehyde. He succeeded in synthesising polythene from ethylene and in defining the reaction conditions that enabled the polymer to be made reproducibly and without explosions. Its commercial development was mainly due to Allen. Given the Oxford input into the discovery and re-discovery of polythene, and its industrial production, it was entirely appropriate that in 1939 Bunn, another Oxonian, revealed its structure using X-ray methods.[80]

5.9 CONCLUSION

At Oxford between the wars there was no post of head or chairman of chemistry, the occupant of which might have tried to secure administrative centralisation and to generate intellectual unity. Contrast Cambridge, where Pope abolished the last of the laboratories run by a men's college; he ruled his staff, including other professors, with a rod of iron; and controlled access to laboratory space and funds. Contrast also University College London, where Donnan, head of department, and Ingold, his professorial colleague, wished to avoid specialisation and division in chemistry so they collaborated in promoting their subject as a homogeneous science.[81]

At Oxford, managerial untidiness persisted and intellectual diversity flourished. Indeed research prospered in a variety of locations, partly as a result of that heightened individualism that was endemic in a polycratic university such as Oxford. A new kind of structural chemistry, X-ray crystallography, was launched by Powell and Hodgkin outside chemistry in the department of mineralogy. Metallurgy was promoted, sometimes against the odds, by Hume-Rothery in the Old Chemistry Department. Generally, the leading sectors of research were organic and physical chemistry, with inorganic chemistry languishing. Organic chemistry was built up by Perkin, who took advantage of internal dissatisfaction, external pressure, and of World War I, to focus without interference on research, an emphasis that was re-enforced by Robinson. Both were adept at securing external funding and staff from outside the University. Dedicated to research in the Dyson Perrins Laboratory, opened in 1916, they ran their own show, mainly ignored the colleges and their chemical fellows, and did not try to 'reform' Oxford.

In contrast, Soddy, the first Dr Lee's professor, did. He denounced the colleges and their chemical fellows, many of whom were physical chemists; he deplored the existence of college laboratories; and he wished to organise chemistry centrally. But he was at loggerheads with the University, suspicious of external funding, and left by Perkin and then Robinson to fight solo. He was therefore unable to secure a new University laboratory for physical chemistry, his intransigence prolonging the lives of the college laboratories, two of which were devoted to physical chemistry and closed only in the 1940s. When Hinshelwood succeeded Soddy, he brought to the chair his effortless superiority as a physical chemist and his affection for the college structure. When money for a new University chemistry laboratory became available in 1937, it was earmarked for physical chemistry because Hinshelwood and the college fellows who were physical chemists used their power to defend the superiority of their branch of chemistry. Inorganic chemistry, for which Hinshelwood was statutorily responsible, was left behind in the Old Chemistry building, which was widely recognised as unsuitable.

A unique feature of Oxford chemistry was Part 2 of the undergraduate degree course. No other British university enabled its chemistry undergraduates to spend a whole year researching at the bench under expert supervision and writing a thesis incorporating the results. Part 2 gave Oxford graduates research skills on top of the different but valuable abilities promoted in tutorials. No wonder that these graduates went on to become influential figures in schools, universities, and industry. Part 2 had a big impact, too, on college fellows and laboratory demonstrators. No longer could research be regarded as a personal luxury: it became an institutional norm because academic staff had to supervise Part 2 projects not just occasionally but year after year. Most fellows and demonstrators elected or appointed after 1918 were effective bench researchers. Some of those who were not, such as Taylor and Parkes, made useful scholarly contributions as editors and revisers of well-known texts written by others. The strong research profile and ethos of Oxford chemistry between the wars ensured that its staff were well prepared to make a notable contribution as boffins during the Second World War.

NOTES AND REFERENCES

1. J. Morrell, *Science at Oxford, 1914–1939: Transforming an Arts University*, Clarendon Press, Oxford, 1977, pp. 65–7.
2. T. S. Moore and J. C. Philip, *The Chemical Society 1841–1941. A Historical Review*, Chemical Society, London, 1947.
3. L. E. Sutton and M. Davies, *The History of the Faraday Society*, Royal Society of Chemistry, London, 1996.
4. On William Henry Perkin (1860–1929), *ODNB*, Waynflete professor of chemistry 1913–1929, J. Greenaway, J. F. Thorpe and R. Robinson, *The Life and Work of William Henry Perkin*, Chemical Society, London, 1932; Robinson, *J. Soc. Chem. Ind.*, 1929, **48**, 1008; J. Morrell, 'W. H. Perkin, Jr. at Manchester and Oxford', *Osiris*, 1993, **8**, 104; Perkin, 'Baeyer Memorial Lecture', *Memorial Lectures delivered before the Chemical Society, 1914–1932*, Chemical Society, London 1933, 47; H. E. Armstrong, 'Perkin', *Nature*, 1929, **124**, 623.
5. Chemistry. Proposed Reorganisation of Department, 1908–9, OUA, UR/SF/CHE/2; G. N. Curzon, *Principles and Methods of University Reform*, Clarendon Press, Oxford, 1909; E. B. Poulton, 'The Reform of Oxford University', *Nature*, 1909, **80**, 311.
6. *OUG*, 1911–12, **42**, 1002; Chemistry and Lee's Professorship, OUA, MR/7/2/7; Professorial Elections, 1898–1938, OUA, UDC/M/41/1; Thomas

180   *Chapter 5*

   Herbert Warren (1853–1930), *ODNB*; William Jackson Pope (1870–1939), *ODNB*.
7. J. Greenaway, "Perkin", 28; W. H. Perkin, *J. Chem. Soc.*, 1916, **109**, 815; W. H. Perkin, 'The Position of the Organic Chemical Industry', *J. Chem. Soc.*, 1915, **107**, 557; W. P. Wynne, 'Universities as Centres of Research', *J. Chem. Soc.*, 1925, **127**, 936.
8. J. C. Smith, *The Development of Organic Chemistry at Oxford*, typescript, no date, is invaluable; New Chemistry Laboratory Papers, OUA, UM/F/4/15; *HCP*, 1921, **118**, 61; *HCP*, 1922, **121**, 87; Charles William Dyson Perrins (1864–1958), *ODNB*.
9. M. Sanderson, *The Universities and British Industry 1850–1970*, Routledge and Kegan Paul, London, 1972, pp. 214–42; H. Gay, *The History of Imperial College London 1907–2007*, Imperial College Press, London, 2007, pp. 121–6.
10. W. H. Perkin, *J. Chem. Soc.*, 1915, **107**, 557.
11. *OUG*, 1915–16, **46**, 539, 546, 1916–17, **47**, 550, 556, 1917–18, **48**, 479, 1918–19, **49**, 602; Frederick Charles Hall (1882–1962); Dalziel Llewellyn Hammick (1887–1966), fellow of Oriel 1921–52, on whom E. J. Bowen, *Biog. Mem. Roy. Soc.*, 1967, **13**, 107; James Ernest Marsh (1860–1938), fellow of Merton 1906–30, on whom J. A. Gardner, *J. Chem. Soc.*, 1938, 1130.
12. *OUG*, 1916–17, **47**, 56; W. J. Reader, *Imperial Chemical Industries: A History, v.1: The Fore-runners 1870–1926*, Oxford University Press, London, 1970, pp. 266–75; Frederick Alfred Mason (1888–1947); Edward Hope (1886–1953), fellow of Magdalen 1919–53, on whom S. G. P. Plant, *J. Chem. Soc.*, 1953, 3730; Joseph Kenyon (1885–1961), *ODNB*, on whom E. E. Turner, *Biog. Mem. Roy. Soc.*, 1962, **8**, 49; George Roger Clemo (1889–1983), on whom B. Lythgoe and G. A. Swan, *Biog. Mem. Roy. Soc.*, 1985, **31**, 165; William Ogilvy Kermack (1898–1970), *ODNB*, on whom J. N. Davidson, F. Yates, and W. H. McCrea, *Biog. Mem. Roy. Soc.*, 1971, **17**, 399.
13. Andrea Angel (1877–1917), on whom *J. Chem. Soc.*, 1917, **111**, 321 and *Ox. Mag.*, 1916–17, **35**, 113; Herbert John George (1893–1939), fellow of Jesus 1923–39, on whom D. L. Chapman, *J. Chem. Soc.*, 1939, 1640; Malcolm Percival Applebey (1884–1957), fellow of St John's 1919–28, on whom J. L. S. Steel, *Proc. Chem. Soc.*, 1957, 214; Cyril Norman Hinshelwood (1897–1967), *ODNB*, fellow of Trinity 1921–37, Dr Lee's professor of chemistry 1937–64, on whom H. W. Thompson, *Biog. Mem. Roy. Soc.*, 1973, **19**, 375.
14. L. F. Haber, *The Poisonous Cloud: Chemical Warfare in the First World War*, Clarendon Press, Oxford, 1986; R. MacLeod, 'The Chemists go to War: The Mobilisation of Civilian Chemists and the British War Effort, 1914–18', *Ann. Sci.*, 1993, **50**, 455; Ogston, 'Hartley'.
15. Robert Benedict Bourdillon (1889–1971), fellow of University 1913–21, on whom E. J. Bowen, *[University] College Record*, 1971, **6**, 16; Humphrey Rivaz Raikes (1891–1955), *ODNB*, fellow of Exeter 1919–27; Edmund

John Bowen (1898–1980), *ODNB*, fellow of University College 1922–65, on whom R. P. Bell, *Biog. Mem. Roy. Soc.*, 1981, **27**, 83.
16. W. H. Perkin, *J. Chem. Soc.*, 1915, **107**, 557; *OUG*, 1915–16, **46**, 328, 341, 350, 423, 448, 458.
17. *OUG*, 1916–17, **47**, 184, 251, 303, 352, 428, 448, 466; W. H. Perkin and E. B. Poulton, 'Proposed Statute for the Encouragement of Oxford Research', *Ox. Mag.*, 1916–17, **35**, 121.
18. George David Parkes (1899–1967), fellow of Keble 1930–65, on whom *Ox. Mag.*, 1967–8, **86**, 106; Dorothy Crowfoot Hodgkin (1910–94), *ODNB*, fellow of Somerville 1935–77, Wolfson research professor 1960–77; Harry Munro Napier Hetherington Irving (1905–93), fellow of St Edmund Hall 1938–51; Ronald Percy Bell (1907–96), *ODNB*, fellow of Balliol 1933–67; Leslie Ernest Sutton (1906–92), *ODNB*, fellow of Magdalen 1936–73; Frederick Ernest King (1905–99), lecturer in organic chemistry, Magdalen 1936–43, Balliol 1937–45; William Kenneth Hutchison (1903–89), *ODNB*; Edward Penley Abraham (1913–99), *ODNB*, letter, 13 July 1987; Edward Francis Hussey Caldin (1914–), letter, 21 August 1987; Wilson Baker (1900–2002), fellow of Queen's 1936–44; John Hulton Wolfenden (1902–89), fellow of Exeter 1928–47.
19. S. Leacock, *My Discovery of England*, McClelland and Stewart, Toronto, 1922, p. 80; Autobiographical Fragment, H. M. Powell Papers, A3, Bodleian Library.
20. R. P. Bell, interview, 16 July 1987; E. J. Bowen, 'Hartley', *Ox. Mag.*, 1972–3, **91**, 7.
21. Keith J. Laidler (1916–2003), transcript interview with C. King, University of Ottawa, 1983, pp. 5, 7, 35.
22. L. E. Sutton, interview, 27 Nov 1987; H. Tizard, 'Sidgwick', 241.
23. Information from E. J. W. Whittaker.
24. Information from Professor S. F. Mason; D. Rooke, 'Hutchison', *Biog. Mem. Roy. Soc.*, 1996, **42**, 207.
25. Harold Warris Thompson (1908–83), *ODNB*, fellow of St John's 1930–75; Frederick Sydney Dainton (1914–97), *ODNB*.
26. E. P. Abraham, letter, 13 July 1987.
27. D. A. Long, 'The Sir Leoline Jenkins Laboratories', *Jesus College Record*, 1989, 17; information from L. A. Moignard.
28. *OUG*, 1918–19, **49**, 283, 408; F. Soddy, "Accommodation for Inorganic and Physical Chemistry. University of Oxford", Oxford, 21 Nov 1919; F. Soddy, evidence to Asquith Commission, 29 Sept 1920; *HCP*, 1919, **114**, 185–6, 1919, **115**, 15; OUA, 15 June 1920, NS/R/1/2; OUA, 15 June 1920, NS/M/1/3; *Asquith Report*, p. 117. Frederick Soddy (1877–1956), *ODNB*, Dr Lee's professor of chemistry, 1919–36.
29. F. Soddy, 'The Profits of Research', *Nature*, 1918, **101**, 343.
30. Alexander Smith Russell (1888–1972), Dr Lee's reader 1919–55, fellow of Christ Church 1920–55.
31. *HCP*, 1922, **121**, 87–93, 1922, **123**, 59, 119, 1923, **124**, 57, 1923, **125**, 157–62, 1924, **127**, 193–6; *OUG*, 1923–4, **54**, 221, 1924–5, **55**, 124; F. Soddy

to Vice-Chancellor, 3 Jan 1924, OUA, MR/7/2/7; Balliol-Trinity Laboratory. Correspondence re Accounts 1920–3 and other College Laboratories 1922, OUA, MR/6/3/34; Correspondence re Chemical Laboratories 1923, OUA MR/7/2/7.

32. H. B. Hartley, 'Schools of Chemistry in Great Britain and Ireland. xvi. The University of Oxford', *J. Roy. Inst. Chem*, 1955, **79**, 116, 176. Hostile accounts are A. Fleck, *Biog. Mem. Roy. Soc.*, 1957, **3**, 203 and A. D. Cruickshank, 'Soddy at Oxford', *Brit. J. Hist. Sci.*, 1979, **12**, 277. More sympathetic are: F. A. Paneth, 'A Tribute to Frederick Soddy', *Nature*, 1957, **180**, 1085; A. S. Russell, *Chem. Ind.*, 1956, 1420; and L. Merricks, *The World Made New: Frederick Soddy, Science, Politics, and Environment*, Oxford University Press, Oxford, 1996.

33. F. Soddy to Vice-Chancellor, 26 May 1923, OUA, MR/7/2/7; F. Soddy, 'Old Universities and New Needs', *Nature*, 1918, **101**, 461.

34. OUA, 1923–4, MR/7/2/7.

35. F. Soddy, Petition to the King's Most Excellent Majesty in Council, 19 Sept 1927; University of Oxford, Reply to the Petition..., 21 March 1928, both in Soddy Papers, B18, Bodleian Library; *HCP*, 1928, **141**, 13; *OUG*, 1919–20, **50**, 781, 1920–1, **51**, 179, 1926–7, **57**, 602, 739; W. Garrod to Vice-Chancellor, 1 Aug 1927, OUA, Lee's Chair of Chemistry: Aldrichian Demonstrator, 1927, U Sol/6/2.

36. R. T. Gunther, "The Daubeny Laboratory Register 1849–1923", Oxford University Press, Oxford, 1904–24.

37. E. M. Walker to Physical Sciences Board, 11 May 1932, OUA, PS/R/1/3; Queen's Laboratory Books, 1900–34, History of Science Museum, Oxford, MS 49; E. P. Abraham to author, 13 July 1987; Eric John Francis James (1909–1992), *ODNB*, High Master Manchester Grammar School 1945–62, first Vice-Chancellor, University of York, 1962–73.

38. K. J. Laidler, 'Chemical Kinetics and College Laboratories', *Arch. Hist. Exact. Sc.*, 1988, **38**, 197 is invaluable; J. N. L. Baker, "Jesus College, Oxford 1571–1971", Jesus College, Oxford, 1971, 128, 135; Leonard Ary Woodward (1903–76), fellow of Jesus 1939–70.

39. L. E. Sutton to R. Robinson, 25 Mar 1934, Sidgwick Papers, 79.

40. *HCP*, 1936, **164**, 141; C. N. Hinshelwood to Margoliouth, 22 Feb 1937, OUA, PS/R/1/5.

41. A. G. Ogston, 'Hartley'; R. P. Bell, 'Bowen'; E. J. Bowen, 'Balliol-Trinity Laboratories'; on J. Warrell, R. F. Barrow and C. J. Danby, "Physical Chemistry Laboratory", pp. 77–8.

42. A. G. Ogston, 'Hartley'; H. B. Hartley, 'The Theodore William Richards Memorial Lecture', "Memorial Lectures delivered before the Chemical Society, 1914–32", Chemical Society, London, 1933, 131; Oliver Gatty (1907–40), fellow of Balliol 1931–3; Josiah Charles Stamp (1880–1941), *ODNB*.

43. Lionel Alfred Kilby Staveley (1914–96), fellow of New College 1939–82; James Dewe Lambert (1912–2001), fellow of Trinity 1938–76; Thomas Weston Johns Taylor (1895–1953), *ODNB*, fellow of Brasenose 1920–46.

44. H. W. Thompson, 'Hinshelwood'; C. N. Hinshelwood, *The Kinetics of Chemical Change in Gaseous Systems*, Clarendon Press, Oxford, 1926, 2nd edn., 1929, 3rd edn., 1933; C. N. Hinshelwood, *The Kinetics of Chemical Change*, Clarendon Press, Oxford, 1940; C. N. Hinshelwood and A. T. Williamson, *The Reaction between Hydrogen and Oxygen*, Clarendon Press, Oxford, 1934; Emyr Alun Moelwyn-Hughes (1905–78); Klaus Clusius (1903–63).
45. H. W. Thompson, *A Course in Chemical Spectroscopy*, Clarendon Press, Oxford, 1938; R. P. Bell, *Acid-base Catalysis*, Clarendon Press, Oxford, 1941; E. J. Bowen, *The Chemical Aspects of Light*, Clarendon Press, Oxford, 1942; Hinshelwood, *The Structure of Physical Chemistry*, Clarendon Press, Oxford, 1951, pp. v, 3, 110.
46. B. G. Cox and J. H. Jones, 'Bell', *Biog. Mem. Roy. Soc.*, 2001, **47**, 19; William Francis Kenrick Wynne-Jones (1903–82).
47. R. E. Richards, 'Thompson', *Biog. Mem. Roy. Soc.*, 1985, **31**, 573.
48. J. D. Lambert, letter, 10 May 1988.
49. L. A. K. Staveley, letters 6 April, 15 May 1988.
50. A. J. Downs, D. A. Long and L. A. K. Staveley, *Essays in Structural Chemistry*, MacMillan, London, 1971.
51. D. H. Whiffen, *Biog. Mem. Roy. Soc.*, 1994, **40**, 369; Sutton, 'The Earlier Studies in Great Britain of the Structure of Molecules in Gases and Vapours by Electron Diffraction, with an epilogue', in P. Goodman (ed), "*Fifty Years of Electron Diffraction*", Reidel, Dordrecht, 1981, pp. 92–100; Sutton, account of life, Sutton Papers, A2, Bodleian Library.
52. *HCP*, 1936, **164**, 141, 221, 1936, **165**, 125; OUA, 13 Nov 1936, UR/SF/CHE/1B; OUA, UDC/M/41/1, 20 Nov 1936, 16 Jan 1937; *Ox. Mag.*, 1936–7, **55**, 218, 244, 307.
53. Development Plan 1936–7, OUA, UM/F/4/18; "Oxford [Appeal]: Special Number February 1937"; *OUG*, 1937–8, **68**, 27; R. F. Barrow and C. J. Danby, "Physical Chemistry Laboratory", p. 76.
54. J. Greenaway, "Perkin", pp. 32–3.
55. Harry Raymond Ing (1899–1974), on whom H. O. Schild and F. L. Rose, *Biog. Mem. Roy. Soc.*, 1976, **22**, 239.
56. John Masson Gulland (1898–1947), demonstrator in Dyson Perrins Laboratory 1924–31, on whom R. D. Haworth, *Obit. Not. Roy. Soc.*, 1948, **6**, 67; Robert Downs Haworth (1898–1990), demonstrator in Dyson Perrins 1925–6, on whom E. R. H. Jones, *Biog. Mem. Roy. Soc.*, 1991, **37**, 265; Wilson Baker (1900–2002), on whom J. F. W. McOmie and D. M. G. Lloyd, *Biog. Mem. Roy. Soc.*, 2003, **49**, 15.
57. Thomas Stevens Stevens (1900–2000), on whom E. Haslam and D. G. Morris, *Biog. Mem. Roy. Soc.*, 2003, **49**, 523; William Davies (1895–1966); Osman Achmatowicz (b 1899); Louis Frederick Fieser (1899–1977); Joseph Blake Koepfli (b 1904); Victor Martin Trikojus (b 1902).
58. P. V. M. Benecke, 'Laurie Magnus: "Herbert Warren of Magdalen"', Magdalen College MS 407, p. 67.
59. Robert Robinson (1886–1975), *ODNB*, Waynflete professor of chemistry 1930–55, on whom A. R. Todd and J. W. Cornforth, *Biog. Mem. Roy. Soc.*,

1976, **22**, 415; R. Robinson, *Memoirs of a Minor Prophet: Seventy Years of Organic Chemistry, Volume I*, Elsevier, London, 1976; T. I. Williams, *Robert Robinson: Chemist Extraordinary*, Oxford University Press, Oxford, 1990; R. Robinson, 'The Perkin Family of Organic Chemists', *Endeavour*, 1956, **15**, 92; A. R. Todd, letter, 17 Nov 1986; M. Tomlinson, letter, 23 Oct 1988; J. C. Smith, "Organic Chemistry at Oxford".
60. Gertrude Maud Robinson (1886–1954), on whom *J. Chem. Soc.*, 1954, 2667.
61. W. H. Brock, *The Fontana History of Chemistry*, Fontana, London, 1992, pp. 522–48.
62. Dyson Perrins Lab, 1929–47, OUA, UR/SF/CHE/5/.
63. Alexander Robertus Todd (1907–97), *ODNB*, on whom A. R. Todd, *A Time to Remember*, Cambridge University Press, Cambridge, 1983, and D. M. Brown and H. Kornberg, *Biog. Mem. Roy. Soc.*, 2000, **46**, 515; Henry Norman Rydon (1912–91); John Warcup Cornforth (1917–); Arthur John Birch (1915–95), on whom R. W. Rickards and J. Cornforth, *Biog. Mem. Roy. Soc.*, 2007, **53**, 21; Harold Douglas Springall (1910–82).
64. Frederick Ernest King (1905–99), demonstrator 1931–48, on whom D. Whiting, *Biog. Mem. Roy. Soc.*, 2003, **49**, 301; John Charles Smith (1900–84), demonstrator 1931–55.
65. A. R. Todd and J. W. Cornforth, 'Robinson', 421, 526; Robinson Papers, A35,
    B67–79, B82–6, Royal Society.
66. Society for the Protection of Science and Learning Papers, 24/8, 208/4, 227/3, 208/3, 226/5, 566/2, Bodleian Library; R. Robinson to registrar, 30 Nov 1838, OUA, UR/SF/RC/1.
67. L. E. Sutton to N. V. Sidgwick, 14 Jan, 15 Jun 1934, to R. Robinson, 25 March 1934, Sidgwick Papers, V, 29, Lincoln College, Oxford; P. J. T. Morris and A. S. Travis, 'The Role of Physical Instrumentation in Structural Organic Chemistry' in J. Krige and D. Pestre, eds, *Companion to Science in the Twentieth Century*, Routledge, London, pp. 715–39.
68. *Times*, 26 Sept 1920; OUA, UR/SF/PHA/3, Sept–Nov 1920; *HCP*, 1921, **120**, 25.
69. F. Soddy, annual reports, *OUG*, 1920–7.
70. *OUG*, 1929–30, **60**, 650; Lee's Chair and Old Chemistry Dept 1927–33, OUA, UC/FF/288; Dept of Inorganic Chemistry 1930–47, OUA, UR/SF/CHE/6.
71. L. Merricks, "Soddy"; F. Soddy, *Science and Life: Aberdeen Addresses*, Murray, London, 1920; F. Soddy, Foreword, A. D. Hall *et al.*, *The Frustration of Science*, Allen and Unwin, London, 1935.
72. E. J. Bowen, 'Lambert'; M. P. Applebey, 'Hartley'; Ing, 'Walden'; Frederick Mason Brewer (1902–63), on whom L. E. Sutton, *Proc. Chem. Soc.*, 1964, 381.
73. R. E. Richards, 'Thompson'; Christopher Frank Kearton (1911–92), *ODNB*; P. Gray and K. J. Ivin, 'Dainton', *Biog. Mem. Roy. Soc.*, 2000, **46**, 85; John Wilfrid Linnett (1913–75), *ODNB*, on whom A. D. Buckingham, *Biog. Mem. Roy. Soc.*, 1977, **23**, 311.

74. William Hume-Rothery (1899–1968), *ODNB*, on whom G. V. Raynor, *Biog. Mem. Roy. Soc.*, 1969, **15**, 109; W. Hume-Rothery, *The Metallic State, Electrical Properties and Theories*, Clarendon Press, Oxford, 1931, and *The Structure of Metals and Alloys*, Institute of Metals, London, 1936; H. T. Tizard, 'The Needs of Oxford Science', *Oxford: Special Number*, 1937, 52; Geoffrey Vincent Raynor (1913–83).
75. Sidgwick Papers; H. T. Tizard, 'Sidgwick'; L. E. Sutton, *Proc. Chem. Soc.*, 1958, 310; C. N. Hinshelwood, *Ox. Mag.*, 1951–2, **70**, 284; R. P. Bell, L. E. Sutton and A. G. Ogston, interviews; Alembic Club Minute Books, 1913–38, Museum of History of Science, Oxford; N. V. Sidgwick, *The Organic Chemistry of Nitrogen*, Clarendon Press, Oxford, 1910; *The Electronic Theory of Valency*, Oxford University Press, London, 1927; *Some Physical Properties of the Covalent Link in Chemistry*, Cornell University Press, Ithaca, 1933; *The Organic Chemistry of Nitrogen. New Edition Revised and Rewritten by T. W. J. Taylor and W. Baker*, Clarendon Press, Oxford, 1937; *The Chemical Elements and their Compounds*, Clarendon Press, Oxford, 1950.
76. Herbert Marcus Powell (1907–91), on whom K. A. McLauchlan, *Biog. Mem. Roy. Soc.*, 2000, **46**, 425; *OUG*, 1930–1, **61**, 699; 1931–2, **62**, 670; 1933–4, **64**, 215; 1935–6, **66**, 218; A. F. Wells, *Structural Inorganic Chemistry*, Clarendon Press, Oxford, 5 editions, 1945–84.
77. G. Dodson, 'Hodgkin', *Biog. Mem. Roy. Soc.*, 2002, **48**, 179; G. Ferry, *Dorothy Hodgkin: A Life*, Granta, London, 1998; D. C. Hodgkin, 'Bernal', *Biog. Mem. Roy. Soc.*, 1980, **26**, 17; A. Brown, *J. D. Bernal: The Sage of Science*, Oxford University Press, Oxford, 2005; Sara Margery Fry (1874–1958), *ODNB*; Thomas Lionel Hodgkin (1910–82), *ODNB*.
78. Christopher Frank Kearton (1911–92), *ODNB*, on whom N. Wooding, *Biog. Mem. Roy. Soc.*, 1995, **41**, 219; Peter Christopher Allen (1905–93), *ODNB*; Sydney Barratt (1898–1975); Michael Perrin (1905–88), *ODNB*; William Kenneth Hutchison (1903–89), *ODNB*, on whom D. Rooke, *Biog. Mem. Roy. Soc.*, 1996, **42**, 205; Geoffrey Gaut (1909–92); Bryan Topley (1901–86); John Donald Rose (1911–76), *ODNB*, on whom A. W. Johnson, *Biog. Mem. Roy. Soc.*, 1977, **23**, 449; Wallace Alan Akers (1888–1951), *ODNB*, on whom Lord Waverley and A. Fleck, *Biog. Mem. Roy. Soc.*, 1955, **1**, 1; Ernest Walls (1881–1961).
79. Reader, "ICI", **1**, pp. 91, 219; **2**, 11, 70–80, 93; Sir Peter Allen, Baron Kearton, private communications; Francis Arthur Freeth (1884–1970), *ODNB*, on whom P. Allen, *Biog. Mem. Roy. Soc.*, 1976, **22**, 105; Kenneth Gordon (1897–1955); Herbert Edwin Cocksedge (1884–1962); William Henry Horner Demuth (b 1898); J.N.L. Baker, "Jesus College, Oxford 1571–1971", Jesus College, Oxford, 1971, pp. 131–2; Joseph Lincoln Spedding Steel (1900–85); William Lutyens (1891–1971); Digby Lawson (d 1947); Charles Ross Prichard (1903–76); Charles William Bunn (1905–90), on whom U. W. Arndt, *Biog. Mem. Roy. Soc.*, 1991, **37**, 71; Walter d'Leny (b 1902).
80. Reader, "ICI", **2**, pp. 349–58; D. G. H. Ballard, 'The Discovery of Polythene' in R. B. Seymour and T. Cheng, eds, *History of Polyolefins*, Reidel,

Dordrecht, 1976, pp. 9–53; M. W. Perrin, 'The Story of Polythene', *Research*, 1953, **6**, 111; R.O. Gibson, E.W. Fawcett, and M. W. Perrin, 'The Effect of Pressure on Reactions in Solution', *Proc. Roy. Soc. A*, 1935, **150**, 223; E. G. Williams, M. W. Perrin, and E. W. Gibson, 'The Effect of Pressure up to 12,000 kg/cm^2 on Reactions in Solution', *Proc. Roy. Soc. A*, 1936, **154**, 684; C. W. Bunn, 'The Crystal Structure of Long-chain Normal Paraffin Hydrocarbons. The Shape of the CH$_2$ Group', *Trans. Far. Soc.*, 1939, **35**, 482.
81. William Jackson Pope (1870–1939), *ODNB*, on whom A. Thackray and M. E. Bowden, 'The Rise and Fall of the 'Papal State'' in M. D. Archer and C. D. Haley, *The 1702 Chair of Chemistry at Cambridge*, Cambridge University Press, Cambridge, 2005, pp. 189–209; G. K. Roberts, 'Physical Chemistry for Industry: the Making of the Chemist at University College London, 1914–1939', *Centaurus*, 1997, **39**, 291; G. K. Roberts, 'C. K. Ingold at University College London: Educator and Department Head', *Brit. J. Hist. Sci.*, 1996, **29**, 65.
82. R. Curtis, C. Leith, J. Nall and J. Jones, *The Dyson Perrins Laboratory and Oxford Organic Chemistry 1916–2004*, Published by John Jones, Balliol College, Oxford, 2008.

CHAPTER 6
Interlude: Chemists at War

JOHN S. ROWLINSON

Department of Chemistry, Physical and Theoretical Chemistry Laboratory, Oxford University, South Parks Road, Oxford, OX1 3QZ

The memory of World War I was still fresh in the minds of many as the approach of World War II became clear after the 'Munich' crisis in September 1938. The loss of Henry Moseley at Gallipoli in 1915 had epitomised the folly of not using effectively the scientific talent available. The Government therefore set up a Central Register of those with "professional, scientific, technical or higher administrative qualifications", and so was prepared on the outbreak of war in September 1939 to call on the services of university departments to help with the solution of the problems of what was going to become an increasingly technological war. It has been said that the first World War was the chemists' war and the second, the physicists', but this is an over-simplification; the Chemistry Departments at Oxford were to be heavily involved in the second.

Robinson (the President of the Chemical Society from 1939 to 1941) was insistent that his team of demonstrators and other research workers be kept intact to deal with whatever problems might arise. From the Dyson Perrins Laboratory only T.W.J. Taylor[1] enlisted for military service, becoming, first, a major in the gas-warfare establishment at Porton, and then becoming a scientific adviser to Lord Mountbatten in the Far East. He was joined there by J.D. Lambert[2] from Hinshelwood's department. James Lambert was the son of Bertram, and had succeeded Hinshelwood as the tutor at Trinity College when Hinshelwood was elected to the chair.

Robinson's own research[3] was directed in part to the sterols on which he continued to publish throughout the war, producing eventually a series of 53 papers on these compounds from 1933 to 1955. He was not generally enthusiastic about the contribution of physical methods to the solution of the problems of organic structures, believing in the greater certainty of classical degradations and syntheses, but he had been impressed with the X-ray studies of sterols by J.D. Bernal at Cambridge in the 1930s and so had helped Dorothy Hodgkin, Bernal's former PhD student, in her attempts to set up similar facilities at Oxford by supporting her applications to ICI and, later, to the Rockefeller Foundation. Robinson and Howard Florey in the Dunn School of Pathology had shared a common interest in the enzyme lysozyme, which had mild antibiotic properties, so he and Hodgkin were therefore both on hand when there came news from the Dunn School of a promising new antibiotic produced by the mould *penicillium notatum*. There, Florey and Ernst Chain, a refugee Jewish chemist from Berlin and now a departmental demonstrator in pathology, had shown on 25 May 1940 that this mould produced an antibiotic that was extraordinarily effective against a streptococcal infection in mice, and six months later, that it worked also in a man.[1] This antibiotic had been discovered in 1928, and named penicillin, by Alexander Fleming at St Mary's Hospital in London, but he had not followed up his discovery. Some later work had seemed to show that it was too unstable to be of clinical use. But with the new experiments in the Dunn School its value in war was now obvious and the resources of the Oxford science area were mobilised.

The chemists needed to know the structure of the antibiotic in order, it was hoped, to be able to synthesise it, and not have to rely on its tedious extraction from the very dilute solutions obtained by fermentation. As well as Chain, there was already another chemist in the Dunn School with the appropriate experience – Edward Abraham, a Queen's College student who had been taught by Chattaway and Wilson Baker in the 1930s and, with the closure of the Queen's laboratory then imminent, had chosen to do his D.Phil. with J.C. Smith in the Dyson Perrins Laboratory. He had recently come back to Oxford after a year in the Biokemska Institut in Stockholm.[6] Chain and Abraham had prepared the barium salt of penicillin at about 50% purity when the Dyson Perrins team became involved. There the two senior members of the team were Robinson and Baker; they were joined by the Australian, John Cornforth, who had taken his first degree in Sydney, had completed his D.Phil. with Robinson in 1941, and who was to receive the Nobel Prize for chemistry in 1975. Others who joined the team included Cornforth's wife Rita, Freddie King and his student Michael Dewar,[7] who went on to have a distinguished career as a theoretical organic chemist, and John Barltrop, who became T.W.J. Taylor's successor as tutor at Brasenose College in 1946 when Taylor went to be the first Principal of the University College of the West Indies.

Progress in determining the structure was slow at first since only degradation products were available. The early samples had been found to contain sulfur, but this element was not detected by the analysts in the Dyson Perrins

---

[i]There are extensive accounts of the work on penicillin at Oxford but most naturally concentrate on the problems of its isolation, production and use.[3,4] This account is restricted to the determination of its chemical structure, for which the principal reference is the book[5] written by a team of those directly involved on both sides of the Atlantic. Robinson was the chairman of the editors.

Laboratory in a later and, it was believed, purer sample. Abraham and Chain at first took this result at face value but in 1943 Baker showed by a simple test that there was a sulfur atom present. It was agreed that there were two rings in the molecule but the chemists could not agree on the structure of these rings.[8] Some, including Robinson, favoured two linked five-membered rings – the oxazalone structure – but Abraham, backed by Chain, suggested a four-membered ring fused to a five-membered ring – the β-lactam structure, which had not previously been found in a natural product. Crystalline salts of the alkali metals were prepared, but the sodium salt was not isomorphous with the potassium and rubidium, and, moreover it was found that there were two different penicillins, differing in their side chains. At this point the crystallographers and spectroscopists were brought in. Hodgkin used X-ray diffraction to study the potassium and rubidium salts, relying on the presence of these two heavy atoms to carry out a Fourier analysis. The sodium salt was studied by C. W. Bunn, a former Exeter College chemist now working at the Winnington laboratory of ICI, who relied on trial-and-error methods backed by optical diffraction patterns of large-scale models of possible structures.[9] The infrared spectra were analysed in Oxford by 'Tommy' Thompson, assisted by Rex Richards, and by G.B.B.M. Sutherland at Cambridge. It was the crystallographers who won the race in January 1945; the β-lactam structure was correct. At first the spectroscopists had backed the oxazalone structure, because they did not then fully appreciate the effects of the crystal field on the infrared vibrational frequencies, particularly of the C=O bond, which was used as the principal diagnostic criterion.[10] The solution of this structural problem by the crystallographers was a notable feat that established Hodgkin's reputation as a leader in this field and was to be a sign of her further successes to come with other natural products and to the receipt of the Nobel prize for chemistry in 1964.[11]

War-time Britain was not the place to take on the difficult and tedious job of preparing penicillin by extraction from fermenting preparations of mould, and much of the technology soon passed to the better-equipped and less-stressed American pharmaceutical industry. The hope that a knowledge of the structure would lead to a convenient synthesis of the antibiotic proved unfounded. By 1945 about a thousand chemists in thirty-nine major laboratories in Britain and America were working on penicillin, but it was 1959 before a complete synthesis was achieved and it is still not an economic way of producing the original drug.

Meanwhile, Hinshelwood and his colleagues had been enrolled in a separate research project for the Ministry of Supply. On succeeding Soddy in 1937 he had moved his office to the Old Chemistry Department that he soon renamed the Inorganic Chemistry Laboratory. Thompson and his junior colleagues J.W. Linnett[12] and L.A.K. Staveley,[13] departmental demonstrators from 1939 and 1940 respectively, were housed there; Leslie Sutton was in the Dyson Perrins laboratory, and the other physical chemists in the Balliol-Trinity and Jesus laboratories. This distribution had long been perceived to be an unsatisfactory arrangement and plans for a new Physical Chemistry Laboratory (Figure 6.1) were underway as early as 1934, even before Soddy had been persuaded to step

Figure 6.1 The Physical Chemistry Laboratory, from the University Almanack of 1942.

down. A site to the east of the Dyson Perrins laboratory was allocated but there was no money available. As so often, it was Lord Nuffield[ii] who came to the aid of the University, with a covenant for £100,000 in November 1937. Work began slowly; the first builder went bankrupt, and after the war started it became difficult to obtain materials. But by early 1941 the laboratory was complete and the staff and their research gear moved in.[14] Sutton had photographed the operation month-by-month on a ciné-camera from his laboratory in the Dyson Perrins and this film is believed to be the only such record of the building of a laboratory.[iii]

The problem that the Ministry put to Hinshelwood was the improvement of the British service respirator.[15] This was a clumsy piece of apparatus and when Linnett went to Porton, soon after the war had started, he came back to say, "My God, they have done nothing since 1918".[14] The main trouble was the inefficiency of the charcoal used as the absorbent in the respirator, and Hinselwood and his colleagues at once started on a study of the methods of preparing effective charcoals from coal, since the better absorbents from coconut shell

---

[ii] William Morris, Lord Nuffield (1877-1963) was the founder of Morris Motors, and a considerable benefactor to Oxford University and hospitals. Nuffield College, a graduate college, was established in 1937.

[iii] Sutton's record of the building of the laboratory, the account by Barrow and Danby[14], and an anonymous supplement to that account, bringing the history up to 2000, are available on the website: http://www.chem.ox.ac.uk/history/

would not be readily available. Most of the physical and inorganic chemists were enrolled: E.J. Bowen, F.M. Brewer, B. Lambert, J.W. Linnett, L.A.K. Staveley, L.E. Sutton, H.W. Thompson, and J.H. Wolfenden. At least a dozen research students were also involved. The financial arrangements for both Robinson's and Hinshelwod's programmes were that the Ministry bought out a fraction, usually a half, of the research workers' time and paid this sum to the university and, if they were tutorial fellows also, to the college, with a payment for overheads to both the university and the college laboratories. Three men were not involved in war work in Oxford: R.P. Bell who was at Chatham House, the independent 'think tank' that then advised the Government, where his knowledge of German and Danish was required, J.D Lambert who was in India, and H.M. Powell, who was away from Oxford and who played little or no part in the crystallographic work on penicillin, but whose war-time work is still a mystery.[16]

The charcoal work went ahead rapidly with this large team and by July 1941 coal charcoals had been prepared that could, in the laboratory, absorb or oxidise the most feared gases. Scaling the treatments up to a production scale was undertaken in collaboration with industrial partners, of whom Sutcliffe Speakman in Lancashire was the most involved. Here, the Oxford contribution was to stress the importance of a uniform and well-controlled time and temperature regime in the activation of the carbon by burning away part of the charge by oxidation with super-heated steam. The effort in Oxford was then cut back in the second half of 1941; Brewer left the team to run the Air Raid Precautions (ARP) programme in Oxford; Wolfenden went to Washington to liaise with the increasing American effort in the charcoal field which was managed by W.A. Noyes Jr at the University of Rochester; Linnett, Staveley, Sutton, and John Danby,[17] later Hinshelwood's 'graduate assistant', worked on fuses and flares; and two ex-students, D.H. Everett and a Canadian Rhodes Scholar, J.G. Davoud, were recruited into a secret group attached to the Special Operations Executive (SOE) where they were to invent offensive devices that could be used by partisan forces on the Continent. This group took in eventually at least another five former Oxford chemists.[18]

Thompson had been part of Hinshelwood's charcoal team at the start of the war but his expertise in infrared spectroscopy was later exploited, not only by the penicillin group, but also in the analysis of aviation fuel from shot-down German aircraft. The sources of the fuel could often be identified from the spectra. Some of the organic chemists, led by J.C. Smith, collaborated in this work by preparing many pure samples of different hydrocarbons to standardise the observations.[19,20] Infrared spectroscopy was then a relatively new branch of physical chemistry and this work was a speculative venture, one of several that were revealed at a meeting of the Faraday Society shortly before the war ended. This meeting was dominated by the contributions from Oxford and Cambridge.[20]

Outside the Chemistry Departments, and so beyond the scope of this account, was the important discovery in the Biochemistry Department of BAL, or 'British Anti-Lewisite' (dimercaptopropanol), an antidote to the vesicant liquid known as Lewisite. BAL was soon found to have other valuable pharmacological properties.[21] The Oxford physicists' contribution to the war effort was mainly in the two fields of radar and the separation, by gaseous diffusion, of the isotope ^{235}U for use in an atomic bomb. This latter project also involved some chemists, who worked in the Jesus laboratory behind closed doors. The diffusing gas, UF_6, is a difficult and reactive substance and it was presumably the problems of handling this gas that were undertaken in Jesus, where the emerging D.Phil. theses all carried the deliberately misleading title, 'Some properties of Tube Alloys', the code name for the British atomic bomb project.[22] This work was supervised by D.L. Chapman and L.A. Woodward. The latter, a student of Sidgwick's, had been elected a fellow of Jesus in 1939 after working for his Ph.D. with Peter Debye in Leipzig.[23]

During the War undergraduates were still admitted to study chemistry and other sciences that were deemed to have 'military' value, and were exempt from conscription as long they remained in good academic standing.[24] The timetable was rearranged to suit the requirements of 'blacking-out' windows, with practical classes in the day, some lectures from 5 to 7 in the evening, and tutorials often from 8 to 10 pm.[19] If the students continued to study for a D.Phil. then they joined one of the war-related programmes for part of their work but could often also study other problems in the remainder of their time. Their theses were commonly in two parts, the first of which was 'classified' but the second of which could be published at once.

When the war was over most of the chemists returned to their pre-war interests, but some embarked on new ventures – Linnett in problems of molecular structure (although retaining his war-time interest in flames and combustion), Woodward in Raman spectroscopy, and Staveley in thermodynamics that he had studied before the War with Klaus Clusius in Munich but had not yet been able to follow up. Hinshelwood published some of the previously classified papers[25] and continued work on the reactions on charcoal surfaces and other kinetic studies, but his heart was not in it as it had been earlier; all that he was now really interested in was the kinetics of bacterial growth and adaptation.

REFERENCES

1. Thomas Weston Johns Taylor (1895–1952), *ODNB*; D.Ll. Hammick, *J. Chem. Soc.*, 1954, 767.
2. James Dewe Lambert (1912–2001), J. L. Houlden, *Trinity College Report*, 2001, 17; [Anon.], *The Times*, 2 November 2001.
3. Lord Todd and J. W. Cornforth, *Biog. Mem. Fell. Roy. Soc.*, 1976, **22**, 415.
4. G. Macfarlane, *Howard Florey: the making of a great scientist*, Oxford University Press, 1979; T. I. Williams, *Howard Florey: penicillin and after*, Oxford University Press, 1984; E. Lax, *The mould in Dr Florey's coat: the remarkable true story of the penicillin miracle*, Little, Brown, London, 2004; V. Quirke, *Collaboration in the pharmaceutical industry*, chap. 3, Routledge, New York, 2007; R. Bud, *Penicillin: triumph and tragedy*, Oxford University Press, 2007.

5. H. T. Clarke, J. R. Johnson, and R. Robinson, ed. *The chemistry of penicillin*, Princeton University Press, 1949.
6. Edward Penley Abraham (1913–1999), *ODNB*; G. Lowe, *The Independent*, 26 February, 1999; Letter from Abraham to J. B. Morrell, 13 July 1987.
7. Michael James Steuart Dewar (1918–1997), J. Murrell, *Bio. Mem. Fell. Roy. Soc.*, 1998, **44**, 129.
8. H. T. Clarke, J. R. Johnson, and R. Robinson, 'Brief history of the chemical study of penicillin', Chap. 1, and J. R. Johnson, R. B. Woodward, and R. Robinson, 'The constitution of the penicillins', Chap. 15 of ref. 5; E. P. Abraham, *Natural Product Rep.*, 1987, **4**, 41; R. Curtis and J. Jones, *J. Peptide Sci*, 2007, **13**, 769.
9. D. Crowfoot [Hodgkin], C. W. Bunn, B. Rogers-Low, and A. Turner-Jones, 'The X-ray crystallographic investigation of the structure of penicillin', Chap. 11 of ref. 5; for Charles William Bunn (1905–1990), see U. W. Arndt, *Biog. Mem. Fell. Roy. Soc.*, 1991, **37**, 71.
10. H. W. Thompson, B. R. Brattain, H. M. Randall, and R. S. Rasmussen, 'Infrared spectroscopic studies on the structure of penicillin', Chap. 13 of ref. 5; R. E. Richards, personal communication, 2008.
11. Dorothy Mary Crowfoot Hodgkin (1910–1994), *ODNB*; G. Dodson, *Biog. Mem. Fell. Roy. Soc.*, 2002, **48**, 179; G. Ferry, *Dorothy Hodgkin: a life*, Granta Publications, London, 1998.
12. John Wilfred Linnett (1913–1975), *ODNB*; A. D. Buckingham, *Biog. Mem. Fell. Roy. Soc.*, 1977, **23**, 311.
13. Lionel Alfred Kilby Staveley (1914–1996), J. C. G. Calado, *J. Chem. Thermodynamics*, 1997, **29**, 247; [Anon.], *New College Record*, 1995[sic], 70.
14. R. F. Barrow and C. J. Danby, *A History of the Physical Chemistry Laboratory, 1941–1991*, published by the Department, 1991.
15. J. S. Rowlinson, *Notes Rec. Roy. Soc.*, 2004, **58**, 161.
16. Herbert Marcus Powell (1906–1991), K. A. McLauchlan, *Biog. Mem. Fell. Roy. Soc.*, 2000, **46**, 427.
17. Clement John Danby (1916–2002), E. W. Gill, *Worcester College Record*, 2003, 78.
18. Douglas Hugh Everett (1916–2002), R. Ottewill, *Biog. Mem. Fell. Roy. Soc.*, 2004, **50**, 95; F. Boyce and D. Everett, *SOE; the scientific secrets*, Sutton, Stroud, 2003.
19. J. C. Smith, *The development of organic chemistry in Oxford, Part 2*, unpublished mimeographed manuscript, no date, Radcliffe Science Library, 1933d.312.
20. Harold Warris Thompson (1908–1983), *ODNB*; R. Richards, *Biog. Mem. Roy. Soc.*, 1985, **31**, 573; General Discussion, 'The application of infra-red spectra to chemical problems', *Trans. Far. Soc.*, 1945, **41**, 171.
21. M. G. Ord and Ll.A. Stocken, *Trends Biochem. Sci.*, 2000, **25**, 253; *The Oxford Biochemistry Department, 1920-2006*, published by the Department, 2006; J. A. Vilensky and K. Redman, *Ann. Emergency Med.*, 2003, **41**, 378.
22. D. A. Long, *Jesus College Record*, 1989, **17**.

23. Leonard Ary Woodward (1903–1976), [Anon.], *The Times*, 7 June 1976; *Nature*, 1976, **262**, 335.
24. Vice-Chancellor's Annual Reports, *OU Gazette*, 16 October 1940, **71**, Suppl. to No. 2306, p. 45; 13 October 1941, **72**, Suppl. to No. 2341, p. 39.
25. C. J. Danby, J. G. Davoud, D. H. Everett, C. N. Hinshelwood, and R. M. Lodge, *J. Chem. Soc.*, 1946, 918; R. F. Barrow, C. J. Danby, J. G. Davoud, C. N. Hinshelwood, and L. A. K. Staveley, *ibid.*, 1947, 401.

CHAPTER 7
Recent Times, 1945–2005: A School of World Renown

ROBERT J.P. WILLIAMS

Inorganic Chemistry Laboratory, Oxford University, South Parks Road, Oxford, OX1 3QR

7.1 GENERAL INTRODUCTION TO THE PERIOD: THE THREE CENTRES OF INFLUENCE

This chapter cannot be like any of those before it since, in the approximately 60 years from 1945 to today, chemistry and the chemistry school at Oxford have undergone very rapid changes and enlargement and even now the eventual condition of the chemistry school (and the university!) is far from finalised. However, while chemistry and its school changed, the major organisation of the university[1] stayed almost unaltered and, as it forms an essential background to our account, we outline in this introduction its bottom-up nature and its relationship with the chemistry school in this period. The central point, as described in the introductory chapter of this book, and that we elaborate here, is that the university and the colleges had reached in 1945, that is after more than 700 years, an uneasy condition of more than thirty independent colleges, see Table 1.1, a central administration body, and it had three separate chemistry laboratories[2] with two departmental professors amongst some ten other independent science departments and their professors. In effect there were then three centres of influence, the university, the colleges and the professors with laboratories, the third of which has been clearly shown to have arisen in the period after 1910, see Chapter 5. It is the changing influence of the three as they affected the teaching and research in chemistry which is given in the next two

---

Chemistry at Oxford: A History from 1600 to 2005
Edited by Robert J.P. Williams, John S. Rowlinson and Allan Chapman
© Royal Society of Chemistry 2009
Published by the Royal Society of Chemistry, www.rsc.org

sections. This section describes the powers of the three within the university and laboratory organisation in 1945 and then in Section 7.1.1 how their internal interaction led to the changes in teaching and research in the chemistry school in three separate periods, Section 7.2 draws attention to the details of selection, duties and the nature of the posts of professor and fellow/ lecturer for they have affected the school's composition and hence its attitudes. Notes on sources of information about the present-day university are included at the end of the chapter. Section 7.3 is devoted to teaching, and later sections to research.

Collectively, the colleges, which each had its own governing body of fellows, whom colleges financed and who had special college privileges, formed the strongest power base. The colleges had their own finances from endowments and fees and the fellows controlled their own activity under the limitations of statutes. In 1945 the colleges, not the university, were still responsible for the election of fellows. Hence the fellows, including the few in chemistry, were very strongly attached to their colleges even if a few of them had taken positions in the university laboratories as demonstrators. The fellows, of whom there were then some 10–20 in a college, not the university, selected (and still select) undergraduates whom they instructed in tutorials. The majority of fellows had always been in the Arts and the elected head of a college was in 1945 and is still usually one of them. The Arts in Oxford are largely college-based, while science is in shifting balance between colleges and laboratories. The second power base, the university, had salaried officials under the Vice-Chancellor but controlled by a Council, with connecting committees,[2] which, until recently, was composed of Arts fellows in the majority and some professors. The committees were not concerned with either what was taught or what research was done. The dominance of Arts in both colleges and the university meant that the scientists views were often poorly represented. The arts/science balance in committees is more even today. The university had its own financial strength, paid lecturers, called demonstrators in science from 1945 to 1964, who, if fellows or not, had university stipends. It paid professors fully. The university controlled the number of professors, demonstrators (later lecturers) but not fellowships at that time, and importantly it financed in part the laboratories that could be looked upon and certainly were also fiefdoms of designated professors in 1945.

The laboratories with their professors, not selected by a college, formed the third power base, and although professors were fellows of allotted colleges (without many privileges), they were more concerned with research and the running of the laboratories and with the central university body than with colleges and teaching. They could also seek funding from outside the university. It was within the laboratories that the professors and fellows decided their independent research projects and general lecture programmes. Here, we shall be interested only in the chemistry professors and their laboratories, which by 1945 had become three very separate departments, inorganic, organic and physical, which acted rather like three different colleges. We turn now to outline the particular way in which the three power bases and their changing natures affected chemistry's development after 1945 noting first the alterations in the ways in which the laboratories were managed.

As we have seen in Chapter 5 it was the setting up of two professorships of chemistry in two separate laboratories, reinforced with a third laboratory in 1941, which created the chemistry professors' influence. The professors were responsible for the finances and the technical staff and could appoint demonstrators (see note 1), personal assistants and research officers in the laboratories but not fellows of colleges, some of whom, those without college laboratories, also became demonstrators and did research in the professors' laboratories. However, it was the fellows of the colleges not the professors' staff who organised and did the bulk of the chemistry teaching in colleges, tutorials, while lectures and demonstrations in the laboratories were shared tasks with demonstrators who had no fellowship. Now, following the closure of the last of the college laboratories, the very last in 1947, all fellows of colleges needed research space in the professors' laboratories. The university decided that they too should have independent space there and as a result there was less coordination of research in a given laboratory with the exception of that under a professor's control. Thus, the power of the professors was considerable but limited. At this time, however, the professors set the general line of research. From 1964 the university changed the structure of elections to both fellowships and demonstratorships (from then on called lectureships, see also note 2) to one of joint, college/university, (fellowship/lectureships) appointments,[3] thereby necessitating consultation, see Section 7.2. The years between 1945 and 1964 form a first period of interest in this chapter, after which time the influence of the colleges and the professors separately was reduced and moved into the hands of laboratory, joint university/college committees.

Between 1965 and 1980 the fellow/lecturers in the three departments could continue to do more or less as they wished in both teaching and research without much financial or administrative stress. During this second period the university, which was better financed,[4] changed little in influence but the professors now had to work within the constraints of departmental committees of themselves with fellow/lecturers. The colleges and these committees had also to cooperate. After 1980 the costs of running chemistry increased greatly and the university became more directly concerned with chemistry departments, introducing financial constraints. This was increasingly inevitable since by 1990 the university income from outside was twice that of all the colleges combined, and questions were being asked of it as to the overall purpose and value of the university as a whole especially its science departments. Of considerable consequence then to the way chemistry laboratory organisation developed between 1945 and 2008 was the way in which the relative financial strengths of the colleges, the professors with the departments, and the university changed and were used in control over appointments, see Section 7.2, and general laboratory activities and they have yet to settle. There is little doubt that the university will have to be more closely involved in the future.

We now need to see how the organisation in the laboratories met with that in the colleges. Given that the majority of fellows on nearly all governing bodies of the colleges and of the university were and are still from the Arts, and that they were and still are college based with all its freedoms, they were and remain inevitably a major barrier to this increased central authority.[1,7] In this respect their wish, different from that of the professors, was that a chemistry fellow was loyal to a college just like one of them. Remember through many centuries the colleges were the places of work, even the homes of fellows and a certain ethos had become built into them. Clearly a decisive point for an individual chemistry fellow was the degree to which he felt more responsible to a college or to a laboratory[5] and hence to the university. Again attitudes have changed with time.

Turning to this division of loyalties that persists for a chemistry fellow/ demonstrator (lecturer), it is necessary to see not just duties, see Section 7.2, but the salary and other benefits from the tutorial centre, the college,[6] and from the university and its laboratories. Of the total salary the university has provided in recent times two thirds, as a demonstrator or lecturer, and the college one third, as a fellow. However, the college has for many years given benefits for example meals and housing allowances of considerable value, which unfortunately are very different in the different colleges. A college can still be the base for social life as it was in 1945 but is not of family concern. It provides a homely room, quite unlike a laboratory office, and every fellow has always been on the democratic governing body of his college. Meanwhile, the laboratory is the research base, where many a chemistry fellow now spends 90% of his working day. Despite this imbalance and speaking generally the college has attracted the more loyalty until recently as it gives the fellow clear general living and social advantages, with a sense of independence and self-importance. At first, it must be remembered, a fellow had little influence on the running of a laboratory, that is before 1964. In passing we note that the colleges could also elect junior research fellows without reference to professors who virtually had to give them laboratory space. More recently The Royal Society selects research fellows who must also be given "independent" space for research in the laboratories. In early times sponsored posts for research fellows were more or less in the gift of the professor.

Increasingly the pull of research in the period from 1965 to today is driving a fellow/lecturer out of the arms of a college and teaching into a closer relationship with the department and the university due to the increased intensity and costs of research. The problems have arisen in fair part because there has been no overall organisation of the finances or the structure of the chemistry school in which college, university and laboratory committees share responsibility. Let us leave these organisational problems for the moment and turn to an outline of the development of teaching[8] and research in the chemistry school itself, which we shall divide into three periods much though there is considerable overlap between them. In the next sub-section we give a brief survey of these periods before more detailed analysis so as to give the reader a perspective of the challenges that a very rapidly expanding chemistry school posed for the collegiate university within which it had to function. Certain themes will be seen to be persistent and the account will then be somewhat repetitive in the different time periods.

7.1.1 The Three Periods 1945 to 1965, 1965 to 1980, 1980 to Today

Change in teaching and research in chemistry in what we shall call the first period from 1945 until 1965 was slow since the aftermath of the war prevented it and financial constraints were considerable. Even so, the university had to adjust to mounting numbers of students, especially in subjects such as chemistry, 40 in 1939, 35 in 1945, 60 in 1950, and over 100 by 1960 in all of four years with relatively small changes in teaching strength or facilities.[2] This expansion was largely driven by *the colleges*. In this period there were increases by 1960 from 15 to only 27 demonstrators, not all fellows, and only relatively small laboratory extensions had been built except for that for inorganic chemistry, which was greatly enlarged.[5] Moreover, the war had created a more serious and purposeful atmosphere amongst students, while as we shall see their teaching course remained poorly organised overall. The dominance of the two very different pre-war attitudes on the content of the subject changed little in this period. We shall call one of them intellectual (mainly physical chemistry) to a large degree inspired by Professors Sidgwick and Hinshelwood in the physical and inorganic laboratories and historically consistent with college teaching traditions, much though Sidgwick retired in 1948. The other, organic chemistry in its laboratory, was more practical as we shall call its concern with analysis and synthesising of compounds, as directed by Professor Robinson, which persisted much as Perkin had initiated it with lesser concern for college interests and despite the fact that Robinson retired in 1955. The word 'practical' is used throughout this chapter not in any derogatory sense but to provide a distinction between chemistry strongly dependent on synthesis and analytical degradation mainly in the organic chemistry laboratory, often based on insight and green-fingered skill, and that based on systematic enquiry into underlying principles of the subject, which we have called 'intellectual' here. This second approach has been mainly associated with the physical and to a smaller degree the inorganic chemistry laboratories both in teaching and in research. The division is not strictly between the laboratories as all experiments require practical effort. The difference leads to a ready connection of the practical rather than the intellectual approach to industry and hence to different sources of funding. In turn, industrial funding gives an independence from government and indeed from a need to serve strictly educational requirements. The absence of overall management allowed the different laboratories to go their own way to a large degree. Quite probably these differing styles altered little especially since the demonstrators/fellows in physical and inorganic chemistry had changed little until after 1965. Several were born and bred loyal college men who naturally inclined to the intellectual style, which, as we have emphasised, was very suitable for giving tutorials. Hinshelwood's influence as well as Sidgwick's was still strong too on new fellows who were largely Oxford graduates, see Table 7.1. Although Jones succeeded Robinson as professor in 1955 and there were several changes in demonstrators in organic chemistry, some even before that time, it too was little altered in style since the new demonstrators, three of whom came from Manchester with Jones, and both old and new college

Table 7.1 Teacher/pupil connected appointments before 1955.

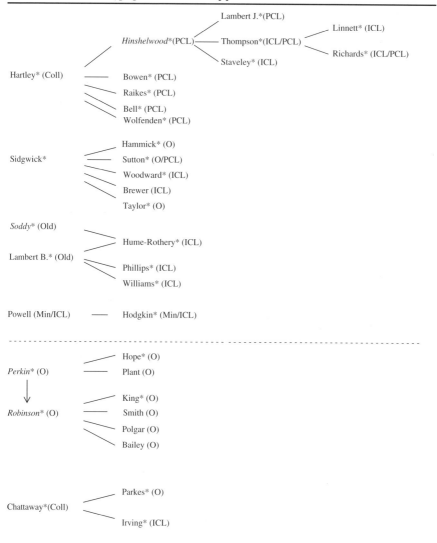

Notes: * = College Fellows; Department Professors are in *italics*; laboratory location in brackets; (Coll) = College. All but Perkin, Robinson, Smith, Polgar and King were Oxford Undergraduates. All five were organic chemists.

fellow/demonstrators were, like Jones himself, mostly synthesis, practical, chemists, not concerned with mechanisms of reactions, which is more easily related to the activity in the other departments. Moreover, there were a considerable number of demonstrators without fellowships and hence with no attachment to colleges. Because of this inertia in the period to 1965, when Hinshelwood retired, we shall devote a considerable part, Section 7.4, of the chapter to the atmosphere that the three 1945 professors created showing how certain of their attitudes, shared in physical and inorganic chemistry especially amongst their colleagues who were college fellows, and to a degree separate from those in organic chemistry, became strongly embedded in the Oxford chemistry school for many years even after 1965 and perhaps even to today. We must stress that, despite the difficulties and differences of this period, by 1965 Oxford chemistry, with its peculiarities, stood out as a leading school in both physical and organic chemistry research worldwide, as we shall show in Section 7.5. However, there was virtually no inorganic chemistry being practised and the physical chemistry had a limited scope. *In this period the major influences were exerted by the professors and the colleges, acting largely separately, with the university playing a tight restrictive financial role on the number of its appointments.*

The second period we shall describe, from 1965 to 1980, was one of rapid expansion in the school in numbers of fellow/lecturers to match the greatly increased but stabilising numbers of undergraduates. This became possible since the *university's* financial position had improved greatly. By 1970 there were close to the final number of about 175 students a year, while it took to around 1980 for the number of fellow/lecturers to reach about 55, the number today. With the expansion and improvement in research possibilities and novel equipment many new research topics arose. There remained for the most part the noticeable separation between the practical attitudes in organic chemistry (in the style of Robinson) and the intellectual style in the physical chemistry department (reminiscent of Sidgwick and Hinshelwood) much though there were many new techniques and programmes. The most considerable changes in subject matter were in the revival of the inorganic chemistry department, which became quite independent from 1963, under a new professor. This professor was, however, a thoughtful as much as a practical man. His style helped the continuation of the election of fellows who were intellectual in style and they were encouraged to keep the personal college-based tutorial education. Hence, the tutorial system went unquestioned in its suitability for chemistry teaching especially in both the inorganic and physical chemistry laboratories.

After the arrival of the now three new professors, Jones, Anderson and Richards, in 1955, 1963 and 1965, respectively, there was less sense of a dominant personality in any of the laboratories. These professors did bring with them special research interests, but the lack of relationship of the research between members of one department, especially in the one, "inorganic", created casually by Hinshelwood, remained. In order to manage the changes in all the laboratories there evolved gradually in this period laboratory organisation in three independent departments more closely resembling that in colleges with the professors chairing committees of lecturers. The effect of the change upon elections will be described in Section 7.2. It was supposed without much thought that this expanding system of independent research workers could be financed by the university, the research councils and industry without too much stress.

As far as undergraduate education was concerned we shall see in Section 7.3 that, by the end of the period, it had settled down as the great changes of understanding of chemistry from 1930 and before 1980 necessitated a required training in a well-established core discipline. The teaching course became better, not yet well organised, and the fellow/student ratio improved so that students were helped more.[8] Throughout these first two periods it will be explained that for fellows the association with an undergraduate college still gave everyday life with its contact with students and members of other faculties, and social comforts of relaxed dining and chatting, much of the style of a rich private home, see section 5.3. Such a life gives a great deal of cosy satisfaction. However, if the increasing associated duties were to be done, as they had to be done, there was the problem that a college as well as the department began to take much time in administration apart from in academic and normal leisurely social exchange. (This could of course be made a reason for doing little research.) Simultaneously, the pressures to do research and the possibility of making very many measurements, giving rise to much data needing interpretation, increased rapidly. *Once again in this second period the university had very little influence on the development of teaching and research in chemistry but its financial control was relaxed. Change depended now on joint college/department committees. The professors had less say and research diversified further.*

As research everywhere gathered pace in the third period, 1980 to today, with more research workers in a group, with more equipment, and more administration and responsibility in the departments, the pressures on the Chemistry fellow/lecturer in the laboratory together with that in the college became felt strongly. Chemistry research had increased year on year so that more research space was needed[5] and, together with more people and equipment, costs rose rapidly. Thus, a Labour government minister, had been led to declare at the beginning of this time that "the party is over" for scientists.[9] From around 1980/85 the Government did not expand funding sufficiently to meet needs and certainly funds were harder to obtain. This greatly increased financial and workload stress hit the running of chemistry both generally in the university and in the individual research groups. Change happened quickly and somewhat unexpectedly to the only half-aware chemist. Quite possibly this was a major cause of the refusal of Oxford University in 1985 to give the Prime Minister at the time an Honorary Degree.[10] It was not really an appropriate response to a cut back and not an effective one. The Prime Minister was Margaret Thatcher, the first woman to hold the post and herself an Oxford chemistry graduate. The insult was not taken lightly and no change of policy followed, "The lady was not for turning." We see the 1980/85 years as an unavoidable watershed after which the very life of an Oxford chemistry fellow/lecturer changed quite quickly from one of relative ease to one of considerable hassle and bustle, even discomfort. The manager of a research group became intensely aware of competition for funding and of review pressures[11,12] to demonstrate not just academic but useful research success. He felt more and more compelled to travel and seek fame to demonstrate his ability while gaining egoistic satisfaction at international conferences. The idea of 'pure' research unsullied by contact with the practical side of society, which had been encouraged by the 'intellectual' stance of many earlier Oxford chemists with their collegiate background, became difficult to maintain in the face of government demands for "national wealth creation." Moreover, both Inorganic and Organic Chemistry now had 'practical' men, Green and Baldwin, respectively, as professors and the intellectual style remained dominant only in the Physical Chemistry Laboratory. The value of research faced judgement not only of this kind but by peer review as Research Assessment Exercises by Research Council Committees[11] were introduced alongside teaching proficiency checks, both of which were linked to Higher Education Funding.[12] The idea of "value for money" was forced upon the vastly increased research activity in chemistry.

The effect of all these internal and external pressures meant that the simple idea of highly individual research workers and teachers loosely cooperating, which was carried over from traditions in colleges to departments, could not be maintained in sciences such as chemistry. We return to the changes in the organisation of the chemistry school. The central authority of the university now had to find ways to impose itself on the Colleges as it held the greater financial responsibility. One considerable change was that teaching fees were channelled through the university not as previously through the colleges, which led to a different way of allocating funds to colleges and laboratories. At the same time in the sciences, financial management had to be exerted on the laboratories through the professors. In 1996 to aid organisation the university placed responsibility on a Chairman of Chemistry linked to a Physical Sciences Board and then created Science, and other, Divisions[13] that had financial control over each subject, including the Arts. Of course by this time the costs of science had increased enormously relative to the Arts and were directly in the control of the university, while the colleges' contribution was proportionately greater in the Arts. Thus, we see the struggles between the three centres of influence unwittingly generated initially in the separation of authority between colleges and the university in the Middle Ages and that had come out in favour of the colleges for many centuries, and that was made complex by the appointment of professors and later their laboratories, swinging toward the central university now including the professors and laboratories with more organised science but not without a fight. For Arts fellows there was no great pressure for change and their attachment to a college, not the university, was made the greater, as, for many, two thirds of their salary still came from it, a quite different position from that of their scientist colleagues. Consequently, resistance to change was and is still very strong much as it is associated with 'academic freedom'.[7] Today, internally, and with the pressures from the requirements of society outside Oxford, the need for compromise is great and forcibly present. Financial strength may still rest to some degree with colleges and can protect Arts fellows but it is not able to protect chemistry departments and the interest of chemistry fellows, who now found that they had to find funding for their own research. A fellow in chemistry is more and more likely to see the college as a residential place for students where he teaches a few tutorial hours a week, say five, has his lunch there and is seen no more, despite the fact that it pays one third of his salary and gives benefits. As we shall see the scientist has become something of a driven researcher with little time for much else. The structure of the chemistry departments within the collegiate university needs reappraisal. We shall describe the rapid changes in this last period to today in Section 7.7.

7.1.2 Summary

It is clear therefore that the period from 1945 to today is very different historically from all previous periods of the study of chemistry in Oxford. Not only was there a great increase in the number of fellows, undergraduates, and graduates, but there was also a greatly improved, Figure 7.1 and enlarged laboratory space, Figures 7.3 to 7.6. The very nature of chemistry changed by 2000 from a small subject in 1945 in which there was only a glimmering of understanding of relatively few facts to a large one with a well-developed unchangeable core of theoretical understanding, with much improved and faster ways of making measurements, and an ever-increasing mountain of facts allied to the vast numbers of new compounds. The impact of the combination of all these factors led to increased research activity in a variety of what became

Figure 7.1 The 1879 old (now Inorganic) teaching Chemistry Laboratory in its somewhat modernised state with fume-cupboards (hoods) at either end, compare Figure 7.6. (Photograph by M. Lodge, published with permission.)

highly specialised parts of chemistry. A search for possible new materials or compounds of applied value became necessary to generate the greatly increased required funding. At the same time it became highly unlikely that any entirely new theory would appear. From a somewhat gentlemanly activity chemistry became a worldwide competitive scramble to uncover useful novelty. To be secure and well financed the chemist also had to publish or perish and to apply regularly for grants, if he had no link to industry, for his work now had to satisfy inspections. For a fellow/lecturer to have ten papers in print a year was not unusual but writing books became rare except in retirement. The changes hit the Oxford fellow/lecturer hard for he had to work within the historical influences of the collegiate structure confused further by departmental divisions. He had to do his research and his college teaching and administration although now the teaching was much reduced and almost routine around a set core of the subject, making individual tutorials of less value. Equally the task of the professors became increasingly difficult with the combination of complex organisation and research. It is this quick surge over fifty to sixty years in the changes of structure and costs with the magnitude of the activity in chemistry that provides one major reason that the description of the subject at Oxford in this chapter has to be so different from that in earlier chapters. While the chemistry school was able to grow from 1945 perhaps even to 1990 in a haphazard disorganised way, in which the influences of colleges, the university, departments and professors varied, as the century ended the size of the activity became such that it clearly had to be managed to a degree as a whole.

It would be wrong to finish this introductory section on a seemingly half-despairing note without drawing attention to the national and international standing of the teaching and research very largely carried out by individual professors and lecturers. It is undoubtedly fair to state that the Oxford chemistry school is of world renown in all its branches. National and international recognition can be estimated in part by the number of Fellows of The Royal Society and of the corresponding foreign academies, by the award of numerous medals by Scientific Societies, and by honorary degrees awarded by outside Universities. In the 1955–1960 period there were at least ten Fellows of The Royal Society and that number has often been exceeded since. The research has attracted a large number of foreign scientists as visitors and many Oxford chemists have spent time abroad as visiting professors. It is the belief of the authors of this book that the school has maintained a standing well into the top ten teaching and research schools worldwide. The most noteworthy achievements will be described in the following sections.

Turning to the style of this chapter the vast increase in people and activity means that we cannot maintain that of other chapters of the book in several further ways. In previous chapters the characters and achievements of individuals could take pride of place and we could give brief biographies. As we shall show it is still useful to follow this descriptive style for the professors since on election they could introduce a novel section of research on a considerable scale. However, following the expansion of staff from some twenty to over fifty we cannot describe the nature and activities of all the individual fellows in chemistry departments, while we note those of a few prominent members. Again many researches are far too much concerned with detailed practice for us to give short simple descriptions. Instead, we stress the way many contribute to the advancement of chemistry in a corporate sense. (Of course individual fellow/lecturers will not necessarily see their personal contribution in this way as Oxford stresses the need to be seen as an individual.) It is impossible too to refer to the literally thousands of their publications and we have had to limit ourselves mainly to books and not even to all of them. Again the number of research grant applications, reports and documents add up to many hundreds so that we cannot provide references for them either. Fortunately, very much detail is now available via the university website that can be used to follow the details of even individual's activities.

Finally, this chapter differs from the others in that it concerns contemporary history. The author has been in Oxford through its period from 1944 to today, successively as student, graduate, postgraduate, fellow/lecturer and professor. It is inevitable that the chapter will contain some remarks that will appear to have a personal bias since any individual has limited experience and vision concerning the times in which he has lived.

7.2 RECRUITMENT AND THE NATURE OF PROFESSORSHIPS AND FELLOWSHIP/LECTURESHIPS

We first go into some essential features of the methods of the appointments and selection of teaching and research staff and their alterations together with their duties from 1945 to today for this will reveal many peculiarities of Oxford Chemistry. It is the involvement of the colleges that is the major source of problems and differences from any other university. Two previous histories of the Organic[14] (see also reference 103) and Physical Chemistry[15] Laboratories have been used extensively in the writing of this and subsequent sections of this chapter but the emphasis here is quite different from them as they are internal reports and as such are naturally favourable to the activities in them.

In 1945 there were three professors of Chemistry, Robinson and Hinshelwood in charge of three chemistry laboratories, one Organic (Robinson), and two others Physical and nominally Inorganic in the Old Laboratory (Hinshelwood). The third professor was Sidgwick who had been appointed *ad hominen*. The election to chairs in Chemistry, the above three in 1945 or five in charge of the departments of chemistry now (Organic, Chemical Biology, Physical, Inorganic and Theoretical) has remained a university committee matter from the middle of the nineteenth century. Electors in earlier times, see Chapters 4 and 5, included the Vice-Chancellor, or his nominee, and say two external and four internal members including two college fellows.[1] Today[2] the committee includes the Vice-Chancellor or his nominee, two persons from the college to which the chair is attached, two persons appointed by Council and four persons appointed by a new Sciences Divisional Board.[13] Although most members of the committee have always been fellows of colleges the college outlook has kept only a limited say and no veto and it influences decision but slightly today, much though it may have done so before 1945. The fellow/lecturers in the department concerned are asked to advise. From the election of Perkin to today the decisive factor in the election of a professor has been research achievement and potential, not teaching, or administrative ability, or interest in the colleges. A professor is made a fellow of a specified college but he receives few benefits from it and he is paid by the university. The list of professors from 1945 to today is given in Table 7.2. It is clear that until very recently organic chemistry has not favoured an internal Oxford graduate candidate and only once has Inorganic Chemistry elected an internal candidate, Green, but even he was not an Oxford graduate. The Physical and Theoretical Chemistry Chairs have a quite different history with the election

Table 7.2 Professorships in Oxford Chemistry School from 1930[2].

Department Heads			Others	
Inorganic	**Physical**	**Organic**	**Specialist or Royal Society**	
1930 Soddy	(Soddy)	Robinson*	1936 Sidgwick (A)	
↓	↓	↓		
1937 (Hinshelwood)	Hinshelwood			
1955		Jones*	1952 Coulson (Theory Maths, Chemistry (1972))	
			1960 Hodgkin (Royal S)	
1963 Anderson*		↓		
1964	Richards		1964 Powell (A), Thompson (A)	
	↓			
1970	Dainton		1967 Waters (A)	
1974 ↓	Rowlinson		1974 Williams (Royal S)	
1976 Goodenough*			1977 March (Theory)	
1978 ↓		Baldwin*	1984 Carrington (Royal S)	
1989 Green*	↓		1985 Turner (A)	
1993	Simons		1989 Lowe (A)	
2000 ↓	Klein		1992 Hill (Bioinorganic) (A)	
2003 Edwards*			1994 Child (Theory)	
2006	↓	Davies	2003 Bayley (Chemical Biology)	
2007 ↓		↓	2007 Logan (Theory)	

A = Ad hominem * Not educated in Oxford or Cambridge for first degree

Notes. After 1996 Lecturers could be entitled "professor" after assessment of the quality of their contributions. However, this meant no change in their duties and so this large group is not included here. From 1972 to 1994 there was a separate department of theoretical chemistry under Coulson 1972 and then March 1977. See Note 9.
Between Coulson's death and March's appointment there were three IBM visiting professors: Rudi Marcus (Illinois, 1975, later Nobel Laureate), Roy Gordon (Harvard, 1976) and Ben Widom (Cornell, 1978).

of Oxford (or Cambridge) graduates. Does this reflect a rather different outlook on the organisation of chemistry within physical chemistry at Oxford? Its elected professors did not have a particularly conservative view of chemistry but their background made for easy acceptance by them of the style of a collegiate university. This could not be said to be true of most of the professors in the other departments. The duties of these professors are clearly defined by regulations[3] and it would appear that they are in charge of the departments, with all the diversity of its tasks, but this is true only in so far as they are actually in charge of accounts, technical and an administrative staff, and such matters as safety and allocation of space. In a general sense the selection of a professor takes little regard for the nature of the management problems in the confused relationship between colleges, the departments and the university administration. The very separate procedures for selection of professors from that of fellows has at least 400 years of history and has created a somewhat strained relationship, both academic and social, between them.

We turn now to consider how the recruitment of other staff members affects the position of a professor. It was the case that in the period from 1945 to 1964 that professors could control the election to university demonstratorships, research officers and assistants but they had little *direct* say in those of college fellowships. The limitation on expansion was the lack of university funds. The professor could and did if he wished exert considerable pressure especially on poorer colleges through this gift of demonstratorships, which then paid a significant portion of a combined salary, to accept his fellowship candidate. The professor could then control much of research direction in principle. However, at that time, and all previous times, college fellows could be selected by a college committee alone, provided the college paid all the salary, and the professor virtually had to accept them in his laboratory with the above financial caveat, that is without a demonstratorship. Often, however, the fellow obtained one after a time. Thus, there could be a strong professorial group with demonstrators, not fellows, elected for research ability and a number of fellow/demonstrators each with a small group doing unconnected research elected with a view to teaching and college administrative as well as research ability. In Chapter 5 it was seen that this could result in outright antagonism between the professor and some fellows. It was not until the university made the post of lecturer (a re-designation in 1964 of demonstrator) a joint appointment with that of a college fellowship[1,3] that a democratic scheme for these elections was devised, which persists. In it the professor still has a strong hand in principle but frequently he has not wished to use it even though the university pays two-thirds of the fellow/lecturer's salary.

The new selection procedure for a joint post has a committee of which the department, two fellows of the college concerned and the professor have membership with one outside representative from a different department. The effect of this change was most apparent in organic chemistry where the separation between those with and without fellowships was most obvious in 1945. As we shall see in Section 7.3 the major drive even to today to increase fellow/demonstrators (lecturers) was from the colleges' need for tutors opposite increasing numbers of chemistry undergraduates. The increase in numbers of fellow/lecturers was only made possible after 1965, however, by the greatly improved financial resources of both the university and the colleges. Colleges retained a strong interest and influence on elections at first, but increasingly it was the joint professor/fellow committee of the laboratory that selected for research potential. The compromise with a given college was not always easily managed. We shall see that in practice these historically based operations have led to great diversity in research especially after 1964 as the committees allowed selection on merit more than on a chosen research direction. It may well be that as central organisation of the university increases, there will be a somewhat more directed sense of purpose again. However, at present, when financial restrictions are strong and while continuation of the given fellowship/university lectureship post may be guaranteed, no new lectureships can be created. (Somewhat curiously there is an effort to overcome the unequal distribution of the limited number of such posts between colleges by opening any vacated post to fellowship bids from all colleges. Hence, a college may lose a post but from its side it may just discontinue it. Thus, a firm sense of continuity in teaching or research is still missing.) Today, there is even a tendency for the laboratory committee to be directly interested in the ability of the candidate to raise money.

In recent years the appointment of more administrators and research officers to look after complex equipment, again selected by committee, has re-introduced social differences as these posts are not associated with a college. In general, it is clear that the university and its chemistry departments are not easily brought into step with colleges over appointments. Each laboratory appoints its own technical and administrative staff.

It is the historical creation of the unique position of a fellow of a college that has brought about this so-called democratic laboratory organisation, something akin to that of a college. As a consequence, the professor's position has been reduced to that of an operational director, a considerable change from earlier times. Sometimes this has led professors to bury themselves in their own research often letting administration be managed virtually by the fellow/lecturers and an administrator. Alternatively, some professors have seen the job as one of management internally of laboratory or university affairs or seeing to it that Oxford is represented on the ever-growing number of external committees linked to funding or national programmes and facilities. It is easy to see that the procedures are biased toward individual 'brilliance' in research whether of a professor or a lecturer, and away from organised strength. A partial way out that is developing is for a laboratory to have a research professor and separately a department head to organise matters.

In order to see the attractiveness of the fellow/lecturer position at Oxford and hence the very strong competition for these posts we give next in some detail the duties and freedoms nominally associated with the dual position in Oxford's collegiate university. The outstanding feature is the high degree of independence of a fellow/lecturer from any authority.

A college fellow/lecturer (demonstrator) has always been required to do tutorial duties and assist in the running of the college under the college head. However, the tutorial load per week has been reduced from ten hours in 1945 to five today and college administration is not so much by academic fellows as professionals. Curiously, as mentioned in the introduction to this chapter, it is the benefits and social nature of a college that frequently held and for some still hold a greater loyalty to the college than to the laboratory. This attachment is helped by functions of the fellow in the college including the selection of the undergraduates whom he teaches (notice this is not a university or faculty function), and a considerable input to the election of other fellow/lecturers in his subject in his own college. A fellow's attachment to students rather naturally leads to a common sense of loyalty that the college encourages as this relationship may well lead subsequent alumni to make gifts to the college rather than the university. (Here lies another difficulty for the overall organisation of the university.) It is very rare for smaller gifts to be made directly to chemistry. The commitment to the tutorial teaching of one college puts the tutor to a degree in competition with other college tutors as to whom can produce the most top class pupils. A great asset of his position, not shared by the professor, arises from this connection to undergraduate students. Many students are of considerable ability and the tutorial contact or through lectures may well lead to them deciding to do 4th-year research with the tutor in his college. It is very unusual for a professor to know a particular undergraduate, he is not asked to give tutorials, and it is then not usual for a Part II student to wish to do research with a professor. The Part II itself often leads to a doctorate for these brighter students so that the research groups of several fellow/lecturers can easily be larger and even now and then more powerful than that of a professor. This was certainly not the intention of Perkin when he set up the Part II scheme during the First World War.

The position of a lecturer in a laboratory is also particularly favourable in Oxford. Duties since 1964 are to give lectures on topics in which he has a considerable say, to teach in the practical classes, much of which is now delegated, to assist the professor by mutual agreement and to do research. The research is not controlled by anyone but the lecturer/fellow himself. He is assisted by departmental funds but today he needs to raise his own research money from Research Councils, charities or industry if he is to run a large group, often of nearer twenty than ten including Part II, doctoral and more senior research personnel. He must publish, almost furiously today, and apparently he must satisfy the professor that he is making a reasonable contribution in all respects. In reality this last requirement is usually met by the fellow/lecturer's own ambition and pride in his research and teaching. Do notice, however, that his ability to do independent research is now linked to raising money for his own work. (If a fellow wishes to remain active in retirement then he may do so if he can cover *all* costs.)

What is very noticeable is that once selection moved away from the professor or from a college, with very different objectives in mind, to a laboratory committee there was no drive to appoint with a particular purpose in mind either in teaching or even in research.

It is clear from the above that a fellow who holds a university lectureship is nominally obliged to assist in both the laboratory and the college, but the separate independences of his fellowship and lectureship positions mean that he can only be asked to do things and not easily told to do them. Resistance to organised activity in teaching, administration and research is not easily overcome. Some, perhaps especially a professor from outside Oxford, might well consider that the history of the collegiate university has created too great an independence of the fellows in chemistry.

A particular feature of the selection procedure of the new fellow/lecturer and the collegiate teaching system was that election to Oxford fellowships of those who had been Oxford (Cambridge) undergraduates was common in 1945 to about 1960, see Table 7.2 and is not uncommon today. The underlying fault in a succession of steps from leaving school for Oxford to becoming a fellow/lecturer can be that Oxford 'in-breeding' can risk insulation from novel changes in research taking place elsewhere, and we shall look at this risk in Sections 7.5 to 7.7. Since 1965 much of this internal linkage has disappeared although as many as a third of appointments to fellowships are still of Oxford graduates. More naturally, many are selected from those who have completed their doctorate at Oxford. They, and others who have been elected, have usually had postdoctoral experience, often abroad, and perhaps under Research Fellowship schemes such as that of The Royal Society.

The overall highly independent position of a fellow/lecturer is seen to have many advantages but it also introduced many problems. Given the more favourable attitude of the university to research, pushed by outside influence, individual fellow/lecturers increased the number of research students dependent upon them, many selected and paid for by them, in an effort to increase the speed at which their research progressed and undoubtedly to increase their prestige. They could do this, as there were no controlling limits on an individual fellow/lecturer's research group. Grave organisational problems arose just from these steep increases in numbers, to which we now return, see Section 7.1. One was laboratory space to house not just first- to third-year undergraduate practical teaching classes but the fourth year or Part II of the course, a research project involving 150 students, and the ever-rising number of graduate research workers not just those studying for a doctorate, approaching some 70 to 80 a year today.[17] There was also the quite considerable number of post-doctoral research workers, many from overseas, and an increased staff of university officers to look after the sophisticated new equipment. Clearly a programme of new buildings or extensions was continuously necessary, see Appendix 1 to this chapter. A second problem was the financing of all this rapidly increasing activity – people, buildings, and equipment, including all the servicing required, and of laboratory non-teaching staff. At the same time the sophistication of methods grew in all departments. The change in costs including inflation rose from requiring some £100 a year for each research worker in 1945 to a sum closer to £10,000 in 2000, see Appendix 2. The university has not seen this expansion in laboratories and costs as its responsibility in the same way as it sees its libraries. The chemistry school has had to seek help not only from Research Councils but also from outside organisations, industry and charitable trusts, but it has also turned to a commercial attitude in the creation of external companies, see Appendix 2. The pressure to do and publish research and to bring in money has become very strong so that internally the activity of the lecturers/professors has changed. Each professor and fellow has now a management job of great weight, which forces him to abandon general chemical knowledge and become very much a research specialist with management skills, and to seek for opportunities to finance the work. As mentioned earlier the magnitude of these tasks and the difficulties in financial control have led the university to introduce control over departments through the creation of a new post, that of Chairman of Chemistry[16] and Science Divisional Boards directly responsible to the university, see note 8.

Putting Sections 7.1 and 7.2 together we see a large shift in the character of the chemistry school over fifty years. In 1950 a lecturer carried out his teaching duties, his research, his college duties as a fellow and still had time for leisurely pursuits, not forgetting socialising with students and teachers from different disciplines, a strength of the college system. A chemistry fellow could readily take on the mantle of that of an Arts man and could manage the complex set of duties imposed by his dual functions. There were few financial constraints. He could have the odd leisurely afternoon and had time to read around. The very late acceptance of the sciences, including chemistry, did not appear to be causing significant problems for the unique collegiate university that was Oxford. Unfortunately there grew to be a larger and larger gap between the detailed work associated with a chemistry fellowship/lecturer's research with the large numbers of people involved in a group as well as the need to raise large sums of money for research, and an Arts fellow's work. The colleges and much of the central organisation of the university, which had always been biased toward the Arts by weight of numbers, did not change much as it attempted to maintain the style of 1945. By way of contrast, as the number of Science fellows increased with the increasing importance of its subjects, including chemistry, and with ever increasing business in the laboratory they have considered, correctly, that they do not have the time to undertake the essential duties in college, laboratory and university. The loyalty to a college is no longer felt strongly, and has been replaced by pride in research and research pupils. Questions must arise not only as to the structure of the chemistry school in the university but as to the style of the university itself. Meanwhile, the role of a professor changed from research toward management both in a department, in the university and on many outside committees.

In summary, the creation of science schools, here we consider chemistry, largely forced upon the university, together with the creation of laboratory-based research but college-based teaching with their historical development, have been modified step by step to change the relation between the university and the colleges. In chemistry the importance of teaching, especially the tutorial, has diminished relative to research, to the degree that today Oxford now has a research rather than a teaching school where collective sub-faculty organisation in the university is becoming more important than that of any college-based structure. This condition has been clear in the case of the appointment of professors from its initiation but the appointment of fellows (demonstrators/lecturers) has moved steadily from within the college orbit to in effect that of the university, although still department based. So far this has had little effect on independence. This movement, of both the nature of professorships and fellow/lectureships, a part of the history of the collegiate university, has gained pace only since 1945, though starting from 1912 with the election of Perkin as seen in Chapter 5. Only recently is it being led into being more centrally organised as we shall see in the following sections. While teaching can still be individually based, research is in danger of becoming linked to the need to raise money, which may require more cooperative effort.

In the next sections we give first an account of undergraduate admissions and teaching and then at greater length the history of research in the period from 1945.

7.2.1 A Note on Women Fellows in Chemistry[2]

There have been very few women appointments in Chemistry. In 1945 there was only D. (Crowfoot) Hodgkin appointed at Somerville, fellow 1935, demonstrator 1948. The next appointments were of M.L. Tomlinson at St Hilda's as a fellow/demonstrator in Organic Chemistry (1948) and of M. Christie also at St Hilda's as a fellow/demonstrator in Inorganic Chemistry (1963). Men could not be appointed to any of the women's colleges until after 1975 and it was not until then that women could be appointed in any of the men's colleges. Somewhat oddly this has not increased the proportion of women on the chemistry staff as men were appointed at women's colleges and no woman in a man's college until very recently. Thus, there has never been more than some 5% women chemistry fellow/lecturers while the number of women undergraduates increased from 10% or below in 1945 to 1970 to over 35% today. The situation is very similar to that in other physical sciences and not dissimilar from that in the higher relative to the lower reaches of the scientific civil service. A question arises as to the wish of women to be involved in the very demanding lifestyle of a chemistry fellow/lecturer or parallel professions. It is the case that more women have been elected in Arts subjects.

7.3 THE UNDERGRADUATE ENTRY INTO OXFORD AND THE CHEMISTRY COURSE

7.3.1 The Butler Education Act 1944

As explained in the previous chapter before the Second World War the chance of coming to Oxford for many students was limited by the English class structure with its educational and financial biases. In general, undergraduates in the 1920–44 period were likely to be from relatively rich families with an education in public (private) schools much of it along more or less conventional arts subject lines. There were not so many from such schools interested in chemistry (stinks) that often appeared to these schools to have too little academic merit. Moreover, many of the students at Oxford were not academic in attitude. There were some, it must be said, of a different ilk, most, not all, from Grammar Schools or other State-aided schools, many of whom had gained scholarships. However, the cost of an Oxford education and of board and lodgings away from home kept most students from middle and lower classes in home-town universities. These schools in contrast with the public (private) schools often had a large science sixth form effectively promoted by local employment opportunity. For students from these schools university education was a more serious pursuit to gain advantage in employment and as a consequence universities such as Manchester[18] and the combined colleges of London had larger undergraduate science schools than Oxford with a considerable number of chemists. (Undeniably Oxford already had the largest number of honours chemistry students, however.) The Butler Education Act of 1944,[19] implemented by the Labour government, changed the situation in that it provided finance for a student away from home at such places as Oxford. The resultant pressure on the colleges to take scientists increased very considerably from both political sources and these financial changes and from the drive in many of the Grammar and Direct Grant Schools to get pupils into Oxford, now more open to them, as can be seen in the immediate jump in numbers in science undergraduates after 1945 and their further increase over the next 20 years. The Grammar Schools were proud of their students who won such scholarships. We note, however, that in 1955 there were per year only 141 entrance scholarships in sciences but 316 in Arts. Cambridge was considerably more attractive to science applicants as the number of scholarships in science was greater.[1] As remarked in section 7.1.1, in the years just prior to 1939 19% of the total Oxford student population were in sciences, some 40 undergraduates per year were reading for an honours degree in chemistry, while in 1945 the number had dropped, it had increased in 1950 when there were 60, (when 35% of all students were scientists) and the numbers increased further in 1955 to 90, and to 160 by 1965, after which the number fluctuated a little around a slightly higher value. Note that less than 10% of its undergraduates were women from 1945 to 1965.[2] By 1966 only one third of the science entry was from public schools compared with almost two thirds of the entry in Arts. It has remained the case that state schools are more heavily represented in the sciences, including chemistry, than in the Arts and it would be interesting to see if this is so amongst teaching staff. By 1950 the Oxford Chemistry School with its four-year course became comfortably the largest in Britain, even perhaps worldwide, but it is no longer so today. With its well-established research it also became by 1960 possibly the best chemistry school in Britain. The increase has to be seen against the changes in other sub-faculties.[2] In major Arts subjects such as Classics and History there was little change from pre-war times but newer subjects such as Philosophy, Politics and Economics also had a considerable jump in student numbers, Table 7.3. There are later parallels of such switches of

Table 7.3 Growth of undergraduate subjects[2].

Year	Chemistry	Physics	History	Lit. Hum.	PPE
1939[a]	45	40	250	120	125
2000[a]	160	150	280	125	325

[a]The numbers are for average years over 4 year periods rounded to the nearest 5.
Note. These numbers could give a false impression of the total change in the University that show that numbers of undergraduates in Science subjects as a whole increased from 20% in 1939 to somewhat below one half of the total in 2000.

interest now in the sciences when biological studies, including biochemistry, increased while those in chemistry and physics remained steady or fell somewhat perhaps due to the impact of the supposed or real adverse effect on the nature of the environment of the chemical industry's products. There also grew to be considerably increased numbers of students at school reading biological subjects after 1970. The drop in students taking A-levels in chemistry has been reversed since 2005, maybe because it is seen that chemistry is able to provide possible solutions to problems such as global warming.

Students were taken in by colleges in 1945 and for some time after on the basis of an entrance examination, linked to scholarship awards, whether or not the college had a science (chemistry) tutor or even sufficient numbers of living rooms in college to match the increased total entry. Such students, paying fixed fees provided by education authorities, were profitable as science (chemistry) tutorial teaching was usually done relatively cheaply much by non-fellows in the university. It has remained the case to this day that students are admitted individually by colleges, though there is no longer an entrance examination, and not by a central scheme with the exception of a few last places that are selected by a university-wide clearing-house system. Nowadays too, the students are well looked after domestically in the colleges, which has not always been so, and the number of tutorial fellows is sufficient for them to do a large percentage of the teaching.

Now, to reach the standard of ability in required subjects for entry into a specific Oxford course via an entrance, scholarship examination, the candidates were restricted by their schools to subjects suitable for Oxford sub-faculty courses describable in terms of Higher School Certificate or Advanced school examinations (A-levels) as follows: fundamental mathematics, with a physical science, only for mathematicians; advanced mathematics and physics for physicists; mathematics (at a lower level), physics and chemistry for chemists. All Arts subjects were virtually outside a science course from age 15 to 18 and no examination results in biological subjects were in general acceptable to these sub-faculties and so were excluded for aspiring school chemists. These restrictions are much less marked today but they characterised attitudes in the chemistry school for many a year. The requirement of having passed a national test in Latin was only removed in 1960.

A very interesting development in the university was the idea of mixed colleges of men and women, unknown in 1945. In the years from 1945 to 1965 there were no more than ten women undergraduates reading chemistry in the strictly female colleges,[2] with the number of men per year rising above a hundred in their separate male colleges. The admission of women to the first batch of 'men's' colleges to change their statutes was in 1975[2] and subsequently all colleges have become mixed. As a consequence the number of women undergraduates increased from about 10% in 1970 to over 30% in each year from 1990 to today. The total number of undergraduates in chemistry has not changed but has fluctuated since 1970 around 170. There is then considerable concern that the number of women who have become research workers in chemistry in the UK, not just in Oxford, has remained small, see Section 7.2.1.

7.3.2 The Structure of the Chemistry Course

The very nature of the restricted subjects studied at school and the subject matter that a student had to take in the Oxford entrance/scholarship examination in 1945, and many years thereafter, meant that the course at Oxford in Chemistry was very intensely just chemistry with no teaching in mathematics, although it was becoming increasingly important in chemistry, nor in geochemistry nor in bio-sciences, which were increasingly chemistry dependent. The course was quite hard work, not at all to be taken in the relaxed manner portrayed in the books of the 1930s. Each day there were two hours of lectures and a practical course, which could take up to several hours of a five-day week. Tutors set an essay a week and expected it to be handed in before the end of the week. Students often had to work over weekends and to do some work in the vacation. Curiously at that time students were expected to monitor their own progress in all three branches of chemistry in which they were given tutorials, inorganic, organic and physical chemistry. They were not examined at all during the course of the first three years, except for a simple test in translation from a German chemistry text, for they had sat what was the nominal first-year examinations (preliminaries) in basic chemistry, physics and mathematics in the summer before they entered the university. This meant that abruptly at the end of the third year the student sat six written papers (one elementary and one advanced paper in each of the three branches) and three 6-hour practicals to finish the Part I course. There followed a fourth year, Part II, which was a nine-month research project. The student selected his own research supervisor for Part II but previous contact with any research activity was so small that in most cases the student chose to do the fourth year with one of his tutors of the Part I course. There then followed an examination of the Part II thesis after which a classified degree was awarded on the combined Part I and Part II performance. The curious nature of the Part II with no taught classes or tutorials arose from its origin not with the colleges or for educational purposes but from the desire of Professor Perkin (1919) to see that students received a practical research training, presumably at that time to be of some use to *his*, Perkin's, research. Some colleges refused to pay tutors for supervising Part II work since it had no tutorials.

Subsequently, the system of examinations, but not most of the basic features of the course, has been changed in steps. Even in the 1950s it was possible to take subsidiary courses in certain topics such as quantum mechanics or pharmacology, which were examined at the end of the first or second year. They did not count in the classification of the degree. Later, the preliminary examinations were changed and made compulsory, in chemistry alone, taken at the end of the first year and a small number of very weak students were then expelled (no more than 2 to 3%). In the more recent preliminary examinations, there is included a compulsory paper in mathematics, taught in classes, and this examination remains today. An end-of-second year examination was introduced in 2003. By this time practical training had been greatly reduced in part due to the over-stringent health and safety rules. There are no practical examinations today but a student is required to complete a limited number of experiments. In the opinion of many the examination of a student three times in three years at university level, once at the end of each year, with tests called collections in all other terms in some colleges, is too much. For a reasonably well-motivated student, the majority, some period should be allowed for browsing especially as the content of the chemistry course has been widened to include the major advances in chemistry over the last 40 years and at least a part of biochemistry. In this view a university should not be a place of over-specialisation such that a student has a limited perspective. Again the chemistry course and its examinations have become more quantitative, problem solving, rather than description oriented. In the outside world although obviously exact solutions to a problem can only be reached by calculation, many problems have to be described clearly and solutions sought that do not lend themselves to exact numerical analysis. In the opinion of some such intensity of numerical effort in the Oxford chemistry course could well be left to the Part II or postgraduate years for those determined to be professional research chemists especially since the majority, say 60%, of those trained in chemistry do not use chemistry greatly ever again after leaving the university.[20] It remains, however, a strongly held belief by those teaching chemistry that the logical structure, at least in fair part based on mathematics, is a very good training of a mind for later life. The question remains as to how good the chemistry graduate is at expressing himself verbally. The final year Part II is a research-only year, which may aggravate the problem but it does require the student to write a clear presentation of the year's work.

It was the pride of Oxford in the years following 1945 that a first class honours degree implied that the recipient was of quite outstanding merit. The classification was in four classes from 1945 to 1973.[2] While 15% were given first-class honours some 25% were classified as thirds or below. This division was more exactingly selective than the classification in say most of the top ten or twelve British universities. Pressure on Oxford from outside effectively forced there to be an increase to more than one third 1st class degrees, the vast majority of the remaining undergraduates to be awarded 2nd class honours but now divided into 2.1 and 2.2. This division, in force today,[2] was made in order to come into line with the requirement of a 2.1 or better in all universities if a student was to continue to a doctorate. The indignity of 3rd class and earlier 4th class honours has almost disappeared.

7.3.3 The Content of the Undergraduate Course

The essence of the course as described already is of tutorials and lectures during three years, Part I, followed by a year of research, Part II. We look at the content of Part I first. In 1945 it was not seen to be necessary to have an ordered framework, a syllabus,[21] for Part I as it was generally thought that chemists knew what chemistry was about. As a consequence the lecture course itself often had a strong flavour of the research interests of the lecturers. The fact that the school had lost touch with the medical/biological sciences, which had so helped its formation, and with geological and environmental science, which had for centuries been a strong part of it, did not seem to concern those who taught the course in 1945 but this is less true today. Tutorial subject matter, which, it might be considered, should have dovetailed with the lectures, was very idiosyncratic, tutor-by-tutor, and often a freewheeling exercise, with a student given a topic for a week, see Chapter 5 for a more thorough description. The topics were very variable in selection between old-hat and modern chemistry depending very much on the age of the tutor and often in no obvious sequence. Later we raise the question as to whether the tutorial is the best way to teach chemistry. Should fellows teach basic chemistry today? After all it is an adapted approach from leisurely culture lessons. In making these remarks we must not overlook the value of Oxford's tutorial teaching. It provides direct individual contact between a fellow/lecturer and a student with all its potential educational and social advantages. It is, however, costly per student and in a fellow's time. Before indicating how matters have changed very considerably it is also useful to have an impression of the lectures themselves in 1945/1950.

In 1945 the outstanding features of the Oxford Chemistry course were the lectures of the three professors, two later Nobel Prize winners, Robinson and Hinshelwood, and Sidgwick, a man of great scholarship. We shall describe these men in some detail in Section 7.4 as their distinction gave to Oxford Chemistry its particular flavour for many years to come, see also Chapter 5. The first two professors lectured once a week for two or three terms to the first-year students. Their manner of presenting material was quite different. Hinshelwood's physical chemistry lectures were based on an intellectual, almost philosophical, command with little reference to practical matters. He concerned himself with the underlying concepts of thermodynamics and kinetics and their application more than with structure or the theory of chemical bonds. The less able found them difficult and were glad that others gave simpler descriptions of the same material. There was little reference to condensed phases or to biological systems. Robinson detailed the classes of organic chemicals and routes to their synthesis. He was clearly teaching classical analysis and synthesis of organic molecules starting from an elementary level. He showed little concern with any underlying theory and he again rarely mentioned biological topics. Sidgwick was different in that his lectures were scholarly in the sense of an Arts man. He ordered the facts of inorganic chemistry following the Periodic Table and, taking elements in turn, he noted their properties and those of their compounds. This could have been a mere catalogue of things but his insight into links between compounds and his historical knowledge saved the lectures. There was little reference to theory and he, like the other two, gave a nod to industrial chemistry only in passing. His lecture course, linked to his book, was remarkable in that it lasted for 2 years, two lectures a week covering the Periodic Table from hydrogen to uranium. A first-year undergraduate could find himself exposed to beginning at hydrogen then Group I (year one) or at Group V, nitrogen (year two). As the material was largely descriptive this was no different from starting English Literature at any one of two periods say Chaucer or Dickens then going forward to others following historical sequence to today before starting at Chaucer again. Surprisingly, the lectures were well attended not just because there was no other decent source of information but due to the pleasantness of style and the clear fascination the man had with knowledge. There is no doubt that the influence of these three men was exceedingly great upon receptive students. Two of the professors, Sidgwick and Hinshelwood, wrote major teaching books on their subjects,[22,23] which from their date of publication were very useful to students, but Robinson only wrote a short research book on natural products.[24]

Apart from the lectures of the three professors there were a diversity of topics, as stated often chosen by individual demonstrators or fellows linked to research interests in the different departments, and with no obvious organisation. There was no indication of the links between the more practical and empirical organic chemistry, on which there were several series of lectures on particular classes of organic compounds, and the treatment by the physical chemists. The main lecturers in organic chemistry were not just given by fellows of colleges, Hope and Taylor, but included those by non-fellow demonstrators, Smith, King and Plant. One additional college fellow, Hammick, a maverick organic chemist, lectured on the phase rule in the Inorganic Chemistry Laboratory. He was made a Fellow of the Royal Society in the year he retired. When asked how he felt, he commented, "Rather like a bride having a baby in Church during the marriage ceremony!" Notice that with the exception of Sidgwick the organic chemists did not write books. The physical chemists in their laboratory gave lectures, in essence on their research topics on which several of them had written books: Valence Theory (Sutton), Photochemistry (Bowen),[25] Spectroscopy, mostly infra-red, (Thompson),[26] Acid/Base Solution Kinetics (Bell).[27,28] All the lectures had a strong theoretical base and little application, quite the opposite to the Organic Chemistry lectures. In addition, Wolfenden wrote a student book on problems in physical chemistry.[29] The lectures given in the so-called Inorganic Chemistry course, were really only called such after the place in which most of them were given, the (Old) Inorganic Chemistry Laboratory. Many were really a part of a physical chemistry course such as those of Staveley (Thermodynamics) Woodward (Electrochemistry), Hammick (the Phase Rule, see above), Linnett (Aspects of Valency)[30] and lectures on crystallography by Hodgkin and Powell. The latter were of surprisingly little appeal to undergraduates as both concentrated on such properties as space groups and symmetry properties. Hardly an undergraduate in chemistry was told of Powell's work on structures of inorganic chemicals except by reference to some papers with Sidgwick. Even less was revealed of the achievements of Hodgkin in solving the structures of cholesterol and later penicillin. The lectures by one or two members of the Inorganic Chemistry Laboratory, who might have qualified as "inorganic" chemists, hardly stimulated any interest. Brewer, deeply involved in local politics (later Lord Mayor of Oxford), and Irving, effectively an organic chemist who had been injected with the closing of college laboratories, did not inspire. In effect there was no real course in inorganic chemistry but for that by Sidgwick, given in Jesus College! Undergraduates were forced back to books; Sidgwick's large two-volume work[22] when it appeared in 1950 under his Lincoln College address, and a book by Emeleus and Anderson, revised in 1950.[31] (Anderson was to become the first genuine inorganic chemistry professor at Oxford in 1963.)

Perhaps the most curious feature of the lecture courses, even of lectures in any one laboratory, was that there appeared to be little connection between them. Although in later years there was more collaboration it often still seems that individual staff members choose to lecture on a topic, linked perhaps to their own inorganic, physical or organic chemistry interests. In Section 7.5 we shall consider how this situation came about against the background of internal struggles between colleges and the university (laboratories) to manage academic matters.

The practical course was superintended by the demonstrators (re-named lecturers in 1964) in the three laboratories, Figures 5.1, 6.1 and 7.2. The organic chemistry experiments were routine minor syntheses that taught the student the need to follow instructions as closely as possible. No work was carried out in fume-cupboards, no eye protective glasses were worn and a student was expected to be able to do all the operations without close supervision. A similar attitude prevailed in the other laboratories. In inorganic chemistry the major tasks were qualitative and quantitative analysis. The poisonous gas, hydrogen sulfide, used in analysis, was the only chemical confined to the one or two fume-cupboards available for a class often of 30 students, Figure 7.1. Physical chemistry involved simple problems in physical properties, mostly kinetics and equilibria, using several basic procedures. Anyone inspecting the course today would declare that most of it broke health and safety rules but very few accidents occurred. As with the lecture course it was considered to be unnecessary to relate one practical course to another and in fact the approaches were quite different in the three laboratories.

In summary, demonstrators and tutors took an attitude to the student assuming he was adult. He swam or drowned, for there was no syllabus. Many found the Radcliffe Science Library a wonderful resource even looking at latest issues of journals, of which there were not vast numbers as there are today, and they were accessible. The great advantage of such an education, based on the assumption of a very rigorous school background and a presumed high intelligence, was that it encouraged independence, or should we say it left little choice but independent effort. Its obvious failings were its random character and lack of relevance to industrial chemistry, biology, geology and the environment. It must be remembered that research topics were just within the reach of undergraduates and at least three undergraduates produced research papers in journals such as *Nature* before their first degrees were awarded. The students joined their tutors at weekly meetings of the Alembic Club after dinner when they listened to experts from far and wide lecturing across the width of chemistry. The Club could not hold a student audience once research became very sophisticated and specialised and the Club had ceased to exist by 1965. Its junior branch was run by undergraduates and lasted somewhat longer in modified form.

The chemistry course in general progressed between 1960 and today in step with knowledge world-wide but with special features through the changes in Oxford appointments. The arrival of Charles Coulson in 1952 pushed forward new emphasis in lectures on the theories of chemical bonding that relatively quickly entered examination papers. In physical chemistry development of theory by Longuet-Higgins and Orgel amongst others, and new instrumental methods, especially in spectroscopy made possible the teaching of a closer understanding of molecules and molecular dynamics (kinetics). (Only when Rowlinson became professor in 1974 was more stress placed on the relationship of gases to condensed phases.) Organic chemistry teaching changed too in that mechanisms of reaction began to play a greater and greater part shortly after 1960 first due to Waters and then to Norman.[32] After 1965 Knowles and Lowe introduced enzymological syntheses into the course. Later, organo-metallic chemistry was also included. Probably the biggest impact on the course came with the publication of the two-volume book on Inorganic Chemistry by Phillips and Williams,[33] both Oxford graduates, in 1965/66. The book was novel, concentrating on intellect-based discussion of as much modern theory as possible (and more than most students could stomach) followed by correlation of data on inorganic compounds in various graphs. It was immediately taken up in Oxford and serves to illustrate how far the dominance of physical (intellectual) chemistry had entered in the inorganic chemistry thinking and teaching at Oxford. It was not a highly successful textbook for undergraduates worldwide as it was too difficult and also unrelated to practical concerns. It was a teacher's or at least a graduate's book of the principles underlying inorganic chemistry, highly suited to Oxford. Much of its novel approach is now incorporated generally in inorganic chemistry courses but the more recent books for undergraduates give the approach of descriptive and synthetic chemistry an appropriate balance. The much more recent book by Atkins (an Oxford fellow/ lecturer not an Oxford graduate) on Inorganic Chemistry[34] is a successful combination of principles and practice. It achieved this objective by being co-authored by chemists from outside Oxford more closely linked to practical inorganic chemistry. Atkins, from the Physical Chemistry Laboratory, had previously written outstanding textbooks on that subject.[35-38] In particular his book on Physical Chemistry (1978) now in its eighth edition has dominated the teaching of this subject in Oxford and elsewhere. At an early period Smith had provided a very useful student book on thermodynamics.[39] In succession, the new inorganic professors following Hinshelwood, Anderson (1963) and Goodenough (1976) brought lectures on condensed matter, under the general heading of solid-state chemistry and Malcolm Green first as a college fellow then as a professor made a considerable impact introducing organo-metallic chemistry. A major change in all parts of chemistry brought about by new methods of measurement was the emphasis on the spatial and electronic properties of compounds in all types of phases.

In a rather desperate effort to interest the young in biological chemistry, Williams gave a set of lectures in the Zoology Department lecture theatre in the 1970s as he had been discouraged from giving them in the Chemistry course and lecture rooms directly. Later, the course was properly incorporated into the Chemistry School by his pupils, amongst others. Williams himself became a tutor in biochemistry in his college (1967) but remained a lecturer in inorganic chemistry before he became a Royal Society Research Professor, see Section 7.7.

If we now move forward from the 1960–1980 period to today the changes in the Chemistry course can be seen to have been considerable. Some change especially most recently was necessary as education in schools slowly lost its intensity and it was possible to study chemistry there in a new "relevant" form with much less mathematics and physics while biological topics could be included. Gradually the tutorial course has had to assume a lower basic knowledge simplified against a background of a huge change in our insight into chemistry. At least in outline what had to be included in the course were the jumps in the theory not only of the binding of atoms to one another in compounds, but also in statistical approaches to the huge array of novel groups of inorganic and organic compounds and in the great advance in instrumental methods leading to new experimental approaches to many traditional topics especially in physical chemistry but also into the properties of old and novel inorganic and organic compounds. On top of these changes interest in biological systems has gained some hold on the Oxford course. The huge advances of knowledge have demanded and altered the relationship between teaching and research. The undergraduate today is now far from the research front and its chemical literature is in hundreds of different journals with a massive weekly output. Even the rows of books, set aside for students for some time in a special part of the Radcliffe Science Library have considerable extent and there is a special series of slim volumes,[40] Oxford Chemistry Primers, especially suited to the course. In this ocean of information the student would drown if left alone. Instead he is guided by a more systematic lecture course and his tutorial work, both based on what, from around 1980, is a secure core of necessary knowledge. They now have to be aimed at a series of university examinations, one at the end of each year, and are closely organised. As we shall stress the new research into the theory of the chemistry of chemical bonding and into structures as well as the advances in physical methods have made the handling of some parts of mathematics an essential part of undergraduate chemistry. There is also a course in the use of computers so very necessary today. The mathematics is one more burden on the student that is particularly difficult for those who have loved chemistry at school for the pleasure of making a chemical substance and doing operations with skill but who are poor at mathematics. In fact many have chosen to do a combination of chemistry with more descriptive A-levels since now-a-days the tight restrictions imposed on doing purely chemistry at Oxford from say 1945 to 1970 have disappeared to some degree though the course is still not very broad. Naturally the student has to have more help than he got in the early period. Perhaps it is unfair to call this present teaching spoon-feeding but it is very different from the course of fifty years ago, see section 5.3. We ask again, would not classes be a better form of teaching than tutorials? The present tutorial system probably rests more on the value of social contact than on the imparting of knowledge.

The practical classes are also unrecognisable. In all chemistry departments, experiments have to be selected within health and safety rules. It is then better to devise experiments to illustrate principles rather than for the teaching of manipulation. Here, the advance of equipment makes for an easy way forward as even very simple materials can be tested by complex equipment, unfortunately in "black boxes". Only organic chemistry practicals can be seen to be related to those of fifty years ago. The downside is that the student is less well prepared for the fourth year, Part II, of research.

Without any doubt the advances in chemistry have made for a course that leaves little room for imaginative adventure. We shall see later how these advances and the business of the whole activity have also come to affect research and even the style of living of those engaged in the subject, reducing the interest in teaching. Perhaps it is not quite true to say that the three-year course, Part I in chemistry, is now a "training of the mind" in logical thinking throwing us right back to the very beginnings of the university eight hundred years ago. In this sense the course has a classical ring without the accompaniment and thrill of active novelty. The numbers now doing research each year exceed the total numbers in the undergraduate school of 1939 and it is here, to some extent in the Part II year of the course, that novelty can be introduced. As mentioned before this implies an interest in research in chemistry yet the average Oxford student reading the subject to the end of Part I cannot be looked upon as a future citizen occupied in advancing chemistry.[20] Much though the fellow/lecturers are very dedicated, even preoccupied with this target, the student is more likely to benefit from learning to treat information with an eye on relevance than on a quantitative view in order to provide understanding.

The Part II of the course, the research year, was devised by Perkin as explained in Chapter 5, and required "presentation of experimental investigations" and could be followed by a study for a doctor's degree by selected individuals in his laboratory. Very soon, however, many chose to go to the physical chemists for their Part II and then a doctorate. Today this wider choice of supervisor is encouraged. The weakness is that there is still no course work in the Oxford Part II, no required lectures and no explanation of the links of one research group to another. In fact a Part II student can be led into the now narrow path of a supervisor's research and although some broadening is provided by departmental seminars there is very little cross-fertilisation between the departments. The student is not encouraged to go to seminars in other subjects. As the Part II decides the future doctorate study its subject may well remain the speciality of the student in further research without an ability to see connections to even related topics. There are clear grounds for giving the research Part II course a very thorough review with an examination of its purpose for over 150 undergraduates a year.

Despite the above criticisms, the quality of the chemistry graduate from Oxford has always been very high by any judgement. In part this owes itself to the quality of the intake plus in fair measure dedication in the tutorial system. There is, however, no reason that the course could not be improved noting the aspirations of students and the likely concerns of tomorrow's society. After Part II, research-minded students do not just do research in Oxford but have entered into research groups in other UK Universities and then proceeded to doctorate and post-doctorate experience all over the world, where many have proved to be very distinguished. Others have entered industry, and also commerce, but very few have become school teachers.

7.3.4 The Graduate School

Oxford has had in recent years a large and increasing graduate school studying chemistry that has risen today to 80 graduates per year studying for doctorates and a succession of post-doctoral students from many parts of the world. For many years after 1945, however, few colleges took much interest in them though they were required to pay college fees. More recently a middle common room has been introduced between the student (junior common room) and senior common room (fellows) in all colleges but only a few colleges make a real effort to provide a comfortable social life for them. In essence the colleges are undergraduate living and teaching places. The graduate students together with the 150 or so Part II students of the undergraduate course[2] in chemistry have no formal teaching either in tutorials or by lectures. There are certainly yearly courses, the Hinshelwood lectures in physical chemistry and the Robinson lectures in organic chemistry, given by outstanding outside professors and there are a variety of special as well as regular weekly seminars but there is no structured course covering all the branches of chemistry, although one or two special topics may be treated more systematically. The instruction in research has grown up in an *ad hoc* manner that needs review. It may well be that the period allowed in which to complete a doctorate (3 years) is so short as to preclude taught courses as a high research output is demanded. Elsewhere, for example in Europe and USA, course work is often compulsory but the doctorate takes longer (5 years). Again, despite these disadvantages the great success of the graduate school cannot be denied not only by the achievements in research but by the immediate successes of the graduates in obtaining post-doctoral positions in Oxford and elsewhere and in the long-term successes in academic life and society generally. We turn next to the nature of the research at Oxford drawing attention first to the dominance of its direction by the three professors of 1945 and then turning successively to the research of all members of staff dividing it into three periods 1945–1965, 1965–1980, 1980 to today.

7.4 THE THREE PROFESSORS AND THE THREE DEPARTMENTS OF 1945

Chapter 5 has described the three major players in the recent history of Oxford chemistry, Sidgwick, Robinson and Hinshelwood, who became professors in the 1918–1939 period. The influence of their styles was still felt strongly after 1945 in all three research branches of chemistry not only while they remained in office but for a considerable time after that. We shall see that they were strong forces in the formation of two disciplines, one in a certain part of physical chemistry (Hinshelwood), strongly intellectual though based on experimental data not just gleaned from the internal work of his laboratory, and with very little of a preparative or analytical approach, and the other in organic chemistry (Robinson) with a different, equally strong, practical approach, using information mainly from his organic chemistry laboratory, with very little physical or theoretical connection. The third discipline of the subject, inorganic chemistry, was almost lost as a laboratory subject much though Hinshelwood was appointed as the Dr Lee's Professor of Chemistry, which was supposed to cover Inorganic Chemistry, but was represented strongly by Sidgwick's lectures. As mentioned previously any connection with medical, biological or geological sciences, the major origins of chemistry, except for Hinshelwood's work on bacteria,[41] was absent. The first two disciplines were carried on in the laboratories named after them, but the third laboratory was not, and never became, simply for inorganic chemistry but was and has remained to some extent a second physical chemistry laboratory for reasons we shall relate. Of course the professors and their teams did not occupy the three laboratories, the Inorganic Chemistry Laboratory, the Organic Chemistry Laboratory, also known as the Dyson Perrins, and the Physical Chemistry Laboratory on their own. The university supported the laboratories financially, and, as of right made it clear that college fellows, with or without university demonstratorships, as lectureships were called from 1939 to 1964, were to be given space in them. There were no college laboratories shortly after the war; that in Jesus College, the last, closed in 1947. Now as seen in Chapters 4 and 5, demonstrators who were not fellows could be selected by a professor and therefore helped to further the chosen direction of his work, as did the senior research officers and the graduate assistants appointed by him. It is very necessary to distinguish this group of laboratory and university-based research workers in small top-down systems, from the independent fellows of colleges who could also hold demonstratorships. There were equal numbers in the first category to fellows in the organic chemistry laboratory in 1945.[2] Several of the college fellows, who were also demonstrators in this laboratory, were not connected to Robinson and they did not contribute greatly to research before 1945, nor until Robinson retired. Even after that only one or two introduced a more physical chemical, mechanistic, approach to organic chemistry before 1965. As a consequence, the organic chemistry laboratory had a strong professor's group leading its research. Quite differently the demonstrators in the other two laboratories were virtually all fellows of colleges, many very active in independent research. Some had come from the college laboratories, several of them were pupils of Sidgwick and Hinshelwood, see Table 7.1, and certain others in the Inorganic Laboratory were left over from Soddy's days. The most recent members from 1945 to 1960 of these two laboratories were effectively selected by Hinshelwood with a college in mind in direct contrast to those fellows in the Organic Chemistry Laboratory who were chosen by colleges and in whose election Robinson mostly showed little interest. Naturally this led to individual research groups many of which dated from earlier, see Chapter 5, and some closely equal in strength to that of the professor, and to dispersal of interest especially in the Inorganic Chemistry Laboratory. Not only was the organisation of the laboratories different but the research styles were differently oriented and have effectively remained so to today. We shall use the distinction between the 'practical' chemistry of the Organic Chemistry Laboratory and the more 'intellectual' chemistry of the Physical and Inorganic Chemistry Laboratories as defined in Section 7.1.1. The second group did not think to take out patents and had few industrial connections. It is against this background that we shall review not only the characters of the three professors but how they used the demonstratorship and fellowship positions as far as they could to advance their particular vision of research in chemistry. Note that we shall refer to fellows of colleges who were also demonstrators as *fellow/demonstrators*, who were highly independent of the professor, and to *demonstrators* as those without a fellowship who owed their position directly to the professor and were therefore selected to further his point of view. We have noted that the bias of a professor could affect considerably even the selection of a fellow, since as noted in Section 7.2 a decent salary could not be provided by the poorer colleges, and a demonstratorship was then required in order for a college to employ a chemistry tutor. We have seen how the changes of 1964, when all appointments were of fellow/lecturer, affected the power of the professors in Section 7.2 but the very fact that selection of new fellows was by a given departmental laboratory, meant that the different attitudes in the laboratories were likely to be persistent. First, we go back to the period immediately following the war to describe the nature and effect of the professors on the departments.

7.4.1 Hinshelwood and Physical Chemistry

As indicated Hinshelwood, a renowned physical chemist by 1945, was in charge of two separate laboratories, that for physical chemistry in the new 1941 laboratory and that for so-called inorganic chemistry in the old 1879, now called the Inorganic Chemistry Laboratory. He had become a very different man from the college tutor described in Chapter 5 and even of the war-time professor as mentioned in Chapter 6. He held himself aloof from most of his colleagues though as a young man he had been good company even for students. It is said that on becoming professor in 1937 he stated that he could no longer be friends with any of his colleagues in chemistry. As a bachelor he lived in Trinity College at first but soon after 1945 he moved to Exeter College. In itself college life had become a lonely one especially in vacations. Outside his college and his laboratory he cut a cold thin figure *en route* between the two. But this loneliness was his wish and his very style of dressing and appearance helped to give the impression of one wanting to be shut off from others. He was of somewhat gaunt looks and, when outside, wore what looked like a black oilskin coat. He walked with head deliberately bent, seemingly contemplating the pavement, his hands locked behind his back and it looked as if he was not really wishing to be seen, never mind engaged in conversation. He had a large office in the laboratory furnished aesthetically and somewhat minimally but with a beautiful Persian rug and one or two ornamental vases of distinction. Undoubtedly he was a man of taste. To enter this office was not possible without permission from a Miss Marjorie Binnie, known by some as 'the dragon', who inhabited an office on the route to 'Hinsh'. Another barrier was his research assistant C. J. Danby, who could act in a high-handed way. A young demonstrator might meet Hinshelwood once or twice a year in the office but the meetings were brief and one-sided. For some years he took tea with younger people but later was more distant though he still had tea now and then with two or three junior lecturers much in the style of many a college Arts don. No one could doubt his intellectual ability as shown in his books[23,41,42] and lectures and he was recognised at the highest level of academic research life world-wide as the 1956 Nobel-prize winning physical chemist.[43] He was President of The Royal Society, 1955–1960 and had been its Foreign Secretary from 1950–55 – both demanding positions. On the Arts side he was an expert on Dante, he was at one time President of the Classical Association, and he painted with considerable skill as shown by his portrait of Harold Hartley. It is undoubtedly the image of the cultured intellectual, which he was, that he wished to project. His distant style created only minor problems in 'his' Physical Chemistry Laboratory for he did not attempt to interfere with the research of the demonstrator/fellows in the department, who had moved in with him in 1941. Mostly they were very distinguished in their own right, see below, and had common purpose in that they were all physical chemists, at first just Bell, Bowen, Sutton and Wolfenden. Their strong sense of independence arose as they were very much college tutors as well as lecturers. Each wrote at least one book,[25-29] and many research papers. These tutors easily attracted undergraduates to work with them in research but Hinshelwood was not able to recruit as many once he became professor and especially later, as he had little contact with students. His team, though it had some Part II students, was more likely to include young men from outside such as Rhodes Scholars. Although the remoteness of his style did not affect the well-established fellows and their science in the Physical Chemistry Laboratory his influence did affect the type of physical chemistry introduced in that laboratory for many subsequent years. Particular importance was placed on gas-phase kinetic properties and on spectroscopy, which have remained strong to this day. His retention of headship of the Inorganic Chemistry Laboratory, however, did cause both immediate and long-lasting problems for the nature of Inorganic Chemistry in Oxford.

The old Inorganic Laboratory was only a couple of hundred metres from that for physical chemistry but Hinshelwood was never seen there although he managed to control who worked in it for some years after his appointment as professor. He also put, under himself, much lesser figures in charge of this laboratory, as we shall see. While there was something of a common theme to much of the work in the Physical Chemistry Laboratory, there was no such theme in the Inorganic Chemistry Department. As remarked earlier, in 1945 and for some years after, it had a miscellany of fellow/demonstrators from the time of Soddy and Sidgwick, for example B. Lambert and Hume-Rothery, together with Brewer and Woodward (Sidgwick's pupils), Irving (an organic chemist) and Hinshelwood's new appointments, Staveley, and before 1942 Thompson and then Thompson's pupil, Linnett. Hinshelwood was said to state that the only subjects worth studying in inorganic chemistry were metallurgy under Hume-Rothery[44,45] and structures by crystallography, the last still under Powell and Hodgkin and in 1945 not yet in the Inorganic Chemistry department but in the Museum. We shall describe the history of crystallography in detail later, noting here that Hinshelwood took into his new laboratory spectroscopy but not structure determination. It was as if he knew which parts of chemistry should be given greatest weight and that was that. We return to the nature and variety of research in this so-called inorganic chemistry laboratory later but as we have remarked before it was largely a second physical chemistry laboratory.

As he grew older his distant style led Hinshelwood to become locked in his own work and no doubt he was distracted by the burden of Royal Society business. He began to ignore others in his own field but also those in an area of biological chemistry, which he had chosen to enter just before the war. In his own field with the help of C.J. Danby he tried to follow molecular fragmentation in gas reactions using a very early mass spectrometer but this was not successful. At the same time he refused to acknowledge the work of Eyring on transition-state kinetic theory.[46] As a purist Hinselwood was correct in describing this theory as inexact, he was not given to inexactitude or qualitative views, but Eyring's theories have been of great practical value in chemical kinetics in solution, but not so greatly in gas-phase reactions, those of Hinshelwood's interest. The author of this chapter (RJPW) experienced this hostility to Eyring when he visited Oxford. Hinselwood refused to meet Eyring on this occasion and it was left to R.P. Bell to entertain him.

Now this aloof style, avoiding the work of probably less able scientists, led Hinshelwood into a serious error in the new science he had chosen to enter just before the war. The error arose from his brave wish to carry over his chemical kinetic schemes into the analysis of rates of bacterial growth. Apart from doing some relatively simple experiments on this topic his group worked on the basic assumption, thoroughly sensible in chemistry, that your starting materials could be well-characterised chemically. This ignored the possibility of diversity and mutants in a population of cells. Despite the fact that the work led to a very interesting book on the algebra of his way of looking at bacterial growth,[41] which was consistent with much data he collected, it was shown by others that mutation could not be ignored. Even 'pure' populations of cells were full of variants so that 'products' were not easily connected to 'reactants.' This makes the algebra of growth exceedingly complex especially when the conditions of growth are varied. Hinshelwood, by ignoring the comments on his work by biologists, especially those knowledgeable in genetics, drew heavy criticism and possibly placed biological science as a subject outside the realm of work for Oxford physical chemists for very many years. It is also rumoured that Hinshelwood with Thompson blocked Molecular Biology from the Faraday Society's subject matter. A somewhat parallel disadvantageous distancing from molecular biology struck Oxford organic chemistry later for quite a different reason.

In university politics Hinshelwood believed he had to fight for the sciences against lasting prejudice against them in the colleges for he was a chemistry professor now and not just a college tutor. We can sense that given the biases of the governing bodies of many Oxford colleges and the university's institutions he was not far wrong here. The earlier history of antagonism between empirical scientists and many on the Arts side is amply illustrated in previous chapters and only in the period from 1920 to 1939 did this attitude begin to change slowly. Even today some scientists feel that chemistry does not receive the full support of a college or even of the university. Hinshelwood was a keen supporter of The Natural Science Dining Club, which had existed since 1850 for the leading scientists at Oxford, and used the Club meetings as a means of finding a consensus view to press for science objectives within the university. Unfortunately for chemistry the respect Hinshelwood craved for science, and perhaps for himself, from the university was only somewhat grudgingly given, partly one suspects through his lack of warmth and his air of superior insight. The author of this chapter (RJPW) was present when the following remark, indicative of an Arts man's view of him, was made by Sir Maurice Bowra, the Warden of Wadham College, a sometime Vice-Chancellor and a classicist. When asked about Hinshelwood's standing Bowra said that he was "the cleverest schoolboy he had ever met" . . . . then, after a short pause, . . . "aged sixty." Hinshelwood, never at a loss for an Oxford-style reposte, described Bowra on a different occasion as "an interesting sociological phenomenon" (see note 10). Another factor in his style throws light on other problems within Chemistry that Hinshelwood's later character helped to create. Robinson's successor as Professor of Organic Chemistry, E.R.H. Jones, remarked that in ten years, 1955–1965, Hinshelwood never came to his, Jones' office, for any meeting between them but expected Jones to go to him. This aloof manner, already apparent in Robinson's time, helped to create the tradition of three very different laboratories – a serious mistake today. The separation was not corrected but worsened when the Inorganic Chemistry Laboratory gained independence as we shall relate. The re-uniting of the three departments, lost in stages between 1910 and 1965, is an on-going battle even today.

A final indication of this extremely able man's latter day inability to see through the eyes of others lies in the creation of the third Chair of Chemistry, and in the separation of an Inorganic Chemistry Department independent from physical chemistry in 1962. While Hinshelwood supported the idea in principle he did not wish to give up his patronage and authority over his 'second' laboratory. To keep his fiefdom he proposed that the new elected professor would share responsibility for the inorganic chemistry laboratory in a troika including himself until he retired in 1965. It did not come about and in 1963 Hinshelwood gave up control of the Inorganic Chemistry Laboratory with the appointment of Professor J.S. Anderson, a forward-looking inorganic chemist.[31,47] It is not clear that Anderson ever saw that he might work with the other departments, or that he exchanged many words with Hinshelwood. There certainly was little precedent for exchange between professors.

Hinshelwood's intellect showed not only in his major book and his research, especially before 1945, but in his lectures, both of which had a considerably philosophical content. As described in Section 7.3 students could not help but be impressed by his controlled lecturing style, given in a manner demanding respect and covering the basic tenets of physical chemistry in thermodynamics and kinetics. The less able found them difficult and were glad that others gave simpler descriptions of the same material but that was not Hinshelwood's style. There was a great contrast in lecturing style between this man and the other two professors and indeed between him and any subsequent professor. As we have said already he and his predecessors and colleagues had by 1945 established a dominance in research in Oxford of certain very essential parts of physical chemistry, kinetics, thermodynamics and spectroscopy, which lasted for a very long period. However, there were weaknesses in this cover that Hinshelwood and his colleagues did not attempt to remedy. Much of the properties of condensed phase and their structural chemistry, where Oxford had leaders in the use of X-ray diffraction in the crystallography department (actually under Hinshelwood but never truly associated with physical chemistry), and of theoretical chemistry including quantum mechanics (only united with physical chemistry in 1994), where Oxford also had many able men but lost them to other universities, were not given due weight by him and the physical chemists. In fact after 1963 solid-state structural and physical chemistry became firmly entrenched in the Inorganic Chemistry Laboratory, while theoretical chemistry had a chequered disconnected history at Oxford, see Sections 7.5.1 and 7.7.3. Hinshelwood's attempt to introduce a physical chemistry approach into biological sciences failed for the reasons already given and the opportunity to develop physical chemistry with its power in this area was also lost. Furthermore his handling of inorganic chemistry in his second department showed his complete lack of interest in the practical and preparative part of this chemistry, and his relationship with organic chemists was close to non-existent. Many of these drawbacks of his attitudes, which his immediate colleagues in two departments may or may not have shared but did not fight hard to remedy, have been only somewhat overcome with time. A certain style had entered Oxford physical chemistry that somewhat inevitably produced new, able, often locally educated young men and who had taken in this style and its limitations with their mother's milk so to speak. It could be said that the department rested on its laurels for several years. To appreciate the feelings of others not in the physical chemistry laboratory (PCL) to the physical chemists' apparent air of superiority we quote a verse of Hammick, a fellow in organic chemistry, referring to (Cyril) Hinshelwood and his colleagues.

> Straight to the throne of God he strode,
> Cyril, first baron South Parks Road,
> Swore that all should go to hell,
> Unless they worked in the PCL.
> But the good Lord would not condemn,
> For 'they' had been got at by 'them'.

The question remains as to why Hinshelwood became so isolated and his research from 1939–1965 was of so relatively little consequence. Did his failure in his approach to living cells hurt him, forcing his withdrawal within himself, unable to admit an error? By the time he retired he was no longer an influential figure in the university, nationally or internationally. To add to his loss of stature he was not permitted to keep his rooms in Exeter College on retirement. (This is a normal Oxford College condition, which has, however, been ignored in special cases.) He left Oxford, perhaps slighted, to live in London and was not seen in Oxford again. He died in 1967.

In Hinshelwood's time the Physical Chemistry Laboratory had many very distinguished demonstrators but there was only one departmental head and professor, himself. Several have been described in Chapter 5 including R.P. Bell, H.W. Thompson, E.J. Bowen, L.E. Sutton, and J.H. Wolfenden (who left in 1947). Their research interests are also mentioned earlier except for Sutton's work using electron diffraction and on dipole measurements. One or two additional appointments were made later including J.D. Lambert (1948, and son of B. Lambert), who was a good tutor and interested in molecular energetics using ultrasonics but this method did not prove very valuable and was superseded when lasers became available; R.F. Barrow (1950) (UV Spectroscopy), and later still R.E. Richards (1959) (NMR Spectroscopy) were further appointments. Five of them were or became Fellows of The Royal Society to which we can add Powell and Hodgkin in crystallography. The contrast with the other two laboratories is considerable for between them they had but three or four such Fellows in this period. We shall turn to the achievements of some of the newer demonstrators after 1965 in the next section.

7.4.2 Robinson and Organic Chemistry

The second professor, Sir Robert Robinson, was not an Oxford University product, as was Hinshelwood, and apparently had little time for its character. His first ten years from 1930 are described in Chapter 5 but his influence on Oxford chemistry was so considerable that it is necessary to repeat and add to the detail given there in order to place Robinson's style in contrast with that of Hinshelwood. As explained in Chapter 5 his origins were Manchester, Liverpool, St Andrew's, Sydney (Australia) and London Universities, but his main background was Manchester, the very different spirit of which place Perkin had brought to Oxford and Robinson reinforced. This background was apparent in his worldly style, hardly changed from that described earlier, for he continued to give the impression of someone who enjoyed living (more or less in his chemistry) with no streak of aestheticism. To go with this he was of considerable height, a good figure of a man with the somewhat rugged face of an experienced life, a practical man, in later life nicknamed by some as 'the grocer', a tradesman in other words and not a true blue Oxford figure (see note 10 and Section 7.1.1). He was undoubtedly a descendant of Perkin, determined to go his practical way. The contrast with Hinshelwood was like that often portrayed as a difference between Cassius, a lean introverted man, and Anthony, more of an extrovert with a fullbody in later life, in Shakespeare's Julius Caesar. Robinson was recognised well before 1939 as another man of extreme ability, but we must see here a quite different ability in chemistry from the intellectual prowess of Hinshelwood. Practical insight such as that of Robinson arises amongst organic and sometimes inorganic chemists and biologists rather than in physical chemistry. It is looked upon as 'intuitive' much though it may be born from an ability to accumulate a vast range of facts and to see patterns in them with the added ability to make practical advances from experience in both. There is no need for mathematical treatment, an adjunct of Hinshelwood's approach. In the laboratory such scientists are called 'Green Fingered' since whatever they touch grows successfully. His fame grew after 1939 to his retirement in 1955. He was one of the greatest organic chemists of the twentieth century. Perhaps the most outstanding organic chemist alive at the time, Woodward of Harvard, never hesitated when asked who (after himself, of course!) was the greatest organic chemist of the 1930–1960 period. "Robinson" he would reply. (We should note though that much of his best work originated before he came to Oxford.)

Robinson's real insight, fully described in Chapter 5, showed in his masterly ability to uncover the natures and then devise the syntheses of complex organic molecules such as alkaloids, steroids and anthocyanins for which he was awarded the Nobel Prize in 1947. Like many an organic chemist his inspiration often came from wonderment at the nature of small but complex natural products, which he isolated, studied their structures and then synthesised. The work also allowed him to formulate schemes of biosynthesis. He had a considerable, almost intuitive, insight in the theory of reaction mechanisms using what is known loosely as the curly arrow notation for electron shifts to describe how organic chemicals react but his mechanistic proposals became controversial as illustrated in the well-documented 'spat' with the London University Professor, C.K. Ingold. If the truth be told, he was not really given to be a chemist interested in principles rather than practice, though he introduced structural features of molecules such as their polarity and conjugation into his approaches. His Nobel Prize lecture shows no sense of electronic or three-dimensional conformational concerns. His small book,

The Structural Relations of Natural Products,[24] gives very concisely his approach to biosynthesis.

Robinson was married, contrast Hinshelwood, for a long period to his first wife, an able chemist, and she saw to it that they ran a hospitable house. He loved the game of chess, had an interest in music, and was a mountain climber in his youth, a hill-walker later. His extrovert character did not really extend to much contact with those laboratory colleagues who had been elected as fellows in Perkin's time or subsequently. Robinson did extend the research buildings of the laboratory around 1940 and he was then able to introduce into the Organic Chemistry Laboratory younger scientists who could follow their own bent and produce novelty, which was the essence of the direction the colleges also wished for them, for example, Todd, Cornforth, Birch, Dewar and King, but none of them became a college fellow, and it did not seem that Robinson wished to direct them to involvement with colleges. For many years to come it would be the colleges themselves that played the leading role of choosing fellows in organic chemistry as we shall see again in the next sections. His wish was for assistants to forward his own programme. He tended to ignore the demands of the university and had no great interest in students, which showed in his lecture course, a mixture of brilliance and casualness. Moreover, administration was very relaxed in his time and this generated a hard task for his successor as the outside world demanded greater and greater accountability in finance and educational style. Robinson, though he could have been a more congenial figure in college and university life than Hinshelwood, never gave much time to either – he was equally but differently dedicated. Of course Robinson too had many engagements outside Oxford for example as President of The Royal Society 1945–50 and of The Society of Chemical Industry 1958/9. Both he and Hinshelwood had numerous medals and honorary degrees that must not go unnoticed. They were both indeed remarkably able men.

There is no doubt that Robinson and his predecessor, Perkin, changed the position of more than chemistry at Oxford. They did so by ignoring the colleges to a great degree, just tolerating college fellows, who had been placed by the university in 'their' laboratories, while making their own departmental appointments. Hence in 1955, when Robinson retired, organic chemists were looked upon as very practical chaps but not very intellectually bright, and not really interested in the style of Oxford academic life into which Sidgwick and Hinshelwood fitted. This attitude plus the remote relationship with physical chemistry, not improved by Hinshelwood, left Robinson to follow Perkin in not seeking to obtain financial aid from the university alone. He relied on industry and especially the Rockefeller Foundation. Here, lies a basic distinction in that organic chemistry readily provides novel compounds for possible use. Once a strong link was established between the university in the form of the professor, not the colleges, and outside funding, organic chemistry could develop without much thought to college wishes if that was what was seen to be for the best by the professor and the university. This remains true today, leading perhaps to the expectation that chemistry should fund itself. The movement of this laboratory away from college influence gave an example to later professors in organic chemistry and to those in many other science departments.[48] The change represented a small part of a slow movement after many centuries toward a possible dominance of the university at least in science (the professors plus the central administration) over college concerns. For example, in 1945 there were 3 or 4 demonstrators or other professorial assistants and only 4 fellow/demonstrators in Robinson's laboratory. This apparent beginning of a top-down local structure has been allowed to continue, though modified and reduced, to today in some sciences but not in chemistry. As we shall see, Robinson's immediate successor could not maintain dominance in the organic chemistry laboratory and it was never exercised in the same way in the other two laboratories. Robinson's and his successor's weakness in relation to the colleges will be shown to be open to an outflanking manoeuvre by colleges using their influence in the university to couple a college appointment with, in the short or longer term, a university demonstratorship or later lectureship. In 1964 this coupling became required and an Oxford professor's chance to dominate was lost. What is best for the development of chemistry is an open question.

Just as in Hinshelwood's case there was weakness in Robinson's approach to organic chemistry. He did not introduce the style of Woodward of Harvard with its ever-increasing use of physical methods in organic chemistry. Much though he could have made room for Dorothy Hodgkin, who assisted his work and whom he assisted and who could have introduced crystallography in his department, he did not do so, despite the success with cholesterol and penicillin structures. Is it not odd that neither Hinshelwood nor Robinson welcomed crystallography? We shall see this reluctance of the chemistry sub-faculty to be deeply involved again close to 1970. He maintained, however, a fine microanalysis laboratory under Weiler and Strauss. A second weakness was that although he was concerned with *small* molecules from living cells he did not direct attention to their larger molecules, though he had thought of doing so before the war. This subject was in course of development in 1950 and onwards in chemistry departments elsewhere for example in Cambridge University under Sir Alex Todd, later Lord Todd, a pupil of Robinson and later Honorary Fellow of Oriel College, (Nobel Prize 1957)[49] and by the chemists who had moved into the biochemistry and physics departments of that university and in the MRC laboratory in the same town, for example under Sanger. (Note that the undergraduate teaching of the Cambridge Tripos system has the great advantage of reducing narrow specialism and allowing easy movement between sciences such as those of physics, chemistry and biosciences.) We shall see that these gaps, often included under Molecular Biology, have taken a long time to fill in Oxford.

As noted in Chapter 6 very important organic chemistry was done in Robinson's time in other Oxford departments; *e.g.* that on British Anti-Lewisite, penicillin and cephalosporin. The Nobel Prize in Physiology or Medicine was awarded in 1945 for the work on penicillin in Oxford. In passing we must note the award of the Nobel Prize to two of Robinson's pupils, Todd, mentioned above, and Cornforth[50] in 1975, yet Robinson did not establish a succession of his people in Oxford. In one sense organic chemistry did not need such a pattern once it was set in its major path of synthesis and not biological chemical investigation.

Robinson retired in 1955. After he left Oxford he did work with Shell as a consultant. He died in 1975. In 1955 Oxford elected as his successor E.R.H. Jones, again from Manchester. We shall add here Jones's initial period until 1965 to that of Robinson as only in the years after 1965 was there much impetus behind a new phase in Oxford organic chemistry. By this remark it is clear that at first the style of organic chemistry research, with its concentration on novel synthesis not mechanism, was little modified by the change in professor. There also remained a stressful relationship with colleges that has remained as an undercurrent to today.

At Oxford, Jones was not dedicated to research to the degree seen in the case of Robinson and his first task, a self-imposed action, was the renovation and then a further extension of the laboratory. He was not a giant in organic chemistry and some thought Barton would have been a better election. Barton, a Nobel Prize winner in 1969,[51] was a founder of conformational analysis and a greatly respected thoughtful man but at the time Barton was extremely young. As a professor, Jones[52] was more open-spirited than either Robinson or Hinshelwood and sought for agreement with the university and colleges rather than confrontation. He could be described as a fair-minded gentleman and there were those in various positions of power, especially college men, who took advantage of what they treated as a weakness in the rough organisational divide between college and professorial authority. Jones was also concerned with the standing of the two British Chemical Societies – one called The Royal Institute of Chemistry and the other simply The Chemical Society. The first was more linked to industry and the second with university chemistry. He worked hard and successfully to unite them, together with the Faraday Society and the Society of Analytical Chemistry, in the Royal Society of Chemistry, of which he became the first President. This care for more general matters outside Oxford was not a forté of Robinson or Hinshelwood, though both served on committees during and after the war. While they might serve as Presidents of Societies their wish to focus on the nitty-gritty of the organisation of chemistry as a whole was small. Jones concerned himself deliberately too with Research Council business with one striking gain for Oxford as we shall relate later. These external matters take much thought and time.

Jones's interest in organic chemistry, his research was on terpenes and acetylene derivatives and later on hydroxylation of steroids, was not really strong in Oxford. It could be said that his best work was done in Manchester, as could have been said of that of Robinson, perhaps that of Perkin too. He did introduce new physical methods and instrumentation but, like Robinson, he did not provide an attack on large biological molecules that had advanced dramatically as mentioned already with the general structural knowledge emanating in Britain from Cambridge University and the MRC laboratory on the outskirts of Cambridge. The leading position of Oxford in organic chemistry research due to Perkin and Robinson was lost. In fact it was junior people in Jones' laboratory who pushed the subject forward in new directions after

1965 as we shall see in later sections. On the very positive side, however, it was an action of Jones with his inside connections in the Science Research Councils that created a major opportunity and not just for organic chemistry in Oxford. Before we describe this opportunity giving rise to the Enzyme Group we turn to a particular difficulty for Jones.

Jones brought with him three of his colleagues from Manchester. (Note that later G. Whitham and J.M. Brown strengthened further the input to Oxford organic chemistry from Manchester.) At that time, 1955, there was no agreement linking posts in colleges to positions supported by the university or by outside finances, so that in one sense his men were, and remained for some time, 'outsiders' in Oxford. These Manchester lecturers had a difficult time as several of the left-over colleagues in the laboratory from Robinson's day had college fellowships with better salaries and social standing. Unlike the other laboratories the number of fellow/demonstrators in the organic chemistry laboratory remained only roughly equal to the number of staff not linked to colleges. While Robinson had almost encouraged dissociation from colleges, Jones just disliked their influence. Furthermore, Jones's decent style allowed the colleges to regain much of the ground lost to Perkin and Robinson as young Oxford men were elected by colleges in preference over his Manchester colleagues although all three did become college fellows after some years. There were no Fellows of The Royal Society in the Organic Chemistry Laboratory except Jones and Waters at this time. The very different style of the laboratory can be seen in that the organic chemists produced very few books. In this early period of Jones's professorship the organic chemists could not be considered as being as distinguished as those in physical chemistry. The next sections, 7.5 and 7.6, of this chapter will show how the balance became more even later.

7.4.3 The Third Professor: Sidgwick

The third professor from 1939 days was Sidgwick, who, as explained in Chapter 5, did not really belong to a laboratory but to a study, a college man. He was not unlike Hinshelwood in that he was scholarly but otherwise very different in style. He too was a bachelor and he too lived in college but he clearly continued to take pleasure from that life. Small, round-headed and bald, he looked benign but could be very sharp. He was careful with his choice of words and his work was always securely based on facts, that is facts based on work of others not from his own work. As described in Chapter 5 he started as an organic chemist associated with Perkin's and then Robinson's laboratory. He underwent a remarkable transformation after the 1920s but he had already written a learned book, *The Organic Chemistry of Nitrogen* in 1910, with a second edition in 1937. This is itself an unusual organic chemistry text as it is interesting through good writing and connections between facts rather than just through fascination with the classes of compounds, their analysis, preparation and properties that was the focus of organic chemistry before mechanistic views held sway. His first considerable step away from organic chemistry was a book on valence, *The*

Electronic Theory of Valency (1927), a masterpiece of 'old' quantum theory ideas before wave mechanics. It may well have been the diversity of chemical bonds that intrigued him and set him off on a more general study of the chemical elements. In the second book, *Some Physical Properties of the Covalent Link in Chemistry* (1933), it was again style rather than mathematical substance or logic that was attractive, but it was an innovative book drawing attention to parallels between carbon bonding and that of 'inorganic' elements, *e.g.* in coordination compounds, but he seems to have missed the power of wave mechanics. (He also took an interest in Powell's crystal structures illustrating the functions of lone pairs of electrons.) These works clearly placed him, apparently an organic chemist, in the intellectual, physical chemistry, Oxford camp and he was definitely not practical. It is said that Lambert, in charge of the inorganic chemistry laboratory, preferred it if Sidgwick did not come in to demonstrate. Then just before and during the war he began to lecture and compile his massive two-volume work on inorganic chemistry, published in 1950.[22] Looking back at this book today shows how intrigued he was by the facts of chemistry, and also with the variation in property of elements and compounds following the sequence of the Periodic Table. He did not show so much interest in the creative practical work that had produced them. A compound was even given its melting and boiling points, its colour and its solubility in a variety of solvents, facts all running one after another before the next compound in a neighbouring part of the Periodic Table was introduced in a manner aiding comparison. The book was referenced in great detail in a fashion reminiscent of the writings of his Arts colleagues, it was truly a scholarly more than an imaginative work, more likely to be a reference source than a book for continuous reading. Sidgwick was not perhaps of so direct an influence on Oxford chemistry as the other two professors before he retired in 1948, but his input was considerable and he placed two or three of his pupils in colleges as fellows, Table 7.1. The combination of Hinshelwood and himself, one more philosophical and the other more simply scholarly, with university insider connections undoubtedly contributed in large part to the way physical and inorganic chemistry evolved almost as one with a strong collegiate link in Oxford and quite separate from that of organic chemistry. It differs in stress on 'thinking about' rather than in just 'doing'. As explained in the introduction this division has coloured Oxford chemistry and not only its chemistry even to today, for it divides major parts of the university, not just the departments of chemistry, from one another. In chemistry it left two practical areas somewhat to one side, the analysis of biological chemicals of great complexity and the preparative side of novel, perhaps useful, compounds based in part or totally on inorganic elements. It could not therefore refer to mineralogy, geology, environmental, or much of bulk industrial chemistry. We shall show how in effect the neglect of these areas in the two major research laboratories, organic and physical chemistry, was partly remedied, largely by chance, after 1963 in the by then enlarged Inorganic Chemistry Laboratory, under a new professor. While it provided or helped to provide several innovative approaches there is no doubt that even then the major approach of this inorganic chemistry laboratory, though deliberately separate, remained more intellectual than practical in the sense of the difference between Hinshelwood's and Robinson's styles. We may question today the desirability of having three chemistry laboratories with such a strong drive to self-identification and during the following years we shall see increasing links between them.

7.4.4 The Acting Heads and Nature of the Third Laboratory of Inorganic Chemistry (1945–1963)

A peculiar situation developed from the appointment of Hinshelwood as the Dr Lee's Professor of Chemistry as a successor to Soddy, who was more of an inorganic than a physical chemist. Hinshelwood, a purely physical chemist, was soon installed in a new physical chemistry laboratory, where he resided, but in addition he kept a hold on the old Inorganic Chemistry Laboratory, though he had no interest in its proper subject matters. As an absent landlord he needed somebody to look after this laboratory, which he never visited after 1941. At first Hinshelwood appointed B. (Bertram) Lambert, who had been an enemy of the previous professor, Soddy, in the 1930s as described in Chapter 5. He had been a demonstrator even before Perkin arrived. (Soddy and Lambert were both Merton College men, Soddy only an undergraduate, and Lambert an undergraduate and a fellow. The extent of their enmity is revealed in that although Merton has portraits, or at least photographs, of many of its fellows, including Lambert, and even of distinguished undergraduates it has no picture of Soddy, a Nobel Laureate in 1921, and he was never made an Honorary Fellow. Quite probably Lambert saw to that.) Lambert had not chosen to follow the rapid changes in chemistry from the 1920 period and as we have seen he and others certainly helped to prevent Soddy from developing his line of inorganic physical chemistry, radioactivity, and the creation of a third laboratory for physical and inorganic chemistry. Lambert's research, such as it still was in 1945, was concerned with ways of determining molecular or even atomic weights by classical physical methods although the mass spectrograph was in use elsewhere. The researches of Lambert required great skill in glassblowing but little deep thought. In fact he was renowned as a glassblower in the university. Lambert did not lecture in the period 1940 to 1948 when he retired. His tutorials had a typical northern character of warm friendliness concerned with student well-being but with not much content, often barely touching on chemistry. He was very much a college man and did not even seem to notice that he was married. This was clearly observed in 1952 when wives of fellows were invited to a dinner in Merton College for the first time. The young wife of a newly elected fellow on her first visit there asked Dr Lambert's wife as they approached the hall for dinner if she could be advised about the form for the occasion. She was astonished by the old lady's reply "Dearie, I fear I am as much at sea as you are for I have hardly been into the college never mind into the Hall in thirty odd years." Wives were not very welcome in Oxford colleges for some years to come. In Lambert's time colleges still had a large influence over selection of demonstrator/fellows. As indicated in Chapter 5 Lambert had helped to place and support Thompson, a pupil of Hinshelwood, Brewer, supported by Sidgwick, and Hume-Rothery in the Inorganic Laboratory before 1939. He, no doubt with the connivance of Hinshelwood, persuaded Merton to elect a very young Merton graduate, C.S.G. Phillips, to succeed him in 1948. Phillips (1953) had no publications, no doctorate and very little research experience but he proved to be an extremely successful tutor. Later, Phillips did some fine research in chromatography, on the advice of Hinshelwood, but research was not really his forté. His wish was to give his pupils intellectual insight, a genuine Oxford ambition, as shown by his book with Williams, described above.[33] Such inside college 'family' elections can be as effective as many an arranged marriage but they need comfortable acceptance on both sides and luck. Phillips had both as well as ability.

When Lambert retired in 1948 Hinshelwood asked F.M. (Freddy) Brewer to take over. Brewer was in one way similar to Lambert in that he did no research of any consequence and for many years he was not elected a fellow of a college. From 1945–1962 he lectured to those bemused undergraduates who still came to his lectures, given in a wandering style, on features of the chemistry of the then seemingly uninteresting (to both the undergraduates and himself) elements such as Ga, In, Si, Ge, Sn and Pb. Much of the lectures remained unchanged year in, year out and perhaps they were formulated before the war when he was at Cornell doing his doctorate. (The doctorate from the USA made him only a Mr in Oxford – only those from Cambridge and Trinity College Dublin were automatically titled Dr!) He was the Oxford University Councillor on the City Council (unelected by the population but selected by the University!) and became Lord Mayor, in which office he died in 1962. None of the three, Hinshelwood, Lambert and Brewer put their mind seriously to the design of the large extension of the Inorganic Chemistry Laboratory of 1957, Figure 7.2, first

Figure 7.2 The front face of the Inorganic Chemistry Laboratory today but from a 1935 drawing. (Reproduced through the courtesy of the University's Estates Directorate.)

mooted in Soddy's time and seemingly nobody else did either and we shall see the consequence when we describe it in Appendix 1.

Given this leadership and Hinshelwood's way of putting scientists into the Inorganic Chemistry Laboratory no matter how little some of their interests were in the subject and now and then removing one to his own department, for example (Sir) Harold Thompson (inorganic chemistry laboratory 1937–1942) and later (Sir) Rex Richards (nominally, inorganic chemistry laboratory 1950–1958) it is no wonder that the laboratory had a more and more diverse set of research workers, extremely individual, and given the selection of those in charge, with next to no management. Two new appointments in the laboratory were R.J.P. Williams (1955), see below, definitely arranged by Hinshelwood and F.J.C. Rossotti (1961)[53] who appears to have been a college choice. As we shall see, it so happened that several, old and young, thrived in this laissez-faire atmosphere. In 1963 the laboratory separated from physical chemistry and was headed by J.S. Anderson, a new professor, but a description of the consequent sharp changes in it will be delayed to the period from 1965 to 1980 in this book. We shall also describe the diverse activity of the other members of the laboratory later. By 1960 it was not without distinction in that Linnett and Hume-Rothery had become Fellows of The Royal Society.

In passing we mention that there was one other somewhat separate 'department', that of chemical crystallography, to be described in Section 7.6.5. Theoretical chemistry did arrive in strength during the reigns of the three professors, it became a department later, but we reserve its history until Sections 7.5.1 and 7.7.3.

In conclusion, by 1965 the pre-eminence of the three professors had set the tone of Oxford chemistry in essence even to today. To a large degree organic chemistry has remained traditional, as it must be, in its stress on a practical approach to synthesis. Its links to theoretical and physical chemistry are slight and somewhat unfortunately its attitude to the biological chemistry of large molecules has been until relatively recently one of distancing itself rather than embracing it. Physical chemistry too was set on its course mainly in kinetics and spectroscopy. As we shall relate it has remained at least until 1975 to some degree as Hinshelwood made it, that is rather removed from biological topics, from much of condensed-matter studies (note especially the absence of X-ray crystallography and electron microscopy structural studies), and for some time theoretical chemistry. A particular attitude to physical chemistry was imposed on the Inorganic Chemistry Laboratory too until 1963, although there were already growing strains with Hinshelwood's approach to the direction of it. After 1963 the Inorganic Chemistry Laboratory has moved away from the limitations of these earlier years to cover a variety of topics of which two major areas, solid-state chemistry and biological inorganic chemistry still keep the intellectual style, while the third major area, inorganic synthesis achieved separate distinction somewhat later and was genuinely 'practical' in the meaning we have given. As we have said, it was the college origins of physical chemistry that especially matched a tutorial style of teaching that led to this particular individualist intellectual tradition. It was then the historical sequence of election of professors and the building of laboratories for them following the earlier independence of college fellows that gave Oxford chemistry a very strong departmental flavour with no overall plan so that each department came to have its own attitudes to teaching and individual research groups with a flavour of the character of the three professors.

Now, as we explore the general development of chemistry in the years 1965 to today we shall see that although there has been general adherence to the major underlying attitudes to chemistry teaching and research set by the three very able professors of some fifty years ago there have been many additional features. While much of the intellectual development has to be seen as tinged with the desire so very apparent in the earlier history of the colleges to remain distant from the outside world, leaving individuals to plough their own academic furrow, there have been recently several direct connections to and attacks on outside targets, see Appendix 2. The university could afford the intellectual style less and less after say 1980 and especially so since 2000 as financial pressures grew. In this sense organic chemistry was and maybe always will be different for it always was a money-spinner. But the need for large sums of money and a style of research more committed to wealth creation is to be found now in all parts of the subject. It is bringing increasing conflict between colleges with their apparent wish to remain outside the practical world and the university authorities and sciences such as chemistry, which cannot take this stance. Chemistry is slowly becoming more centrally organised as is the university. We shall follow the progress of this conflict as it affects chemistry research and as it comes to a head toward the end of this chapter.

7.5 RESEARCH 1945 TO 1965

We have already seen that research in chemistry in 1945 to 1965 had two strongly different parts; that in the organic chemistry laboratory tackled the nature of small but complex molecules by synthesis and degradation and that in the physical chemistry and the Old (inorganic) chemistry laboratories concentrated mainly upon the study of kinetics, stability, and spectroscopic properties of simple chemical molecules. We have called the first part 'practical', especially as it had links to industry, and the second 'intellectual', as it was academic. Outside these two major sectors were the researches on diversity of topics including metallurgy, X-ray diffraction for structure determination, organic molecules for analytical determination of metal ions in solution, all in the inorganic chemistry laboratory or associated parts of the Museum. Much of the professors' researches in this period have been described in the previous section with that of research initiated before 1945 by other members of staff. Here, we give an account of research by staff recruited in the period from 1945 to 1965. As explained, the changes in organic chemistry in the fifteen or twenty years from 1945 were mainly in people and not so much in the direction of research though a variety of new molecules were analysed and synthesised. One or two remained at first even from much earlier times; E. Hope, never really well, and S.G.P. Plant, who were demonstrators with Perkin's group but Plant did not become a fellow, and T.W.J. Taylor and D.L. Hammick who were independent of Perkin and fellow/demonstrators. Additionally, there were those appointed in Robinson's group before 1945; J.C. Smith and F.E. King, both demonstrators, and G. Parkes who had to move to the laboratory on the closure of the Queen's College laboratory and became a college fellow. There were also several men in the laboratory for a shorter time, several of whom became very distinguished elsewhere, see Section 7.4.2. There was little novel work until after 1955 except in Robinson's group and little movement to the use of sophisticated equipment, or to an interest in large biological molecules, or toward mechanistic studies with one or two exceptional pieces of research. W.A. Waters (1947) and later his pupil R.O.C. Norman (1960), both fellows of colleges, had become engaged in mechanistic analysis, the first using electron spin resonance spectroscopy. In a sense Waters and Norman were more of the 'intellectual' type of chemist, as is shown by their books.[32,54] (Note. The dates in brackets here and elsewhere refer to the year of appointment as a University demonstrator, later called lecturer, as a research officer or a professor's assistant as given in the Oxford University Calendar.[2]) Overall, no great change followed from the appointment of demonstrators N. Polgar (1950) and A.S. Bailey (1952), of graduate assistants L.J. Goldsworthy (1950) and F.B. Strauss (1953) all directly by Robinson. Fellow/demonstrators appointed during this time were J.A. Barltrop (1946), Muriel L. Tomlinson (1948), G.T. Young (1952) and B.R. Brown (1955), and they too were concerned with syntheses[55] though not necessarily related to those of Robinson's group. It would appear that college influence was strong in these elections and that Robinson had given up trying to secure College positions for his staff by 1950 but it must be added that, if he ever did try, he did not seem to try hard, although he had assisted in Waters' election. The arrival of E.R.H. Jones, again from Manchester, on the retirement of Robinson in 1955 brought new directions in synthesis, see Section 7.4.2, but, as mentioned earlier it also generated further appointment problems as three of his lecturers came with him. They were T.G. Halsall, G.D. Meakins and M.C. Whiting, appointed in 1957 as Senior Research Officers and V. Thaller (1975) appointed as a Graduate Assistant. Of them, Whiting was best known as he had been an author of the first paper on ferrocenes with Woodward and Wilkinson. Jones' efforts to place his men in colleges failed at first, while colleges managed to persuade (or bully) him into taking their selected candidates, as we shall see. The situation in the other two laboratories was quite different.

Physical chemistry too saw very few changes in personnel or in research topics until the closing years of Hinshelwood's time, 1960–65, which are treated in the next section. As described earlier, he continued his work in kinetics of gas reactions and bacterial growth. He had assistants, C.J. Danby (1945), A.R.C. Dean (1955) and B.A. Coles (1960) but he did not exert great pressure for new appointments especially after 1955. H.W. Thompson and L.E. Sutton, also mentioned earlier had been moved to the 1941 laboratory from inorganic chemistry and from organic chemistry, respectively, but their dates of appointment to demonstratorships appear to be 1943 and 1945, respectively. Two new appointments in spectroscopy were R.F. Barrow and R.E. Richards (at first in inorganic chemistry), both as fellow/demonstrators in 1950. The research work and lecture courses of both of them have been described earlier but we note especially Richards' very successfully nuclear magnetic resonance studies of which more will be said later. The old style of the Balliol-Trinity Laboratory was maintained particularly by Bowen and Bell. The degree to which they had got used to almost primitive equipment was shown by Bowen's eye for waste tin biscuit boxes and string for his energy-transfer apparatus and Bell's refusal to use the glass electrode pH meter, a curious demonstration of a mind set on underlying theoretical advance rather than on the discovery of new facts using sophisticated equipment. Note that during this period some parts of pre-war physical chemistry research were to be found in the inorganic chemistry laboratory. For example, L.A. Woodward worked on Raman spectroscopy, L.A.K. Staveley on the thermodynamics of liquid solutions, and J. Linnett on the kinetics of catalysed gas-phase reactions.

In this 'Inorganic' Chemistry Laboratory, greatly extended in 1957, Figure 7.2, some novelty was introduced when C.S.G. Phillips (1953) took up research in chromatography, suggested to him by Hinshelwood based on his war-time studies of absorption of gases on charcoals. The research in metallurgy, a branch of solid-state chemistry, by W. Hume-Rothery was of a very high calibre especially on alloy phase composition as illustrated in his books.[44,45] His career has been outlined in Chapter 5. The subject was lost to chemistry in 1957 as Hume-Rothery had secured for this subject a new laboratory some distance away. His type of study is part of what is today often called Material Science. It was not associated with chemistry from 1957 until close to the year 2000 but recent years have seen the growth of strong links between the two laboratories. A notable advance in physical/inorganic chemistry in the inorganic chemistry laboratory in this period was that in 1948 when H.M.N.H. Irving with his pupil, R.J.P. Williams, transformed the study of the use of a particular organic reagent, called dithizone, for metal ion analysis into a general study of the strengths of binding of metal ions to all kinds of organic molecules in water.[56] The general treatment applied to the precipitation and extraction of the resultant complexes into organic solvents, a thermodynamic study with wide implications for the analysis of environmental and biological systems as well as allowing an understanding of the principles of the separation of metal ions in living cells. To this day, the Irving–Williams series of stability constants for binding of certain metal ions to organic molecules in complexes is to be found in all textbooks.[33,34] Further study by Williams, based on this work, was used from its beginnings in model studies for the understanding of biological metal-ion chemistry. It has been the source of a flood of research activity to the present time in the subject now called biological inorganic or bio-inorganic chemistry that has developed world-wide. A quite different, now, for once, a practical preparative area, which was started late in the period in the Inorganic Chemistry Laboratory, was the study of the compounds formed between metal ions or atoms with organic molecules, which are soluble in organic solvents. This subject, called organo-metallic chemistry, came to Oxford in appointment steps. One just before 1960 was of M. Venanzi, but most strongly, later, was that of M.L.H. Green (1965). Both were college choices not really influenced by Hinshelwood. We shall treat this topic as belonging to the period after 1965 as it grew greatly in strength then. We also set on one side the beginnings of solid-state chemistry by J.S. Anderson as belonging to the next period 1965–1985, although he came to Oxford in 1963. It is clear too, see next section, that in the very last part of Hinshelwood's professorial headship of the laboratory he had little influence on the selection of demonstrator/fellows in this inorganic chemistry laboratory. Even so, some new obviously physical chemists were put into posts here but probably not on his say so. They included P.D. Dickens (1964), concerned with surface catalysis (a pupil of Linnett), A.D. Buckingham, a theoretician interested in optical properties (1958), I. Beattie, a matrix isolation spectroscopist (1964), M. Christie, a kineticist, the third woman in the chemistry laboratories (1963) and F.J.C. Rossotti,[53] a pupil of Phillips and Irving (1963). This selection of an almost random group of new people caused the variety of chemistry studied in the Inorganic Chemistry Laboratory to grow in a novel but haphazard way. After this time the selection to fellowship/lectureships could never again be dominated by professors or colleges separately for as we have seen in Section 7.2 elections became based on joint committees of professors, departments and college members.

Another big, but disconnected, advance came from the X-ray structure studies of Hodgkin and to a lesser extent of Powell who had moved from the Museum to the Inorganic Chemistry Laboratory when its large extension was completed in 1957. Hodgkin had uncovered the structures of several important molecules extracted from living systems. We shall describe the history of X-ray crystallographic studies separately in Section 7.6.5, as its history is very complex. A very different timetable introduced rapid advances in Theoretical Chemistry to which we devote a separate paragraph as this work was hardly connected to any laboratory.

7.5.1 Theory and Mathematical Research

A considerable change in chemistry after 1945 was due to the advance of theory generally. To a small degree this had already begun in Oxford with the increasingly quantitative treatment of kinetics and thermodynamics as can be seen in the late nineteenth and the first part of the twentieth centuries. The most striking change was, however, in the adaptation of quantum and wave mechanics to the treatment of molecules after its application in the 1930–1940 period to the electronic energies in atoms, mainly in Germany and USA. In Oxford the subject was advanced at first in depth by C.A. Coulson[57] as an ICI fellow (1945–47) and his pupils, H.C. Longuet-Higgins and W.E. Moffitt being the most notable. He left for King's College, London but returned in 1952 as the Rouse-Ball Professor of Applied Mathematics. Coulson, who then had no appointment in the Chemistry School, was not only an innovator but he was a fine teacher and many a research student benefited from his weekly seminars in chemistry. Others, a little earlier, had introduced some of the modern theory in specialised research, for example R.P. Bell on the theory of hydrogen ion and atom transfer and, with H.C. Longuet-Higgins, on hydrogen bonding in boron compounds. There were also lectures by Sutton largely based on Pauling's book *The Nature of the Chemical Bond* (1941),[58] before Coulson returned to Oxford (1952) after a spell away and there was of course the dramatic impact of one year of Linus Pauling himself as a visiting Eastman Professor (1947–48), (Nobel Prize, 1954). In this context we must not forget the theoretical work of Hume-Rothery on alloy phases but not much notice of it was taken in Oxford.[45] Although all these scientists had an influence they did not have the same impact as Coulson. He saw and presented very clearly the general implications of the new theory of chemical bonding, including his own contributions, especially in the description of the simplest molecules, *e.g.* H_2 and CH_4. Coulson was not just a learned professor but he was a man of a giving, warm nature with a principled outlook. He considered that he should teach his mathematical knowledge of chemical bonds in the chemistry school as well as meet his obligations to mathematics, he was elected by that faculty not by chemistry, and to his religious convictions. His seminars in chemistry were an eye-opener for many chemists as we have described earlier. As a Methodist lay preacher he took on many tasks associated with his beliefs, he was a founder member of the well-known charity, Oxfam, and he worked for the cause of the anti-apartheid movement in South Africa (In passing he did not appear to have any common motivation with Dorothy Hodgkin and her husband who supported socialist (even communist) politics in Northern African countries.) This said, Coulson never mixed his science with religion but he would have given any modern-day Dawkins a very difficult time. His conviction in matters concerning chemistry or general life was that his abilities were given to him so that he himself might give, and he did just that in chemistry, in lectures and seminars, while developing the theory of chemical bonds. He was always willing to help others and had a lasting influence on theoretical chemistry and its applications in Oxford.

Stimulated by all these beginnings and work in physics, L.E. Orgel[59] introduced with great erudition the crystal- and ligand-field theories of the physicists to inorganic chemistry in 1950. At the same time M.J.S. Dewar[60] drew attention to the value of wave-mechanical calculations in organic chemistry. By 1960 it was very clear that without a basic knowledge of mathematics chemistry would remain outside the full rational treatments that were becoming so common in physics. We have indicated above its effect on the student chemistry course but mathematical analysis now became essential in much of research, less so in organic chemistry.

It is not clear why Oxford failed to keep the momentum of the Coulson school. Longuet-Higgins left for Manchester, Orgel went to Cambridge and Dewar left for the USA before 1960 and we describe the formation and history of the resultant minor department of theoretical chemistry formed after 1970 later, Sections 7.6.6 and 7.7.3. Some believe that Hinshelwood was not enthusiastic about such theoretical approaches to chemistry, apparently believing that progress in chemistry depended on experiment and that the mathematics required did not need such emphasis, so that he made no lectureships available in the subject. Much though his view may have some truth, theory does coordinate observations, can guide thinking and give rise to *possible* novelty open to test. Its acceptance or rejection by experiment helps progress greatly even if it has not such a powerful standing on its own in chemistry. In this respect it must not be thought that theory is only applicable to calculation of bonding. In Oxford, and in fact everywhere, it began to be and has been applied consistently to statistical studies of molecular interactions in the gaseous and condensed phases as well as to the dynamics, energy distribution, in small and large molecules, amongst many other subject matters especially since 1970. It is very possible that theoretical valence theory has not had the local attention recently that it had in 1950–1970 due to the presentational abilities of Coulson – a hard act to follow. Mathematical approaches to chemistry from the 1930s to the 1960s quickly invaded all physical and inorganic chemistry helping in fact to fortify the difference between the 'intellectual' (thinking) approach and the 'practical' (degradation and synthesis) chemistry. We note that the election of A.D. Buckingham in 1958 to a fellowship/lectureship at Christ Church and placed in the *Inorganic* Chemistry Laboratory with his interest in the theory of the optical properties of molecules, mentioned above, is an example.

7.5.2 Summary

As a closing summary of the changes in the Chemistry School to just before 1965 we observe that the major change in the school was the slowly increasing size of the teaching (research) staff from 1945, needed in response to both the large change in the undergraduate numbers, and to the fact that more undergraduates now went on to take a doctorate. In Physical Chemistry the number of fellow/demonstrators rose from 4 before 1945 to 6 in 1955, 3 recruited from other chemistry departments, and then to 10 with the title fellow/lecturer in 1965. Mostly, the new lecturers kept the centre of the subject on kinetics (largely gas phase) and spectroscopy. Absent were theoretical chemistry appointments, Coulson was an 'outsider' when he led the very considerable change in the interest in mathematical approaches. In Organic Chemistry the number of staff had changed from 1945 to 1965 from 4 fellow/demonstrators plus 4 demonstrators and others without colleges to 9 fellow/lecturers plus 4 non-fellows. There had been a replacement of all but 2 of the original 8 demonstrators. The major direction of the laboratory was synthesis but a few had begun to stress mechanisms of reactions. There had been no thorough-going move toward biological chemistry. Inorganic Chemistry had 7 fellow/demonstrators and one demonstrator without a college attachment in 1945 and it had changed little by 1955 but by 1965 it had 10 fellow/lecturers, only 1 remaining from 1945. Several of the changes were due to retirement but two left with the change in the professor. A striking feature here was the diversity of interests by

1965. As outlined above major new topics arose in theory, much due to Orgel,[59] and in two novel branches of organic/metal ion chemistry. An analysis of the interests of these new faces when added to those already described in the physical and organic chemistry laboratories shows very clearly how the two ways of viewing the importance of different parts of chemistry remained. As described earlier they had been established by the intellectual approach of Hinshelwood/Sidgwick (the understanding of chemical systems) and by the more practical approach of Perkin/Robinson (the analysis of organic chemicals, their synthesis and the general finding of new facts). The styles had become firmly built into Oxford chemistry as almost different subjects, well separated in departments. Undoubtedly, the separation was fortified by the origin of these professors, the first two from Oxford colleges and the second two from outside Oxford (Manchester), and from 1945 to 1965, by the internal election of Oxford graduates to fellowships aided by the first two, which remained well above 50% of the total. Missing to a very large degree were condensed-phase chemistry, traditional inorganic chemistry and the chemistry of molecular biology.

The next period 1965–80 will see very substantial changes but the very nature of elections to lecturers, now closely linked to college fellowships and controlled in part by those already in post in them, would see to it that, although there was change and new recruitment, the essence of the duality of Oxford chemistry would not be too seriously altered much, though a distinct inorganic chemistry department was created. Individual effort would remain above that of coordinated research by fellows. Once such a style has arisen in a Collegiate organisation such as Oxford and that has had six or seven centuries to congeal, it is very difficult to change. In 1965 many colleges could still have had a notice on their doors, "Tradesmen not welcome", implying a lack of concern with the outside and industrial society so closely linked to chemistry. The chemistry fellows themselves kept biological and geochemical topics off their agenda. Despite these internal problems the Oxford chemistry school was of the highest international standing.

7.6 RESEARCH 1965 TO 1980

7.6.1 The Revival of Inorganic Chemistry

This period of twenty years was full of novel chemistry worldwide evolving from the fundamental discoveries of the previous 40 years. In Oxford the activities of the expanding number of independent research groups under Jones and two new professors, Anderson and Richards, who were very different in style from Hinshelwood, were much to the fore. As alluded to earlier the school was possibly the largest in the world with around 150 honours undergraduates and more than 50 postgraduates per year and at the end of the period close to 50 fellow/lecturers. Some of the fellows were young replacements from Oxford or elsewhere for those described earlier who had retired or left while others were in additional posts. The biggest changes and greatest novelty in chemistry naturally came in the Inorganic Department since from 1963 it had been given a separate identity and could free itself at least in part from the image of being largely a second physical chemistry laboratory. The new professor, J.S. Anderson, set the tone by introducing a new area of chemistry for Oxford, that of the solid state. In a limited sense he filled a gap created by the loss of Hume-Rothery and a connection to metallurgy but Anderson's studies were not of metals and alloys but of the properties of oxides. (Later Hume-Rothery's department of Metallurgy (1957) became that of Metallurgy and Materials Science (1975) under (Sir) P.B. Hirsch (1966) giving it many overlapping interests with the work of Anderson and his successors in these solid-state studies.) Stuart Anderson,[61] an Australian, came from the National Chemical Laboratory at Teddington and had a fine reputation not only through his work on oxides but through his book with H.J. Emeleus, *Modern Aspects of Inorganic Chemistry*.[31] Nobody knows why he was selected in preference to Linnett and Irving, considered to be strong candidates for the chair who were already in the Inorganic Chemistry Laboratory, but it is thought that Hinshelwood played a large part: he was an elector of his successor! As a consequence Linnett and Irving felt slighted and left for chairs elsewhere, the first at Cambridge and the second at Leeds. Neither of them really had a strong claim to the inorganic chair as Linnett was certainly a physical chemist and Irving was really an organic chemist.

Anderson could not have been a new broom sweeping away established fellow/lecturers not connected in any way to inorganic chemistry even if he had wished to do so but he could and did provide fresh air with a sense of appropriate purpose. Apart from his own studies and bringing B.E.F. Fender to Oxford, fellow/lecturer (1965), to work in a parallel field, Anderson frequently helped others in the laboratory. For example he used his dowry to re-equip the laboratory, see below. While he developed his research, in which he was a worldwide leader particularly in the field of the electron microscopy of defect structures,[47] he was not a good communicator or lecturer. It is fair to say also that he was not given to social life. Moreover, he did not cultivate college connections and did not interfere much with college/fellowship elections, which, as we shall see, produced further fellow/lecturers with unrelated interests as in earlier periods. He made little contact inside or outside his laboratory. He let Oxford get on with most matters in its own way – there was no longer a Hinshelwood seeking patronage in the Inorganic Chemistry Laboratory. Without going into further detail the study of the solid state became a major part of inorganic chemistry research. In fact the next professor J.B. Goodenough was chosen in fair part to maintain the impetus in the very field Anderson initiated.

At about the same time as Anderson came to Oxford, R.J.P. Williams introduced a different study of the solid state, that of the electron-transfer properties of mixed-valent metal-ion-containing compounds. The studies were intended as models for electron transfer in cell membranes that, with proton transfer, Williams had linked to biological energy transduction (see refs. 92–95). He had amongst others two very able pupils, P.D. Braterman, later a professor in the USA, and then P. Day.[62,63] Day became successively a fellow/lecturer in the Inorganic Chemistry Laboratory (1968), the director of the Von Laue–Langevin Institute at Grenoble, and then Director of the Royal Institution. London. We return to his career in the next section. These studies could well have been developed in the Physical Chemistry Laboratory.

The appointment of Anderson, and the growing interest in solid-state chemistry, could be looked upon in a way particular to Oxford Chemistry. We have stressed that Hinshelwood and his colleagues never encouraged the chemistry of the solid state in the Physical Chemistry Laboratory. They had not shown much enthusiasm for Chemical Crystallography either, which was moved in 1957 into the Inorganic Chemistry Laboratory, although Hinshelwood and his successors remained responsible for it, or electron microscopy and diffraction, initially an interest of Sutton's in Hinshelwood's laboratory. Both these methods were central to Anderson's work and that of others who worked on the study of the solid state. Thus, in a sense the Inorganic Chemistry Laboratory not only looked at novel syntheses of solids but it filled one gap in academic physical chemistry research. Meanwhile another gap, that of the physical/chemical study of biological materials, was also being filled inadvertently and in an unusual way in the Inorganic Chemistry Laboratory, see bioinorganic chemistry described in Section 7.5, and a little later by several departments working within The Enzyme Group. The first followed from a continuation of metal-ion coordination chemistry in water, F. J. C. Rossotti and Williams, with a strong connection to biological protein chemistry.

The concern of Anderson to re-equip the Inorganic Chemistry Laboratory quickly gave a very considerable impetus to all the work in the laboratory. After a discussion with others he decided that both electron spin resonance and nuclear magnetic resonance spectroscopies were the required new methods, both of which were under study in one or another of the other two laboratories. Unfortunately it appeared that there were not sufficient funds in his dowry for both. A chance observation led to the finding that a Japanese company, JEOL, was marketing both spectrometers and wished to break into the European market. It was decided to send one of Williams' pupils with a suitable material out to Japan to test the equipment. Williams had already started to use both techniques in the study of coordination compounds of biological relevance. His pupil Allen (H.A.O.) Hill,[62] fellow/lecturer (1967), took a sample of vitamin B_{12} to Japan. He returned with remarkably good spectra from both spectrometers and negotiations then led to the purchase of both for the price of one. Not only did this remarkably generous gesture of Anderson give a great push to the chemical work on vitamin B_{12} but it was instrumental in providing background knowledge for the subsequent development of several of the researches in the laboratory and to a degree of the Oxford Enzyme Group. (A slight disadvantage was that the JEOL technicians sent to Oxford spoke very little English.) Note how methods closely initiated or being improved in the physical or even the organic chemistry laboratories were introduced into inorganic chemistry with little or no coordination, a defect of the departmental structure. We shall make this comment again.

As mentioned, Anderson allowed the election of fellows to be based on college decisions, and we have noted that the changes to joint college/laboratory appointments of fellow/lecturers had in any case lessened the ability of a professor to control them. There were introduced interests somewhat at random but frequently of pupils of older fellows in the laboratory. Fellows elected were M.L.H. Green (1965), organo-metal chemistry; H.A.O. Hill[62] (1967), studying metallo-enzymes; P. Day[62,63] (1968) analysing mixed-valent conductors and A.F. Orchard (1970), concerned with surface spectroscopy (of which the last three were pupils of Williams); A.J. Downs (1971), Raman spectroscopy replaced Woodward; R.G. Denning (1971), Venanzi's pupil, who returned from the USA to study actinide spectroscopy in 1971 and D.M.P. Mingos[64] (1976) with an interest in the theory of organo-metal bonding. With these elections not only did a huge variety of topics appear in this laboratory between 1960 and 1976 but many of those elected could well have been called physical chemists and in fact their work often was directly related to that in the physical chemistry department. As stated the continuing peculiarity of Oxford was that as with other aspects of physical chemistry noted above, the separation of the three branches of chemistry in different buildings with different traditions resulted in very poor exchange between them no matter how similar their interests.

Goodenough,[65] who succeeded Anderson as professor from 1976–1989, came from Massachusetts Institute of Technology, Boston, USA, where he had worked with great success both academically and with industry especially on the magnetic and electronic properties of ferrites. His time as professor is included here, although it runs well into the third period, 1980 to today, because it clearly had a direct connection with that of Anderson. He was elected in preference to Professor Sir Geoffrey Wilkinson of Imperial College, London, who was a 'practical' organo-metallic chemist linked to industry. Wilkinson was in some ways the more obvious candidate as he was British, but he had a tough, rough reputation, more American than Oxford. He had worked in California and Harvard and had been a professor at King's College, London. He was for a time at Massachusetts Institute of Technology where, with an American, Albert Cotton, an even rougher and tougher man, he had written a very good inorganic chemistry textbook.[66] The book was not at all in the style of Phillips and Williams' book of 1965 (mentioned in Section 7.3.3), was less valuable for many of the more intellectual Oxford undergraduates with their Oxford teaching, but much more valuable world-wide for the majority of more practical chemistry students. Wilkinson might have fitted equally into the organic chemistry department, and we shall see that Jones's successor in that laboratory, Jack Baldwin, had a similar American background and a no-nonsense personality to those of Wilkinson, not really 'Oxford types'. Goodenough with a strong knowledge of theory, was much more suitable in this respect, more of an intellectual. In this election we see the continuing value judgements of Oxford in physical and inorganic chemistry in not favouring the 'practical' approach, this time electing even a physicist. Note that Goodenough could have been elected to a chair in physical or theoretical chemistry, or even in physics. One further reason for not electing Wilkinson was that his pupil Malcolm Green, fellow/lecturer (1965), of whom more will be said later, was building his career in Inorganic Chemistry and had already a strong reputation in organo-metallic chemistry and rivalry between the two was distinctly possible.

The election of Goodenough caused rumblings throughout British inorganic chemistry for outside Oxford (and Cambridge) sympathy is often with the 'practical' man and Goodenough also had two considerable faults. As pointed out he was firstly a physicist and though studying obviously inorganic chemical solids, many oxides, especially mobility of ions in them, he found it very difficult to communicate with chemists, especially undergraduates. Secondly, politically he was naïve in that he thought he could wake up the university and indeed the UK to the importance of science based on his USA experience where science's standing has always been higher. He could not succeed for the permitted style in Oxford, never mind the UK, is that a lot of gentle persuasion is needed to get a little done. Nor could he manage the problems associated with college fellowship/lectureship elections. He let such matters take their course, much as Anderson had done, for example in the elections of C.M. Dobson[67] (1980), S.R. Cooper (1984), P.A. Cox[68] (1984), and A. Hamnett (1985), see the next section for their interests, which if anything further increased diversity. He should have been elected a Fellow of The Royal Society (although a US citizen, he could have been since he had worked in UK for long enough and, undoubtedly, ill-will stood in his way). The very fact that he was not so recognised over a period of seven years (the time permitted for candidate inspection) was a major factor that drove him to return to the USA in 1989. Later he was awarded the Japan Prize, a clear proof of his outstanding ability. The study of solid-state chemistry is now a strong theme in Oxford Chemistry although the Anderson–Goodenough line of research was broken for more than 10 years at the professorship level. It was taken up, however, by several fellow/lecturers during the Anderson/Goodenough period, notably by B.E.F. Fender and P. Day (see above), and has been continued since by them or their successors, whose work falls in the period after 1980. Recently, close to 2000, the study has been augmented by elections to professorships in different aspects of the solid-state chemistry in both the inorganic and physical chemistry laboratories, see the most recent period of research, 1980 to 2005. Once again this shows the tendency of Oxford physical chemistry to be complementary in two departments while staying clearly apart, while the inorganic laboratory also developed an interest in its own conventional inorganic preparative studies. Meanwhile, organic chemistry remained very much separate from both although one major new area to be described next could have developed in the inorganic or organic chemistry laboratories. Mostly at first it gave an additional impetus to the revival of inorganic chemistry. When it did develop in both laboratories it again did so with little or no exchange!

7.6.2 Organo-Metallic Chemistry

A major change in general chemistry that was initiated strongly outside Oxford and to which we refer above is the way organic molecular fragments can be attached to metals, in a manner very different from that in more conventional coordination chemistry in water solutions. This innovative chemistry is rightly attributed to the work of Fischer in Germany and Wilkinson in the USA and Britain who shared the 1973 Nobel Prize.[69] In earlier work direct metal attachment to carbon by conventional bonds as in for example lead tetraethyl, $Pb(C_2H_5)_4$, was well known, but the demonstration that the face of an aromatic molecule, such as benzene, could hold on to a metal atom such as chromium was a great surprise. Earlier it had been realised that a range of more conventional compounds more closely linked to traditional coordination chemistry could be made with phosphorus-based ligands instead of those coordinating via oxygen, nitrogen and sulfur, see Venanzi in Section 7.5. Both groups of new compounds are soluble in organic solvents, and do not exchange units attached to metal atoms readily, making them similar in kind to thermodynamically unstable organic compounds. From both kinds of this novel practical chemistry came a range of catalysts with peculiar bindings of metal atoms even to unsaturated carbon reactants such as ethylene and benzene or directly to hydrogen of saturated carbon. As mentioned before work on the synthesis of these classes of compound and their catalytic properties was introduced to Oxford by M.L.H. Green[70,71] (1965), a pupil of Wilkinson, in the Inorganic Chemistry Laboratory, of which he became professor in 1987. We shall describe Green later in some detail but he was a green-fingered practical chemist of considerable originality. He quickly became a leading figure in organo-metallic chemistry in Britain and well-regarded worldwide. Green was only the second chemist (after Venanzi) in either the inorganic or the physical chemistry laboratories with considerable intuitive skills in the synthesis of small volatile molecules, which until their time were to be found virtually only in the organic chemistry department. It is immediately clear that this particular branch of chemistry is equally suited to an organic or an inorganic chemistry department as it contains a metal element bound directly to carbon of an organic compound. Later, A.J. Downs (1971) came to Oxford to replace Woodward as a spectroscopist and he was also interested in synthesis but of some small but more conventional inorganic molecules. It is noticeable that Venanzi, Green and Downs, the practical chemists, were not Oxford graduates and unlike in approach most of the new appointments given in Section 7.6.1.

We turn next to the complementary novel work in the organic chemistry department based on organo-metallic chemistry although it made greatest progress in the next period, 1980 to today. The laboratory took up this area through J.M. Brown who came from Warwick in 1974 and S.G. Davies, an Oxford product (1980). Both have become extremely well known internationally through their studies of the catalysis of organic reactions. Their emphasis, unlike that of Green, was, and still is, strongly on the stereochemical mechanisms of reactions that have to be controlled, especially asymmetric catalysis, for drug production. Davies in particular has been very active in collaborating with industry on specific processes, which have great potential in pharmaceutical chemical synthesis. In this he falls in line with the traditions of Oxford organic chemistry professors. Considerably later his connection with industry helped to persuade the university to make him the chairman of all the chemistry departments, compare Perkin in Chapter 5, now unified in principle but still largely in three separate laboratories, as well as the titular head of organic chemistry in 2006. We describe this reorganisation of chemistry later but for the sake of continuity of description we shall refer to the different chemistry departments as if they remained separate.

In concluding this section we note how Oxford played a strong part in the revival of inorganic chemistry in three major fields, solid state, organo-metallic and biological inorganic chemistries.

7.6.3 Traditional Organic Chemistry

Before we describe the innovations in traditional organic chemistry we mention again the development from Waters and Norman of studies into mechanism of organic reaction using new methods. Of particular interest was Norman and Coxon's book *Principles of Organic Synthesis*.[32] The impact was to direct attention to synthesis by thoughtful mechanistic approach rather than by intuition and trial and error. This interest in mechanism led through Norman's research pupils to work on enzymes by J.R. Knowles, fellow/lecturer (1966) and G.K. Radda, fellow/lecturer in Chemistry, but later in Biochemistry (1966), together with the direct study of mechanisms of organic reactions on the intake of other new fellow/lecturers. We must also observe that Knowles and W.J. Albery (1963), of physical chemistry, were tutorial pupils of R.P. (Ronnie) Bell and the three met frequently to discuss solution kinetics before Bell departed to Stirling in 1967. Knowles and Albery also had a very fruitful collaboration in work on mechanistic aspects of enzyme kinetics. Because the mechanistic work on enzymes was mostly conducted at least initially when Knowles was a member of the Enzyme Group, a cross-department body, it is better to describe the enzyme work under that heading. We therefore turn to traditional organic chemistry synthesis.

There is some difficulty in writing about outstanding developments in organic chemistry in one period when methods seem to develop gradually and step by step with the use of novel reagents, producing novel synthetic pathways and mechanistic approaches. In this work the use of metal and indeed other elements than C, N, O, and S increased, making again for an overlap between organic and inorganic chemistry, though in Oxford this was not seen as of inter-departmental concern. Looking at syntheses in the Oxford organic chemistry department in the period of the later 1960s and the 1970s the most noticeable change from the classical approach of characterising novel organic chemicals from organisms or by man's creative synthesis was the more direct attack on biological processes in addition to that described above. Jones himself, apart from continuing work on acetylenes, studied biological hydroxylation, Knowles introduced photoaffinity labelling of antibodies with azido-groups and G. Lowe (1965) analysed the active site of the enzyme papain. G. Young and J.H. Jones (1970)[72] turned attention to peptide synthesis, especially the steps of racemisation that are an inherent problem. Toward the end of the period Knowles, who succeeded Woodward in the chair of organic chemistry in Harvard and later became Harvard's Dean, was replaced by J.M. Brown (1974), another Manchester product, who introduced organo-metallic chemical catalysis to the laboratory as already mentioned. We note that while there were other new appointments in synthesis chemistry including those of G.H. Whitham (1965), A.C. Day (1968) and M.J. Robinson (1969) it is not possible to refer to their work here through the complexity of the molecules studied. There were very few retirements of fellows from the earlier period and their work has been mentioned previously. In passing it is worth noting again that two Nobel Prize Laureates of an earlier period had been trained in the Oxford organic chemistry laboratory, Todd,[49] and Cornforth.[50]

Once again it is important to see that organic chemistry in Oxford was not confined to the department with this name any more than physical or inorganic chemistry was so limited. An outstanding development in organic chemistry was the work of Abraham and Newton in the Sir William Dunn School of Pathology on the structure of Cephalosporin C and the related antibiotics, see also Chapter 6. Of course, much research related to the enzyme chemistry of the organic department was carried out in biochemistry and other biological departments and this is shown by the composition of The Enzyme Group, see below. Apart from the more biologically oriented work described above this period of fifteen years saw a switch from very descriptive to more mechanistic approaches to organic chemistry.

7.6.4 Physical Chemistry

In physical chemistry the extreme changes of direction of research did not happen in this period and maybe there was little need for some programmes such as those underway in kinetics and spectroscopy were of high quality and could easily be expanded. The new professor, previously an Oxford undergraduate and tutor, was (Sir) Rex Richards[73] and he had the advantage of a great enlargement of the laboratory in 1966. As already stated he had a deep interest in NMR spectroscopy and he saw its promise in the analysis of a diversity of chemicals. However, he realised that only by constructing NMR instruments, with superconducting high-field magnets and Fourier transform probes and consoles, within his group, could the method be extended to the determination of structures of a wider range of molecules.[74] The spectrometers he built led to their later industrial construction and a worldwide market. He had excellent pupils for example R. Freeman, R.A. Dwek and I.D. Campbell and he managed to assemble an equally good technical team, not an easy achievement in Oxford. The only fault of the group, if fault it is, is that it did not, and possibly had not the time to, look at many detailed applications. It might well be said that it is fundamental development not application that is in fact the duty of this department. Practical application did come about, however, after he resigned and when he was invited to join and chair The Enzyme Group in which there were chemists given to more biological and chemical rather than just physical studies.[74] We will continue with this part of Richard's career under that heading but we note here that for his highly successful research he was knighted in 1977.

Richards, much like Anderson and unlike Jones, did not try to do other than respect the nature of the collegiate university. His direct influence on appointments was relatively small. He also allowed the democratic government of the laboratory to become the reference point for decisions. This habit is now general in all the departments but, unlike the organisation of some departments, Richards encouraged this democracy. In other departments the professors, all from outside Oxford, accepted or put up with 'democracy', previously associated with colleges and much of the university but not so obviously in chemistry, or even attempted to reduce its importance and certainly did not give the impression that they approved of it. Note again that Richards, unlike Jones and Anderson, was an Oxford product and had been a fellow/lecturer.

After Richards resigned in 1970, to be Warden (head) of Merton College, Sir Frederick Dainton (later Lord Dainton), previously Vice-Chancellor at Nottingham University became professor of physical chemistry.[75,76] It is possible that this was a stop-gap measure as the electors sorted out their wishes. Dainton lasted only until 1974 and had ceased being active in chemistry for some years before his appointment. We describe his successor in the next section.

We turn to the nature of the research by the physical chemistry lecturers noting in passing that Sutton retired in 1973 and Thompson and Lambert in 1974. Before 1965, that is in the Hinshelwood era, the stresses were on bulk gas kinetics and spectroscopy and there was additionally the work of Bell on solution reaction kinetics. There was little or no study of either condensed phases or of biological systems. The most noticeable development in the subsequent period was an effort at a much more detailed understanding of how individual molecules react, a continuation of previous work but a much deeper analysis, part of it theoretical chemistry. This requires an intensive exploration of energy distribution in the different bonds of a molecule. Some of the work applied to molecular interactions with surfaces. There was also a diversification to the use of new spectroscopic techniques including photo-electron spectroscopy by D.W. Turner (1968, professor 1985), and C.J. Danby and J.H.D. Eland (1983) (see also J.C. Green and A.F. Orchard in the Inorganic Chemistry Laboratory), electron spin resonance by K.A. McLauchlan[77] (1965), who developed the experimental method while he collaborated with P.W. Atkins (1965) on theoretical aspects of spin chemistry, and the general use of laser spectroscopy for example by G. Hancock (1976). Wayne (1967)[78] had turned attention to the important gases in the atmosphere and their reactions and properties. The previous use by White (1964) of neutron diffraction of surfaces, was continued by his pupil, R.K. Thomas (1978), while their catalytic properties were studied by R.P.H. Gasser (1962). The traditional study of thermodynamics involving inter-molecular forces was by E.B. Smith (1963), later Master of St Catherine's College and Vice-Chancellor of Cardiff University, who wrote a notable book on this topic.[39] A little while later Smith was directing attention to the use of computer studies of inter-molecular forces and he also had a strong research interest in the action of anaesthetics with the professor of pharmacology, W.B. Paton, and in certain aspects of spectroscopy.[79] Bell's studies of solution reactions had been turned to their electrochemistry in the hands of his pupil W.J. Albery (1965). In 1973 Freeman returned from the USA to Oxford to rejuvenate NMR studies in the laboratory and he referred to the period before he moved to Cambridge as his most successful. Finally, there was continuation of work on bacterial growth by Dean but the interest in the work drained away after Hinshelwood retired. It is readily seen that the physical chemistry's outstanding contributions were in spectroscopy in this period, and a broadening of interest in a wide range of chemical systems did not occur until later. It might be said that the shadow of Hinshelwood was still a considerable influence.

Before leaving the subject of chemistry itself we have to apologise to our more general reader that we have often just listed activities in chemistry especially perhaps in organic and physical chemistry. It is difficult to give a simple explanation of the methods and fundamental properties often of single substances studied in these activities. All we aim to achieve here is a general impression of intense individual researches. This sophistication is of the nature of physics and chemistry today but understanding must not be thought of as only an academic pursuit but enlightens much of application, often after a considerable time. We have not been able to list even all the books written or edited by members of the departments.

7.6.5 Chemical Crystallography and Biophysics

An important part of chemistry, which was never incorporated into physical chemistry, though for many years it was under Hinshelwood, was Chemical Crystallography, that evolved after the abolition of the professorship of Mineralogy associated with the Museum from 1860 to 1940. The history of the separation of crystallography is one of wrangling from 1939 to 1944, see Vincent's book.[80] Initially, it remained housed in the Museum with H.M. Powell ('Tiny') in charge as a demonstrator then Reader, 1944. It was Powell who introduced X-ray crystallography to Oxford in 1931. He was without a college fellowship for many years. In 1964 he became FRS, and, long after it was due, a college fellow, and a professor of the sub-department still under Hinshelwood. The department was only transferred to Anderson, the professor in the Inorganic Chemistry Laboratory in 1974, where it had been housed from 1957. Dorothy Crowfoot Hodgkin, a pupil of Powell, worked first in Oxford 1931–32 as related in previous Chapters 5 and 6, and then for a period in Cambridge with Bernal. She came back to Oxford in 1935, financed by a college fellowship, unlike Powell.[81] Now, while Powell continued research into the structures of inorganic substances, later inclusion compounds and clathrates, and was joined by a succession of younger scientists including C.K. Prout (1958) who later headed the sub-department, Hodgkin started a quite independent study of proteins, following Bernal, and also of small biologically important organic molecules aided by Robinson. She had a high degree of independence through her college association and Robinson's support with Rockefeller money and soon had much the greater prominence. With great skill coupled with much determination she elucidated the structures of cholesterol and penicillin,[77] as already described in Chapter 6. Tiny Powell, by way of contrast, was a laid-back character enjoying a liquid pub lunch with a few similarly minded colleagues, including Hammick, mentioned above, and the Magdalen College physics fellow, later President, J.H.E. Griffith, in the snug (no women) of the Kings Arms. He was a quiet, sophisticated man with many interests. Hodgkin's independence coupled to Powell's style led to a lack of coherence in the department, especially since her remarkable achievements helped to set her apart. She was made FRS in 1947 and was awarded the Nobel Prize in 1964 not for protein crystallography but for the solution of the structures of cholesterol, penicillin and vitamin B_{12}.[81] In fact she failed to solve the structure of the protein of her interest, insulin, until Guy Dobson assisted her in the 1970s. Many distinguished visitors worked with her but not so much with Powell and to a large degree, especially after she was made a Royal Society Research Professor (1960), her research became in effect in a separate unit first in physical and then in inorganic chemistry. Quite peculiarly by the end of this period the *inorganic* chemistry department had two Royal Society professors, Hodgkin and Williams, studying *biological* chemical problems.

Now, around 1965 it became obvious that the University and especially the crystallography department should be involved with molecular biology especially using X-ray diffraction analysis. What then happened is confusing and unfortunate since although the chemical crystallography department wished to house it, no space for its study was found in Chemistry. It was Professor J.W.S. Pringle of the *Zoology department* who came to the rescue, supplying space and creating a department of Molecular Biophysics separate from Chemistry. No example could be more clear of the lack of organisation in Oxford Chemistry. The new professor, David Phillips (later Lord Phillips) was a well-known protein crystallographer who had determined the first structure of an enzyme, lysozyme.[82] He brought with him a very active group from The Royal Institution. It became clear that it was not his wish to be too closely associated with Dorothy Hodgkin studying the X-ray crystal structure of the protein, insulin. To provide her with equal facilities she was moved from Inorganic Chemistry into the new *Psychology Laboratory* adjoining Zoology. The future of Molecular Biophysics and Phillip's group was later associated with the Enzyme Group and then with the Biochemistry Department but Dorothy Hodgkin was not connected to either. This left Chemical Crystallography detached from large biological molecule studies and it became, under C.K. Prout (1965), A.K. Cheetham (1969)[63] and D.J. Watkin (2002), a laboratory for the study of some small organic and inorganic molecules and solid compounds produced in inorganic chemistry. They introduced more general use of the electron microscope to this sub-department. Another area in this department was the study of biological minerals by Williams but this topic was only continued outside

Oxford.[83] The decline of the department was clear when in 2002 there was no fellow/lecturer associated directly with it. Although there is a plaque on the front of the Inorganic Chemistry Laboratory celebrating the work in Chemistry of Dorothy Hodgkin the connection between Chemistry and the study of structures of large molecules, both by X-ray Crystallography and by NMR, moved from Chemistry to Biochemistry with the result that structural Biological Chemistry remained from the 1970s for a long period out of its traditional connection with Oxford chemistry. We shall see that only in recent years has it become involved in the main chemistry departments as a sub-department, called Chemical Biology. Another loss of a facility was that of electron microscopy, which declined when Cheetham left for USA. Today electron microscopy is practiced in the Material Science Department.

7.6.6 Theoretical Chemistry and its Short-Lived Department

In 1952 Charles Coulson was elected to the Rouse-Ball Chair in applied mathematics as mentioned in Section 7.5.1. His deep interest was in chemical bonding, which through the development of wave mechanics had become in effect a branch of mathematics. It did not appear, however, that Hinshelwood considered the intrusion of this type of mathematics into chemistry to be of great consequence, see Section 7.5.1. In 1972 a chair in a separate theoretical chemistry department was made available for Coulson and after his death in 1974 it became known as the Coulson Chair of Theoretical Chemistry. M.S. Child (1966) and D.B. Abraham (1972) were appointed earlier to lectureships in this department. Coulson was succeeded by N.H. March (1975) and as a consequence at the end of the next section we shall relate how the theoretical chemistry chair with it a distinct department were for a while almost lost to chemistry and were then incorporated into physical chemistry in 1994. However, there were several theoretical chemists appointed in the physical chemistry laboratory well before 1980.

7.6.7 The Enzyme Group[84]

In 1969 at a meeting of the officials of the Science Research Council (SRC) the chemists present made a complaint that they were under-funded especially for major, expensive, modern pieces of equipment. The response they received from physicists was to the effect that if indeed there was a demand then it had to be demonstrated and given the costs of equipment any application for a grant would have to be made by a large group of researchers. E.R.H. Jones from the Dyson Perrins Laboratory, who attended the meeting, promised the committee that he would look into possibilities. On his return to Oxford he approached J.R. Knowles enquiring as to needs for costly equipment in chemistry. In turn Knowles put the question to his college colleague, R.J.P. Williams, who had been working for some three years on protein nuclear magnetic resonance using the somewhat unsatisfactory Varian 220 MHz NMR carrier wave spectrometer (not a Fourier transform instrument) based, very inconveniently, for a while at Harwell and then at ICI near Runcorn. (This work developed from the introduction of NMR in the Inorganic Chemistry Laboratory, see section 7.6.1, quite separately from the work of R.E. Richards in Physical Chemistry.) Williams suggested that a high-field NMR spectrometer in Oxford would change the prospect of a lot of research and especially in the study of proteins. A meeting, which included G.K. Radda[85] who had started rather different biological phosphorus NMR studies with Richards in the Biochemistry department, was called of all those lecturers with a possible interest in enzymes or related matters from any Oxford department. Some eight department representatives came to the meeting and a general theme, the study of enzymes, was agreed with NMR as the major new (expensive) tool to be requested. A total personnel of well over 100 (20 senior scientists) came together in the Group, some of whom were NMR experts and some chemists, not just from Chemistry departments, who prepared and investigated enzyme samples of many kinds. An enzyme preparative laboratory, several more minor pieces of equipment and several post-doctoral positions were requested also. Considering that many senior scientists had ten research workers, it was a very large undertaking. The group was also joined by D.C. Phillips' unit studying enzymes by X-ray crystallography. (See above for an account of his laboratory then in Zoology.) The project was taken back to SRC for outline approval. Once this was obtained Jones stated that the Group needed an impartial disinterested chairman and it was decided that Sir Rex Richards, recently retired from his professorship and now Warden of Merton College, and very much an NMR expert as described above, should be invited. This proved to be of great advantage for he helped to design, with Oxford Instruments and Bruker, the first Fourier Transform NMR spectrometer of sufficiently high field to resolve NMR spectra of proteins. (Actually these spectrometers have proved to have universal value in all branches of chemistry worldwide.) It was the combination of several projects from different departments most of which could benefit from use of the spectrometer that persuaded SRC to make a quite unusually large grant to Oxford. The Enzyme Group was founded. Protein NMR studies by a group including Williams, Campbell and Dobson, were not the only target and lysozyme was far from the only enzyme chosen by those with different interests. Knowles and Lowe (organic chemistry department) studied phosphorus NMR in certain enzyme/substrate interactions and showed how phosphate was transferred between substrates while Phillips undertook the X-ray structure determination of one such enzyme. Radda and Dwek (a pupil of Richards as mentioned before) formed a team from Biochemistry using NMR for studies on larger proteins involved in phosphorus chemistry, and Dwek also examined antibodies. Later, Radda branched out into biological tissue NMR and Dwek introduced glycobiology. An outstanding achievement was the early spectroscopic (NMR) resolution of signals from individual amino acids in proteins, their distribution in space, and their mobility.[74] The Group held fortnightly meetings with discussion after dinner. Each project was open to criticism and programmes and grant proposals were adjusted. The work proved that indeed it was possible to do protein structures by NMR. In one sense the Enzyme Group failed to reach its major goal, protein structure determination in solution, in part due to the loss early in the project of a person with computer skills. This achievement went to Wüthrich in Switzerland, for which he was awarded the Nobel Prize 2002. On the other hand, the Oxford research developed spectrometers and methods. The work revealed many kinds of internal mobility in proteins that were found to be extremely important for function. In another sense it showed the great advantage that could be gained in university multi-disciplinary chemistry with a large group with no departmental barriers. Later it became the Oxford Centre for Molecular Sciences under the chairmanship of J. Baldwin and C.M. Dobson but it never had the same spirit of cooperative enterprise. It is very difficult especially for a group of independent lecturers of distinction in an Oxford environment not to drive forward their own interests. Slowly the members split apart in different science roles. Radda, later head of the Medical Research Council, on *in vivo* NMR, Phillips, later Lord Phillips, became a general science adviser to the government, Williams (who became a Royal Society Research Professor and had a separate team on metal ions in biological systems), Dobson (who became interested in protein folding, left temporarily for Harvard and then became a fellow/lecturer in Oxford, a Professor in Cambridge and then head of St John's College there), Knowles (later Professor at Harvard and then their Dean of Arts and Sciences), Dwek (subsequently a leader in glycobiology and head of Oxford's biochemistry department), Campbell (also later a biochemistry professor) split apart. If such a Group is to be successful again it will require an underlying group of young people assisted by several older members who have considerable achievement behind them and consequently influence, all of whom are willing to give time to it and to share effort. It must also have a genuine big target. At present, effort in Oxford chemistry is in danger of being too dispersed, as can be seen by any reading of its development of research particularly after 1980, which will make it increasingly difficult to find adequate funding. A particularly important point is that the Enzyme Group came about through the selfless action of Jones with his dedication to Oxford chemistry who saw that the University was represented on National as well as local committees. We return to this attitude later for it is bound to distract from the professor's research.

The question may be raised as to why the major novel features of the Enzyme Group, the study of proteins especially by NMR together with the X-ray crystallographic group of Phillips, became lodged not in the chemistry but in the biochemistry department. We shall see that especially with the recent separation of chemistry from biochemistry in different university divisions this robbed chemistry of the very important study of the structure of large biologically important molecules. Despite the fact that the major players in the Enzyme Group were chemists the first innovative NMR spectrometer was allowed to be placed in the Biochemistry department through the kindness of Rodney Porter who had no interest in it. Maybe Rex Richards, obeying a sound Oxford tradition, that a retiring professor does not continue to do work in what used to be his department, decided he could not put the spectrometer, designed in fair part by him, in the Physical Chemistry Laboratory.

Later a lectureship nominally designated for the Enzyme Group, and which could have been in chemistry, placed Iain Campbell in biochemistry and a similar lectureship was given to Raymond Dwek also in biochemistry. A major section of the Group plus all new spectrometers were housed finally in a new building for biochemistry, The Rex Richards Laboratory, which came to house also the X-ray diffraction studies of Phillips. Later, Oxford Chemistry lost two very prominent folk of great value to these studies: Freeman and Dobson to Cambridge. Cambridge has the great advantage of being able to elect professors independent from department headships. Strong evidence of the success of the Enzyme Group, not only directly in enzyme chemistry, lies in the number of departments worldwide apart from in UK that have hired research workers from it, but it is sadly to the loss of Oxford chemistry that nothing permanent in the way of collaborative effort resulted from it. Could it be that the setting up of strongly independent departments was becoming detrimental to chemistry, which today needs overall organisation?

If we included the researches in the Enzyme Group many based on NMR and the achievements in Crystallography as well as the developments of other techniques described above as being in Physical Chemistry it is clear that it had an extremely fine record of achievement in these years in addition to those mentioned above. It is probably fair to state that at the end of this period, 1980, the department that had been the first to be of world-class, organic, was now the least impressive of the three.

7.6.8 Life in Oxford, 1945–1980

Before proceeding to the times after 1980 the attractiveness of an Oxford fellow's life from 1945 until that time needs to be emphasised. As a bachelor, a fellow lived in college and was looked after in every respect. Food and drink were plentiful at the college High Table, where the fellows dined. There was rarely a shortage of lively conversation, most of it not academic it must be admitted. A married man had a living out allowance and he also had access to free meals in so far that family ties permitted. Fellows exchanged dining visits to other colleges and their homes. They had Gaudies for old members and special Founders dinners to which the 'great' amongst outsiders were invited. In these ways the colleges kept good contact with politicians and civil servants, not to forget those with large sums of money at their disposal, to ensure Oxford's independent collegiate future. Often these men were Oxford graduates, though mostly not chemists. As well as junketing now and then most fellows worked hard either in academic pursuits or in college management. They selected their own pupils whom they taught and met socially and their total teaching loads were not thought to be excessive, variably 10 to 15 hours a week for three eight-week terms. As mentioned earlier they were often loyal to their college but if they were chemists, took little interest in university affairs. The independent research groups of the chemists were mostly not very large and did not demand much administration or large sums of money until around 1980. There was time too for other activities, sporting or otherwise. Before 1960 Dr (later Professor Sir) 'Tommy' Thompson effectively managed the combined University Football Club, Pegasus, which won the amateur cup, and B.R. Brown was an international standard goalkeeper but these extreme examples of chemistry fellows' achievements could hardly be expected later. However, the combined fellows could have put out very reasonable cricket, tennis and golf teams in the 1965–1980 period and several were excellent mountaineers. (They even went on an expedition to the Himalayas.) Life for a fellow was modestly (mostly) purposeful but clearly not stressed, it being supposed that every one would do his measured best, both in teaching and research without outside pressure and little overall organisation. There can be no doubt that Oxford provided very enviable academic circumstances. No matter what one's view of this style many an Arts fellow especially remains deeply committed to maintain it as the best of all possible worlds in which to teach and research. In all of this collegiate activity and laboratory teaching, the professors were never really engaged. It was almost as if they had a caretaker laboratory role while doing their own work. As we shall see very quickly this manner of living and working for a chemistry fellow, adapted from a developed historical Arts tradition, became less and less possible after 1980. It was overtaken by demands to provide 'results' in worldwide competitive, intense activity needing much money, while hopefully gaining personal international recognition. We turn to the life-style of a chemistry fellow today toward the end of the chapter but note that there was still a sense of cooperative venture in a college in this middle period though not so in the laboratories except within the professors' orbits and the Enzyme Group. The history of fellowships (college appointments) and separately professorships (university appointments) had accidentally created social, college by college, as well as research interest divisions within chemistry, which though moderated have remained to this day. The final line must be that no matter what is thought of this style of teaching and research with its individual freedoms, the school remained one of the finest chemistry schools worldwide in all branches of chemistry.

7.7 RESEARCH: 1980 TO 2005

7.7.1 Introduction

The changes in the Oxford Chemistry School were not as considerable in this period as they were between 1950 and 1980 when the major basic features of chemistry had been uncovered by a worldwide effort. We shall see that the three Oxford Chemistry Departments often came to investigate different aspects of these features, while other related departments, for example biochemistry and material science, viewed them with a somewhat different emphasis. It is probable that chemistry itself had become a subject in which principles had been established but that many ramifications remained to be explored. This exploration now led to possible applications and hence to greatly increased activity everywhere. At the same time research had become increasingly expensive both in manpower and equipment and due to the need for purpose-built laboratories. Inevitably the cost led to questions as to the quality of the research, which especially after 1992 became subject to review both internally by the university and by external bodies such as those of the Higher Education Funding Council and the Research Councils.[11,12] Nationally those chemistry departments that were given low scores in these reviews were subsequently poorly funded from government sources with the consequence that several universities closed their chemistry departments: an action easily managed by their top-down organisations. The pressure on Oxford Chemistry was to obtain the highest scores, which it always did. An alternative mode of finance was through industry, which meant directing research to useful rather than just academic objectives. In part this could still be called idealistic as medical, biological and environmental problems were tackled as well as those that were thought to give rise to possibly harmful products. We shall see that Oxford chemical research often moved in these 'idealistic' directions, no matter whether or not the major motivation was financial. Before we turn to this research we note that we have given a brief summary of changes in the undergraduate school and its teaching staff in this period in Section 7.3.

In describing the recent research in chemistry in Oxford we again cannot refer to the work of all the 50–60 fellow/lecturers individually. We have decided to treat separately only the work of the departmental and other professors, see note 9, often grouping the studies of the many others in a department according to common subject interest. However, we shall attempt to mention as many individuals by name as is possible. To some extent the descriptions will be seen not to be limited in departmental location of the researchers, much as this still serves as a useful way of dividing them, as there became increasing and considerable common ground between departments. It is unfortunate that space does not allow any reference to the large number of technicians in workshops and laboratories, several of whom are experts in particular technical matters, and the equally important members of the administration and secretarial staff. Without them a modern laboratory cannot function but they too and their facilities have to be included in the total running costs.

7.7.2 Physical Chemistry

We start the description of each laboratory by reference to the holders of the chairs beginning with the Dr Lee's Professor of Physical Chemistry. As we have seen after Sir Rex Richards resigned to become the Warden (head) of Merton College in 1970 and subsequently Vice-Chancellor of the University, 1977–1981, Sir Frederick (later Lord) Dainton, who had been Vice-Chancellor of Nottingham University was elected (1970–1973). He was not able to accomplish much in his short stay. He was followed from 1974–93 by (Sir)

John Rowlinson[86] (author of two chapters in this book). Rowlinson was an Oxford graduate, but never an Oxford fellow/lecturer, with experience as a chemistry lecturer in Manchester and as a Professor of Chemical Technology in Imperial College, London 1961–1973. His special interests in which he had a very strong international reputation were in the theories of the properties of gaseous and liquid molecular phases. In particular he wrote one very well received book with B. Widom on *The Molecular Theory of Capillarity*[87] and a second on cohesion.[88] As a man not preoccupied with practical laboratory concerns but a theoretician (a true intellectual in our sense of the word), he had the time as well as the skill from his experience to manage the Physical Chemistry Laboratory and to look after the affairs of chemistry in the university in a dedicated manner. Later he showed the same skills as Physical Secretary of The Royal Society. He was knighted in 2000. Lately he has been prominent in writing historical records of scientists.

On his appointment Rowlinson has stated that he considered that the physical chemistry laboratory had remained too associated with spectroscopic tools especially linked to gas-phase kinetics. We have described earlier the missing areas including much condensed-phase chemistry, noting that the solid-state chemistry was by now entrenched in the Inorganic Chemistry Laboratory, but that there had been some new initiatives in the 1965–1980 period especially in the physical chemistry of surfaces as described in Section 7.6.4. Unfortunately, for a few years all science departments suffered from financial restraints imposed by the university. Even so, by 1985 the number of fellow/lecturers in physical chemistry was 16 and had risen from 11 in 1973, 8 of whom were still in post. We shall see later that certainly by 2005, when there were still 16, the interests of the laboratory were very much broader, and we describe below this new work some of it introduced in Rowlinson's time. Some of the difficulties in attempting to elect lecturers with new interests arise from departmental attitudes, others from teaching, including college, needs. It will happen sometimes that the best candidate overall meets somewhat parochial views while those who might introduce novelty fail to match these local needs.

J. Simons,[89] who later studied the photochemistry of biologically interesting molecules, followed Rowlinson as professor from 1993 to 1999. He developed especially a novel way of examining conformations of molecules in the gas phase. He kept the tradition of the laboratory, running it with due consideration for fellows' and colleges' interests. The incumbent to 2007 was J. Klein[90] who had previously been at the Weizmann Institute in Israel. He found the peculiarities of Oxford and particularly the departmental democratic organisation and the collegiate system confusing despite his education both in an English school and at Cambridge University, and he gave them little of his time. His major interest, a considerable departure for the laboratory, was in condensed-phase soft matter, under which heading falls many a material studied in the other two laboratories. After seven years he resigned from the headship of the laboratory to become a research professor and finally at the end of 2007 he gave up his professorship. The head of the laboratory, not the Dr Lee's Professor, from 2006 is G. Hancock who is interested in Fourier transform infra-red spectroscopy and has initiated a Medical Diagnostics spin-off company, see Appendix 2. It is considered today, following this example and parallel cases in other departments, that the notion that the professor can be both a research leader and the administrative head of a large department, with all the baggage that has come to be carried by both, is no longer viable. Research and administration need to be separated, especially as they require different skills and this is now recognised by the university. The election of the next Dr. Lee's professor may well have to keep this in mind.

Turning to the research in physical chemistry by fellow/lecturers rather than by professors we see that there is still a strong section of work on new spectroscopic techniques. We have described above the earlier NMR studies of R.E. Richards. Some of his pupils became members of the department or other departments, of whom those of special note mentioned before were R. Freeman, later a professor at Cambridge, R.D. Dwek the professor and head of the Oxford Biochemistry Department (1988), and I.D. Campbell also a professor in Biochemistry (1992). All three of them became Fellows of The Royal Society. Most recently P.J. Hore (1983), who is still a fellow/lecturer in physical chemistry, has made considerable advances in the NMR technique. A somewhat parallel spectroscopy, EPR, continued to be developed by K.A. McLauchlan[77] and he also tackled the effects of magnetic fields on chemical reaction rates. He too is a Fellow of The Royal Society. The latest chemistry development in EPR is in the Inorganic Chemistry Department partly led by his research pupil C. Timmel (2006), who with P.P. Edwards and others, gained a very large research grant for work in this field in 2006. The research that J.W. White started in earlier years on neutron diffraction has been continued by R.K. Thomas and here we note the association with the work at the Rutherford Laboratory especially that of J. Penfold. Photoelectron spectroscopy also continued under D.W. Turner, who proved to be an exceptionally original experimentalist and could well have shared with his London colleague Prof. W.C. Price in the Nobel Prize awarded in physics to Prof. Siegbahn in 1981. Major work in this type of spectroscopy has gained further strength in the Inorganic Chemistry Laboratory under J.C. Green (1990) and (the late) A.F. Orchard (1970) and R.G. Egdell (1990). Electrochemistry, for a time under W.J. Albery, became strong in the Inorganic Chemistry Laboratory too under A. Hamnett, H.A.O. Hill and F.A. Armstrong. This series of observations illustrate again that the spectroscopic studies in the Physical Chemistry Laboratory were quickly taken up especially in the Inorganic Chemistry Laboratory showing the complementary nature of the two departments, much though they hardly shared their experiences. Other continuing work included the use of shock waves by C.J.S.M. Simpson (1969) and the more extensive research using lasers by G. Hancock (1976). What was lacking, perhaps because of the diversity of techniques in some fifteen different and deliberately independent groups, was purposeful analyses of particular parts of chemistry or the chemistry of biological systems. To see the diversification of research topics we give here a list of many of the people in the Physical Chemistry Laboratory in the period.

Gas and solution kinetics	(Pilling), (Simpson), Wayne, Eland, Brouard, Hancock, Simon, Manolopoulos*
Gas phase spectroscopy	(Turner), Eland, Howard, Softley, Brown, Child*
NMR and EPR spectroscopy	(Freeman), McLauchlan, Hore, Atkins*
Electrochemistry	(Albery), (Coles), Compton
Solid, liquids and surfaces	(White), (Simpson), Klein, Foord, Thomas, Bain, Rowlinson*, Madden*, Logan*
Biological studies	(Dean), (Smith), Richards (G.),*

*Theoretical study

Names in brackets had left by 2000; not all those who retired shortly after 1980 are included. Softley's doctorate supervisor at Southampton, A. Carrington, a Royal Society Research Professor, was in the laboratory 1984–87 and continued in this period his work on gas-phase spectroscopy and kinetics.

The experimental bias in the laboratory remained toward kinetics and spectroscopies in the gas phase but now less so and there is clearly a wider concern. We describe later some of the theoretical topics included above as at first some of them were in a separate department. Only by putting this laboratory with the inorganic chemistry laboratory can physical chemistry be fully covered.

Before leaving physical chemistry, the appointment of an additional new professor, J. Hagan P. Bayley, with research space in the new Research Laboratory in 2003 has brought a new dimension to physical chemistry although his is a separate sub-department of Chemical Biology. In fact the techniques employed there to investigate membranes and single proteins are those being developed by what are generally called biophysicists, *e.g.* Hagan Bayley himself, M. Wallace, a physical chemist (2004) and J. S. Davis of the Inorganic Chemistry Laboratory. It is notable that since 2000, see the listing above, the laboratory has moved very noticeably toward condensed-matter research and this trend has continued in the most recent years.

7.7.3 Theoretical Chemistry Department

As mentioned in Section 7.6.6 the theoretical department became effectively isolated from the rest of Chemistry under Professor March, 1976–1994, who was more of a physicist than a chemist. It had M.S. Child and D.B. Abraham both appointed as lecturers in the department before 1975 but no more appointments were made. As it remained a very small organisation it was transferred to physical chemistry in 1994. This was unexceptional as this laboratory had itself several theoreticians including Rowlinson, described above, P.A. Madden, D.E. Logan (1986), D.E. Manolopoulos (1995), D.C. Clary (2002) and J.P.K. Doye (2006), where Clary had special interests mainly in chemical dynamics in gases, Madden and Logan in solids and surfaces, and Doye studied protein interactions. The chair remained as a statutory appointment held from 1994 to 2006 by M.S. Child with an interest in chemical dynamics of molecules. In 2006 Logan was appointed professor. It can be argued that the amalgamation of the physical and theoretical departments was enforced by the lack of appointment of new fellow/lecturers in the theoretical department from its inception. If it had been an active separate department it could perhaps have been more involved with the theoretical problems of all the departments.

This is a suitable place to remember the role of computational methods in chemistry. The first appointment was of J.S. Rollett (1958) attached to the computing laboratory but assisting greatly with Chemical Crystallography. In the chemistry departments themselves Smith (1963), working on computational approaches to statistical mechanics, and then W.G. Richards (1969), especially concerned with the fitting of small molecules to protein surfaces, continued to be major exponents. Very shortly thereafter computing began to play a major role in assisting rapid calculation and in searching for possible solutions to many rather intractable mathematical analyses of chemical problems.

7.7.4 Organic Chemistry

Turning to the Organic Chemistry Department, following Sir Ewart Jones, Jack Baldwin[91] was elected professor, remaining in office for almost the next twenty years, 1978–2006. He had held positions in American Universities including MIT where it is reputed that the students elected him to the position of the toughest professor in the sciences. He had previously been professor of organic chemistry at King's College, London. His reputation as a strong-willed, impatient man, was maintained in Oxford. He showed little appreciation of the colleges' or the university's style and made himself dependent on outside, often industrial, money to build up a large group. This was a return to the style of Perkin and Robinson, not of his immediate predecessor, Jones, and in the minds of many an Arts fellow confirmed the view that practical men (tradesmen) were best avoided as people not interested in their view of academic matters. ('Tradesmen' reflected the connection to industry, not so much said as an insult as an indication of belonging to a different set, see note 10.) As a practitioner, Baldwin was undoubtedly a first-class, very successful experimentalist and designer of biomimetic synthesis pathways of potentially biological interesting small molecules, especially β-lactam antibiotics of the penicillin family. This is how he wished to be known and certainly not particularly as an administrator or educator. He may well have believed as others have before him that his field could only make striking progress in a large coordinated group, an attitude often accepted elsewhere but not in Oxford. He had several fellow/lecturers working on parallel synthesis projects including R.M. Adlington (1990), M.J. Moloney (1990), J. Robertson (1992) and T.J. Donohoe (2001). Although connected to Baldwin they were by appointment quite independent. Baldwin was knighted in 1997. Later following the work on penicillin oxidation he began examination of the parallel oxidation by iron enzymes from living organisms. The oxidation researches have been continued by C.J. Schofield (1990) who has obtained structures of enzymes involved in these and similar oxidations together with knowledge of their mechanisms and controls. In his laboratory Baldwin also had several independent chemists, those already mentioned included J.M. Brown and S. G. Davies (1980) concerned with a continuation of the synthesis and use of organo-metallic catalysts but we now stress those involved in sugar chemistry. Undoubtedly, under Baldwin the organic chemistry's standing was raised considerably.

An interesting development in organic chemistry has been the growth in the studies of saccharide (sugar) chemistry to put beside the more conventional carbon/nitrogen framework chemistry that has a sustained history in the laboratory from Perkin's time to the present day. One of the first to take up the new area was G.W.J. Fleet (1980) but it has developed so quickly that he is now one of nine members of the Oxford Glycochemistry Centre, which is headed by B.G. Davis. The coverage here is of oligosaccharide syntheses (some of biological carbohydrate mimics), of the complement of different sugars in a cell, (the glycome parallel of the genome and proteome), and of carbohydrate processing enzymes. The last is also in part associated with the above-mentioned Chemical Biology Unit involving especially B.G. Davis and C.J. Schofield. The glycome studies are linked to research initiated by R.A. Dwek in biochemistry that has a special Glycobiology laboratory which has been heavily supported by Monsanto. The Oxford Glycochemistry Centre has been greatly aided by both physical and biology Research Councils and industrial grants.

One of the notable features of this and other organic chemistry research projects has been its dependence on industrial funding for nearly 100 years, while this has only become true of projects in the other two departments relatively recently, much of it due to overlapping interests with organic chemistry in catalysis, materials, biosensors, and drugs. In the context of general change from local organic departmental interests to work more obviously crossing departmental boundaries are (a) the above research of Baldwin and Schofield on oxidative enzymic reactions in that these enzymes contained metal ions, see the work on metal ions in biological systems of R.J.P. Williams, H.A.O. Hill, L.L. Wong, J.R. Dilworth, F.A. Armstrong, and J.J. Davis in Inorganic Chemistry, and (b) that of J.M. Brown, S.G. Davies, D.M. Hodgson (1997) and V. Gouverneur (1998) on organo-metallic compounds that closely overlaps work in the Inorganic Chemistry Laboratory, see especially M. Green in Section 7.6.2. Again the synthesis of conducting polymers by H.L. Anderson and P.L. Burn in the Organic Chemistry Laboratory has a direct link with the studies and analysis of electron transfer by groups in the Inorganic Chemistry Laboratory, see below. This stresses again the need to forge a stronger link between the departments for groups rarely exchanged and even have separate seminars associated with similar work. The contrast of the still somewhat disjointed activities of the departments with the Enzyme Group is marked. A final exceptional development in the organic chemistry department was in mass spectroscopy, especially of very large molecules, by C.V. Robinson (1990) who moved to Cambridge in 2002. In order to have a general idea of research in organic chemistry the following is a view of the main lines of individuals.

Organo-metallic	Davies (S.G.), Brown, Hodgson, Peach
Enzyme-related	Baldwin, (Lowe), Schofield, Fleet
Novel syntheses	Baldwin, Adlington, Gouverneur (F-chemistry), Moloney, Robertson, Donohoe, Jones, Robinson, Jones (J.H.), Peach
Saccharides	Fleet, Fairbanks, Davis (B.G.), Gouverneur,
Materials	Anderson (H.), Burn

There were several retirements after 1980 and before 2000 not recorded here.

In 2004 just before Baldwin retired a new laboratory, called The Chemical Research Laboratory, Figure 7.3, was built as the Dyson Perrins organic chemistry had become condemned by Health and Safety regulations. Perhaps it is a good moment to draw attention to the ever-increasing interest in biological chemistry in this new laboratory in part associated with the new department of Chemical Biology.

As mentioned earlier, 2007 saw the election of a new Waynflete professor, S.G. Davies, the first Oxford graduate to hold the post since Brodie, whose work has been described already. Very quickly the post of Chairman of Chemistry was added to his position and P.J. Donohoe was put in charge of the detailed administration of the laboratory.

Figure 7.3 The South Parks Road side of the new 2004 Research Laboratory. (Photograph by K. Harrison, published with permission.)

7.7.5 Inorganic Chemistry

Turning to the third laboratory, that of inorganic chemistry, M.L.H. Green[92] already described in an earlier period[70,71] as the man who introduced organo-metallic chemistry and its catalytic value to Oxford, was elected professor in 1989 and chosen in preference to P. Day. This was an election that went to the practical (synthesis) man rather than the obvious intellectual for the first time outside the organic chemistry laboratory. Green was an unconventional man, full of energy, humour (not always harmless), and green-fingered originality. Partly his unusual manner arose, we may suppose, from his unusual career, a polytechnic degree but then developing rapidly at Imperial College with Wilkinson before a short period at Cambridge. Today he, just like Wilkinson, could be nearly as strong a candidate for a chair in organic as in inorganic chemistry. He proved to be very sociable, greatly helped by his wife J.C. Green, an Oxford college fellow and spectroscopist in the Inorganic Chemistry Laboratory. P. Day, his rival for the chair, was to some degree a natural successor to Goodenough. While with R.J.P. Williams he became interested in photo-conductivity and the properties of mixed oxidation-state conducting solids. He did excellent, quite independent, work later on these topics and on optical and magnetic properties of both inorganic and organic solids. His career has been described earlier.[62,63] He may have lost preference at Oxford in that it could have been considered that it was time for a change of emphasis at the top of the laboratory. However, Day would have undoubtedly brought credit to Oxford Chemistry but in a very different chemistry field from that of Green. No matter what the cause of the selection the dice fell in favour of Green who continued very successful research on synthesis and especially the catalytic properties of organo-metallic chemicals in which he was a world-wide leader. His period as professor was one of increasing financial problems but he had no real interest in such concerns or gifts in such matters or in administration or in networking within and outside the university. He should perhaps have been a research professor for he often puzzled other members of university and other committees on which he sat with his offbeat and almost flippant comments. For several years he was greatly assisted by A.J. Downs who was an able help with administration. As a professor Green helped to forward the careers of several very able young men both in the department (D.M. O'Hare (1990), P. Mountford (1998) and L.L. Wong (1993)) and elsewhere. A further aspect of Green's somewhat unorthodox style was his willingness to research into areas different from that of organo-metallic chemistry, in which he had gained his reputation. Around the time he retired he formed a catalyst group that worked on solid-state rather than homogeneous catalysts and almost simultaneously he combined with material chemistry departments in the study of a section of nano-chemistry, namely the isolation of minute particles of compounds inside the confined volume of a Bucky-tube, a carbon framework tube of less than 50 Å internal diameter. In both these fields he was again successful, patenting catalysts of value in the first and demonstrating the effects of small-scale chemistry in the second. His continuation in research, after retirement in the department where he had been professor, showed a marked change for and challenge to tradition and leads to many questions for the university to solve.

A second professor in the Inorganic Chemistry Laboratory during Green's period was R.J.P. Williams,[93] the Napier Royal Society Research Professor, 1974–1991, who was the 'grandfather' of biological inorganic chemistry, and his main research remained in this area. In this period and within the Enzyme Group,[84] for which he wrote the research proposals, he concentrated on an examination of the NMR spectroscopy of proteins, with a sustained major interest in those containing metal ions such as iron, calcium and copper. The work, as mentioned earlier, was with, and then independently by, H.A.O. Hill, I.D. Campbell, C.M. Dobson and S. Mann (all now professors, see Section 7.7.2) amongst those who at one time were college fellows and who became very well known. See also the account of the Oxford Centre for Molecular Sciences in Section 7.6.7. On leaving Williams, so as to lead their independent careers in and outside Oxford, all these men have shown their considerable individual abilities in a range of subjects. All are Fellows of The Royal Society. However, his own major purpose remained in the biological chemistry of the elements. After retirement he had time to write four influential books on this topic with J.J.R. Fraústo da Silva.[94–97]

Green was succeeded in 2003 by Prof. P.P. Edwards[98–100] whose work represents to a limited degree a return to the metallic-state chemistry of Hume-Rothery, on metals and alloys, before 1960 and with a special additional interest in alkali metal/liquid ammonia solutions employing electron-spin resonance spectroscopy. It is this that gives a link to the work on conducting polymers in the organic chemistry laboratory. He is also greatly concerned with the hydrogen storage problem, associated with the possible use of hydrogen as a replacement for carbon-based fuels. Unlike any of his predecessors in inorganic chemistry he has been and is a member of many outside bodies concerned with general matters within science and its management and with long-term problems of society, for example with energy sources. His laboratory time is thus limited. Once again there is presented to the professor a number of requirements not all of which can be managed by one person. As we commented earlier Oxford needs to change the structure of its chemistry departments. Edwards met a second problem on coming to Oxford, much as did Jones in an earlier period. He brought with him three of his senior research assistants in Birmingham. The Oxford appointment system is now strictly governed by committee and there are no positions open to professorial gift so once again a group of research workers was isolated from colleges. This applies especially to fellow/lectureships so that in Edward's first five years the six new Oxford appointments in his laboratory have been of lecturers in a variety of topics, mainly solid state and catalysis, but none of his own group and none directly related to his own research were made fellows although Timmel (2006) was much involved with him in the EPR application, mentioned above. A continuation of solid-state chemistry from the earlier time is found in the groups of D.M. O'Hare (1990), P.D. Battle (1989), R.G. Egdell (1990), and S.J. Clarke (2000), all appointed in Green's time as professor, much of it connected to

Oxford material science, and for a short period M. Rosseinsky (1992), one of Day's pupils, elected to a chair at Liverpool in 1998.

The group interested in biological inorganic chemistry has changed hands in the inorganic chemistry laboratory and now has F.A. Armstrong (1993) and J.J. Davis (2003), pupils of Allen Hill, L.L. Wong (1994) a pupil of Malcolm Green, and Jon Dilworth (1998) who had worked on vitamin B_{12} in R.J.P. Williams' group. These fellow/lecturers are in part associated with the development of an Oxford centre for medical imaging (Dilworth) and in part with experiments leading to bio-sensors (Hill, Wong). The somewhat amusing feature is the slow return to a link between medical science and chemistry, already present in the organic chemistry laboratory but more so in biological departments, which was one of the cornerstones of the beginnings of chemistry but that was seemingly lost from Oxford chemistry in the 1945 to 1980 period. (In passing, note that such a change presents funding opportunities in today's climate.) There are somewhat isolated related biochemical studies in the Oxford Biochemistry Department[99] but that department is now more closely linked to physiology and genetics. Only a few stray members of that department remain interested in enzymes or membrane processes and surely they should be in the chemistry sub-faculty with Prof. Hagan Bayley, see above. Another overlap is the synthesis of organic molecules of complex structure that also hold metal ions by P.D. Beer (1994). Some degree of integration has been achieved as the work of Beer and Bayley, some parts of physical chemistry and all the organic chemistry laboratory are housed in the new Chemical Research Laboratory. (It was the intention of the design stage of this laboratory to remove departmental divisions.) It takes little imagination to see that today much of the chemistry in any one department interacts strongly with that in the other departments, but unfortunately their immediate contact is still inhibited by the laboratory structure, for example the larger part of the inorganic department remains in the extended Old Laboratory. To give a parallel impression to that provided for the other two laboratories we summarize the activities in the inorganic laboratory in 2000 as follows:

Solid-state chemistry	O'Hare, Egdell, Battle, Clark, Mountford, Edwards, (Orchard), (Green (M.)), Cox (Denning), (Prout)
Biological inorganic and coordination compounds	Armstrong, Dilworth, Wong, (Hill), (Williams), Davis, (Rossotti)
Organo-metallic and molecular synthesis and catalysis	(Green (M.)), (Cooper), Beer, Mountford
Spectroscopy	(Downs), (Dobson), Green (J.), (Denning), Heyes, Smith, (Orchard)
Crystallography	(Prout) Watkin

Names in brackets are of those who had recently retired or left by 2000 or shortly after. Not all those present in 1980 are included.

7.7.6 Oxford Chemistry Today, 2008[101]

In the years from 2005 to today the chemistry school at the teaching level has changed very little. It maintains its style of tutorials, lectures and examinations in Part I and with a Part II research year. The content of the courses has changed little too, perhaps even in the last twenty years, but there is a slowly growing sense of the need to widen the course and to coordinate all teaching as much as is possible, while leaving a considerable amount of choice of exact topic to the fellow/lecturer. The overall decision-making processes concerning the administration of the laboratories and elections also remain much as described above, but we have already noted the new importance of the Chairman of Chemistry and the Divisional Board in global matters. Surely each laboratory should have a representative on every election committee to any new post. Given the number of recent appointments in chemistry and the obvious fact that it takes time for any new study to show its importance we shall not give the names or subject matters that have been introduced very recently.[101] We have included much of on-going work started before 2005 in earlier sections. Many of the new appointments in all three departments have been made to increase the skills in the core areas, such as catalysis, biological and solid-state chemistry, which we have described already rather than to introduce great novelty. Any research must have in mind today possible, even if long-term, value to society since the costs of the work continue to rise, and at the same time research must meet the primary purpose of the investigation of areas where knowledge is lacking. There is undoubtedly ever-growing pressure on the school to demonstrate effectiveness and efficiency in all activities. It is clear that larger groups will be the most successful and the school needs to look at the provision of technical staff in research for modern instrumentation needs new technical staff possibly more than new fellows with new research topics. A major problem for any university is how to finance the now very expensive science subjects including chemistry, see Appendix 2.

Problems for the school, which face chemistry schools everywhere and that many have already faced, are which areas of science should chemistry cover. Historically its strong links with biological and mineral sciences are most obviously those that were lost from and are not yet firmly re-established in the present Oxford school. Elsewhere there has been a merging of at least biochemistry and chemistry in many universities. The importance of these connections are made the greater by the way schools teach science on a broad base today and the obvious need for those trained in chemistry to assist at both the research and the political level in the problems that society will face shortly. Looking back we can see how the present school came about, arising through haphazard, often college and/or personal interest, with no overall vision or organisation, but it is not so easy to see how to make it more equipped to meet modern day needs.

With these matters in mind we turn to a glance at future internal problems. Toward the end of the last century and increasingly rapidly now are both financial and organisational difficulties and stresses in universities in Britain.

Coming immediately to the way they affect the chemistry school in Oxford, the size of the school, greatly increased from 1945, and the great expense of running its teaching but mainly its research have led to the need to raise very large sums of money through the university or directly from outside bodies, Research Councils, industry and charities, see Appendix 2. The colleges now have little or no say in these matters either directly or through university bodies. Outside money has come with requirements, some of which apply to the university as a whole, but many are directed at chemistry. In the financial appendix we see that chemical laboratories and research are already coupled strongly to industry. This is the very coupling the university avoided for centuries. At the same time the university seems to see chemistry as a possible funding source for some of its own general needs, in which case increase in application cannot be avoided. The extent of activity, the numbers in research, the search for money and response to enquiries of many kinds all take organisation and time. Such pressure on chemistry demands dedicated effort in research and must reduce the loyalty of fellow/lecturers to colleges and teaching. The style of Oxford chemistry research with 50 or more independent fellow/lecturer units arising largely from college teaching needs is doomed to raise questions as to its general efficiency. So far its very considerable successes have saved the school from asking itself could it be better organised? Change may well be necessary and can only come about if individual freedoms are limited, which returns us to the question posed at the beginning of the book. Is a whole-hearted democratic system the best way to run a university and in particular a very large chemistry department today? Let us remember again that the whole nature of Oxford has evolved accidentally not rationally as many separate colleges with independent fellows and several laboratories with a few professors who have now relatively little control over their departments. The advantage of a contrasting well-coordinated institute in research is seen in the top-down Max Planck laboratories in Germany and the dedicated MRC unit in Cambridge. It may be that success at the level of winning a Nobel Prize now depends on a considerable size of a research group and its finance. Sir Hans Krebs, a Nobel Prize winner and an Oxford professor in biochemistry, believed that it was almost impossible for an Oxford Research scientist to win a Nobel Prize as the structure of the collegiate university does not allow the growth of a particular area in sufficient strength no matter whether a fellow/lecturer or a professor is in charge. If this is the case then in chemistry it is not so much individual originality as large-scale effort that will bring success. In fact, everywhere in the university we find a re-examination of the best way of running affairs in the future in, if possible, a bottom-up system that has arisen close to uniquely in Oxford. Re-examination is needed then in several respects of the chemistry school. What is the purpose of its undergraduate course and how broad (or specialised) should it be? Who should teach it and in what mode? Has the departmental structure outlasted usefulness? Can research carry on in such a multi-faceted, individual way with the present lack of coordination? What is the appropriate connection to the colleges and how should they be run? Many colleges are now effectively run by administrators, not by academic fellows, and fellows have a much decreased involvement and hence interest in their college. Should the sub-faculty be the real centre of control over the future through a chairman? The very great success of the school may well have hidden the need to compare the changes that have taken place in other universities, especially those of equally great success, for example Harvard, with the organisation we have, after searching its strengths and weaknesses. We hope this history of the development of the Chemistry School at Oxford has revealed that much of it now needs reappraisal in the light of knowledge that its present structure is still largely a consequence of the history of the collegiate university. At the same time Oxford Chemistry can be extremely proud of its past and present standing.

APPENDIX 1

The Laboratories (see Table 7.4[101] and Figure 7.4)

The earliest chemistry laboratories in Oxford have been described in the several chapters of this book including the basement of the old Ashmolean Museum,

Table 7.4 The Chemistry Laboratories.

Date	Laboratory	Departmental Topic
1683	Old Ashmolean Basement	Minerals; Fossils; Simple exploration of nature of gases and other substances, closed for research, 1860 (now a museum)
1767	Christ Church	Used at first with anatomy, only for chemistry after 1860 (closed 1941)
1848	Daubeny Laboratory (Magdalen)	General chemistry (closed 1923)
1853	First of Balliol Laboratories, second 1879 and joined with Trinity 1897	Became home of the study of chemical kinetics (closed 1941)
1860	Museum Abbot's Kitchen	General Chemistry (still used in part)
1879	Considerable extension of Museum Laboratory	General Chemistry later associated with Inorganic Chemistry (still used)
1900	Queen's Laboratory	Organic Chemistry (closed 1934)
1907	Jesus Laboratory	Kinetics (closed 1947)
1916–1922	Dyson Perrins	Organic Chemistry (abandoned 2003)
1941	Physical Chemistry	Mainly kinetics and spectroscopy (still used)
1950	Museum House	Used as overflow of Inorganic Chemistry for 5 years (closed 1957)
1957	Large extension of old Inorganic Chemistry	Inorganic Chemistry (still used)
1990	New Laboratory (Pharmacology)	This was a temporary research space and acted as an overflow for all three major laboratories (largely vacated 2004)
2003	Research Laboratory	Mainly Organic but something of a mixture

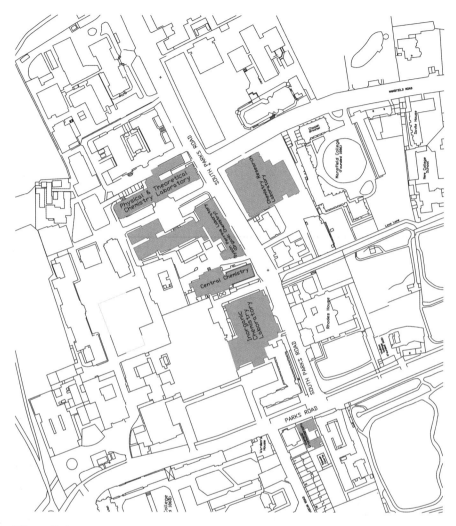

Figure 7.4 A map of the positions of the chemistry laboratories today. In addition, note the house on Parks Road that contained the crystallography department. Much of the plan was devised as early as 1935. Parks Road runs due North to the left (Published through the courtesy of the University's Estates Directorate.)

1683; the Daubeny Laboratory, 1848; the Abbot's Kitchen, 1860; the old Chemistry Laboratory, 1879. In all these four, chemistry was treated in a very open manner without college teaching considerations and often not truly separately from other sciences. The Dyson Perrins laboratory, 1912, and the Physical Chemistry Laboratory, 1941, saw the development of specialisation and the splitting of chemistry into departments. There were also a variety of specialist laboratories built in colleges within two centuries from 1750 but all were closed by 1947. Here, we collect together the further building from 1945. The old Inorganic Chemistry Laboratory was given a large extension in 1957 and it has had further smaller extensions later; the Dyson Perrins was extended in 1940, 1959 and 1971 (this laboratory was abandoned for research chemistry in 2003 as it no longer met health and safety regulations); the Physical Chemistry Laboratory had an extra east wing in 1959, was increased in capacity by extra floors in 1966 and small extensions to it have been made subsequently; and finally a large new Chemical Research Laboratory mostly for organic chemistry opened in 2004. The Physical Chemistry Laboratory was linked to the Rex Richards Building associated with biochemistry in 1984. At various times other buildings have been used for short periods while the above new buildings were being built. In the early 1950s Museum House, seen on the far right of Figure 4.1, and in Figure 5.3, and the Old Physiology Laboratory housed parts of inorganic chemistry and the Pharmacology building acted as a general overflow laboratory from 1995 to 2004. A further new laboratory is planned in part to re-house inorganic chemistry. A new laboratory is still required for the undergraduate practical courses of all three departments. Only aborted attempts have been made to integrate the three departments of chemistry in one laboratory. They remain largely separate and do not yet sit easily with the inter-disciplinary nature of the chemistry of today. The spread of interest in chemistry generally has seen sections of the subject separate and grow in new departments and buildings. Theoretical chemistry split away for a while and was settled in houses close to the chemistry buildings in South Parks Road. It was united with physical chemistry in 1994. Material Science, at first Metallurgy, became divorced from chemistry in 1957 in a building remote from chemistry and is now treated as a separate subject. Mineralogy, an early Museum Science associated with chemistry, contained Chemical Crystallography before 1940 but crystallography became a separate sub-department of physical chemistry in the Museum in 1944 but did not move into the Inorganic Chemistry Laboratory until 1957. Later, after incorporation in Inorganic Chemistry for several years, it was located for many years in a rather beautiful large house on Parks Road quite close to chemistry, see Figure 7.4, which had been the premises of Soil Science and is next to a brown building that had a Latin tag for "Rural Economy". It returned to the new Chemistry Research Laboratory in 2004. Note that the part associated with Dorothy Hodgkin was for a while in the Inorganic Chemistry Laboratory and then the Psychology Laboratory. Major interest in biochemistry grew from 1950 in the chemistry laboratories but a separate Chemical Biology unit only came into being in the new Chemistry Research Laboratory in 2004. The sub-faculty of Biochemistry itself had originated, somewhat strangely, from Physiology[102] as a separate new department in its own building from 1923 and Biochemistry's major activities have remained in laboratories dissociated from chemistry ever since, see note 12. It is even in a division, separate from chemistry, of the university's activity! There has never been an effort to re-examine thoroughly what to all intents and purposes has been the random growth of separated topics and laboratories related to chemistry. Today, chemistry has strong connections to material science and biochemistry but major barriers remain in part due to the rather distant relationship of the buildings.

Looked at from the viewpoint of the open style of development of many modern university campuses with laboratories well separated from one another by green spaces and easy access the jumble of tightly placed Oxford science departments, Figures 1.1 and 7.4, must seem very strange. The problem has arisen from the ownership of land. Colleges owned at one time or another large areas of North Oxford that were developed for houses, or otherwise left as green areas for sports fields and parks. In particular, Merton College owned the land from South Parks Road to Norham Gardens that, at the time the Natural History Museum was built on it, 1860, had become the University Parks. A section of it bordering South Parks Road was seen to be the best possible site for science laboratories shortly thereafter. Slowly, building and infilling made it obvious that the site was too restricted, Figure 7.4, and with nowhere else set aside, it became a crowded ill-organised science area. Further laboratories have now to go elsewhere. Various nearby sites have been considered and some allocations were made for the use of land encroaching on nearby sports grounds and houses. As a result, related sciences will often remain not in juxtaposition, which hinders collaboration. A consequence for chemistry is that it has been forced to build its most recent laboratory, the Chemical Research Laboratory, on the south side of South Parks Road. The laboratory mainly for organic chemistry also includes about a third of inorganic chemistry and a small part of physical chemistry plus a sub-department of Chemical Biology. It is hoped to remedy this clumsy arrangement with further building for chemistry on the same south side of South Parks Road.

The external and internal designs of a chemistry laboratory have shown remarkable changes that illustrate the very fast alteration in the whole nature of the science. At first, laboratories were hidden behind facades of presumed architectural interest for example the basement of the classical old Ashmolean Museum, the Daubeny Laboratory and the mock-Gothic Abbot's Kitchen of the Natural History Museum and its extension. A major feature of the Abbot's Kitchen, with Victorian stonework, Figures 4.1 and 5.3, was that there was ventilation for removing noxious fumes via four external chimneys and a high central vent. The floor was open space left for "benches". The first step, it was a big step, toward a totally new concept of a laboratory came with the building of the 1879 extension of the Abbot's Kitchen that was still of Gothic revival external architecture but its interior was now light with large windows on two floors. The ground floor was a series of small rooms, Figure 4.2, and it had later a raked lecture room, while the upper floor was a large barn-like single room for teaching, Figure 7.1. This combination provided a very common way of isolating research from teaching space. Despite the façade it was for its time a modern laboratory and much of it remains in use today, though it is likely that it will become part of the Museum. The next building for chemistry, the Dyson Perrins Laboratory in 1923, had a front of huge windows on two storeys to gain light, Figure 5.1. The ground-floor windows stretched below the level of the street so that passers-by could gaze on activity within. Inside there were benches, properly serviced, both in the ground floor research rooms and the upper-floor teaching laboratories, a lecture theatre and a small library. There were far too few fume-cupboards (hoods). Its large extension was of simple brick but it had a lecture theatre on stilts of architectural note. The front of the Dyson Perrins Laboratory is now a listed building but the whole laboratory has been abandoned by research chemistry and undergraduate teaching laboratories are to be moved as soon as possible.

The Physical Chemistry Laboratory of 1941 is the first chemistry laboratory with an exterior designed purely around function, Figure 6.1. It is oblong with strict rectangular brick sides indicative of the full use of space. The windows are clearly designed to give good functional light. The laboratory is divided by a simple entrance onto a small foyer with a large upstairs lecture theatre. One wing of the oblong shape is devoted to research rooms and the other to a large teaching room on the upper floor and research rooms below. It has had several extensions. Architects must have thought out carefully the design with Hinshelwood and his colleagues, in marked contrast to the extension of the Old Inorganic Chemistry Laboratory.

At the time of the construction of this extension Brewer was head of the Inorganic Chemistry Laboratory. Apparently with no thought, a plan, little altered from its conception at least as early as 1935, when Lanchester and Lodge were the architects, was used. It is based on a quadrangle of a semi-Georgian style on three new sides with the old mock Gothic laboratory of 1879 on its western side, forming a quadrangle. We can only presume that in 1935 this mock Georgian style was thought to be suitable for a building so prominently placed opposite Rhodes House and next to the Science Library, Figure 7.2. The old laboratory of 1879 had floor levels separated by four metres, typical of the Museum of Natural History after which it had been fashioned and clearly the room heights of the 1957 building have been matched in this extension. It has to this day a narrow archway entrance in front and to the road, indicated by two lamps on obelisk-like stands by the roadside that are referred to kindly as phallic symbols. This archway is too narrow for heavy goods vehicles, which are forced to approach by a narrow side road to the back. The original grass centre of the quadrangle (now a tarmac car park) indicated further that the archway was not a tradesman's entrance and the whole design clearly reflects that of an Oxford college, Figure 7.5. A very interesting historical problem arises as the Lanchester and Lodge layout of laboratories in 1935 shows both the ICL (old laboratory) extension and the PCL design of the 1941 building, Figure 6.1, had been prepared simultaneously. Now this date, 1935, is before Soddy resigned yet the university resisted any attempt by him to build any new inorganic or physical chemistry laboratory. As a consequence the physical chemistry laboratory was built first in 1941 to suit the new professor, Hinshelwood, and all development of the old laboratory was delayed!

An entirely separate question concerns the suitability of the laboratories' internal fittings. There were essential requirements for a large lecture theatre for about 200, student practical laboratories and many research suites in each of the three departments. Each laboratory should also have had a foyer for

Figure 7.5 The inner quadrangle of the Inorganic Chemistry Laboratory of 1957, now a car park, taken from the front arch on South Parks Road, see Figure 7.2. (Reproduced through the courtesy of the University's Estates Directorate.)

immediate ease of entering and leaving its lecture theatre. The research space should have been flexible and provided with separated offices. Laboratories today must also have fume-cupboards (hoods) in which to do experiments, and supply and drainage services required in addition to water, gas and electricity (simple household needs) for access to different gas supplies, nitrogen and say pure oxygen, refrigerants such as liquid nitrogen, and high voltage electricity and helium especially for high-field magnets. The modern Oxford Research Laboratory of 2004 is an excellent example of such a well-designed building and contrasts now with what was in 1923 a fine design for organic chemistry, the Dyson Perrins Laboratory.

As we have mentioned, all the three departmental buildings underwent extensions between 1960 and 1990. Around 1995 it was revealed that the Dyson Perrins Laboratory definitely, and the Inorganic Chemistry Laboratory possibly, did not meet health and safety regulations. A new building was essential and it had to be of the highest technical quality. The problem facing such a development was that the university did not consider laboratory building to be

Figure 7.6 One aspect of the working space, well separated from office space, in the Research Laboratory, contrast Figure 7.1. (Photograph by K. Harrison, published with permission.)

of equal financial responsibility to that of libraries. To be able to build, it was deemed to be essential to secure much funding from outside the university, see Appendix 2 on finances. The building is designed to be strictly useful for research. It is in today's style of unbroken flat-face glass frontages on the two sides facing roads giving light and reducing the sense of size, Figure 7.3. Unfortunately, but of necessity, to make the most use of space its entrance is hidden away from the road. One feature of the interior is a large atrium stretching the full height of the building with a sufficient ground floor space to serve as a place for lunch. There are large open spaces in the basement for instruments but there is neither a large lecture room nor a teaching laboratory. Internally, there is a separation between laboratory and desk spaces and each laboratory is fitted with multiple supply systems and hoods, Figure 7.6. The whole is continuously ventilated. The overall consequence of lack of land and, in a sense for the necessary long-term planning for research only, is that two problems await solution. There is still a need to integrate the departments, which will require a new building next to and hopefully connected to this one and that together with it will house at least the total complement of the old organic and inorganic laboratories but with these labels removed. Furthermore, the teaching laboratories and lecture theatres will still need relocating. As we said at the beginning of this chapter there have been great changes in the Chemistry School but it has not yet settled into a final condition even in its buildings. A further major necessity before further planning is for the university to see the relationship of chemistry to other subjects such as biological, material, environmental and geological sciences, when considering further changes in the science area building for such considerations will be vital not only for research but for teaching in the future.

ACKNOWLEDGEMENT

Ms Joyce Thompson and Mr Colin George of the University Estates Directorate, and Mr Matthew Lodge, Mr Karl Harrison and Mr Keith Waters of the Inorganic Chemistry Department provided the photographs for this appendix.

APPENDIX 2

The Chemistry School Finances

We divide the finances into capital and running costs while also referring to income specifically to chemistry from research grants. Capital costs are for new buildings, extensions, or for general refurbishment and for major items of equipment. Full information is published by the university but is easily obtainable only on request. Since 1950 there have been running battles over the cost of chemistry both in the university of professors, lectureships, laboratories and their research, and in colleges over support for college fellowships linked to university lectureships. These battles, due to the early divisions of the collegiate university, are described in Chapters 2 to 5. It took a long time for Oxford to awaken to the importance of chemistry and much blame goes to the mutual relationship with public schools and educational values agreed with Oxford. In earlier chapters the importance of the financing of the setting up of the chemistry school, its laboratories, chairs and other staff positions have been described and the cost to the colleges in their support of chemistry has been outlined. The huge flourishing of chemistry after 1945 required financing on quite a different scale and had a much greater and increasing magnitude in university costs leading to the situation described at the end of Chapter 7. We shall therefore cover only the period from 1950 to 2007 here.

Turning first to capital costs the university has had to budget for a continual series of extensions of all three laboratories, which are cited in the accounts and that we do not detail here.[1] A major problem did not arise until it was realised that at least two of the chemistry laboratories, those for organic, particularly, and inorganic chemistry no longer met modern standards. The costs of a new building to meet a proportion of these needs and including a complete replacement of the organic chemistry research laboratories was such that the university saw no easy way forward. It is greatly to the credit of the then chairman of chemistry, W.G. Richards 1996–2007, that he managed to find a way to obtain £64 million from outside sources. Part of the capital cost, £20 million, came through an agreement with an outside London Commercial Institution whereby for fifteen years the institution would get half of the university's equity stake in any spin-out company. In this way the university abandoned its independence of the outside world in a piecemeal way, seemingly holding fast to its general principles.

The increase in running costs have to be taken against the falling value of the pound and must be seen against the need to observe Health and Safety

Regulations. In the last 50 years the pound has fallen some 25 fold. The cost of running chemistry has increased from some £50,000 a year in 1950 to around £25 million in 2007. Thus, in real terms the university has faced a twenty-fold increase in costs corrected for the value of the pound. Some 10 million is offset today by research money that has been recovered from science research councils and other bodies. A major factor here is the success of Oxford's response to the RAE and then to HEFCE. The picture is gloomy but it must be remembered that the chemistry department now actively encourages exploitation of intellectual property, hence providing sources of both income and capital to the university.

As far as initiatives in setting up companies are concerned the chemistry department has been involved with a variety of enterprises including those listed in Table 7.5. The department has then been able to contribute to the university a sum close to £100 million over the last ten years. The effort has been greatly aided by Oxford's technology transfer office, Isis Innovation Ltd.

Finally, it is important to observe that the department does not believe that its academic work has been overly affected, but undoubtedly it is aware of and involved with the outside world in a way quite unacceptable to the university in 1950. The largest change is as recent as from 1992 and was forced by the rising costs but much of the university seems barely aware of the problems entering all sections of academic life. Through financial need and then university financial strength the colleges are rapidly losing control over many chemistry matters, except perhaps tutorial teaching itself.

ACKNOWLEDGEMENT

Prof. W.G. Richards and the University Archivist, Mr Simon Bailey, have been of considerable assistance with this appendix.

NOTES ON OXFORD UNIVERSITY

Information about Oxford University is available in several publications, see notes 1 to 7 and also www.ox.ac.uk

1. The Oxford University Calendars give detailed information about the composition of the University and membership of Boards and Committees, fellows of colleges, lecturers and other university appointments. They also give full lists of undergraduates in the colleges and degrees awarded.
2. The Statutes, Decrees and Regulations give the rules under which the different branches and positions in the university are managed.
3. The Financial Reports of the university are published in outline in the Oxford University Gazette but details for the years after 1970 are only available on request. Details are also held in the University Archive.
4. The Oxford University Gazette is an official publication of the University and is published frequently throughout the year. The Gazette contains

Table 7.5 Oxford chemistry spin-out companies.

Year	Company	Description
1987	**Medisense**	Based on the work of Allen Hill using electrochemical methods to measure glucose. Became part of Abbott Co. USA.
1989	**Oxford Molecular**	Computer-aided molecular design software from Graham Richards' group. IPO in 1994. Now part of Accelrys Inc.
1992	**Oxford Asymmetry**	Chiral chemistry of Steve Davies. IPO in 1997? Now part of EvotechOAI.
1997	**Opsys**	Light emitting molecules from Paul Burn. Merged with Cambridge Display Technology.
2000	**Oxford Biosensors**	Biosensor technology from Allen Hill, Luet Wong and Jason Davis. www.oxford-biosensors.com
2001	**Inhibox**	Based on the screen saver project of Graham Richards for drug lead discovery. www.inhibox.com
2002	**Glycoform**	Glycochemistry from the research of Ben Davis and Anthony Fairbanks. www.glycoform.co.uk
2002	**Zyentia**	Structural modification of proteins derived from Chris Dobson's research. www.zyentia.com
2002	**Pharminox**	Small molecule oncology originating with Gordon Lowe. www.pharminox.com
2002	**ReOx**	Drug discovery based on controlling activity of hypoxia inducible factor, incorporating work of Chris Schofield. www.reox.co.uk
2003	**VASTox**	Drug discovery from Steve Davies using chemical genomics. IPO 2004. www.vastox.com
2004	**Oxford Medical Diagnostics**	Laser-based analytical techniques from Gus Hancock.
2005	**Oxford Catalysts**	Catalysts for clean energy from the research of Malcolm Green and Tiancun Xiao. IPO 2006. www.oxfordcatalysts.com
2005	**Oxford Nanolabs**	Single-molecule detection based on research of Hagan Bayley. www.oxfordnanolabs.com
2006	**Oxford Advanced Surfaces**	Coatings using chemistry of Mark Moloney.
2006	**OxTox**	Electrochemical detection based on research of Richard Compton.

This table was kindly provided by W.G. Richards. IPO, Initial Public Offering.

notices of all actions and proposed actions of the university open to scrutiny by university members. Matters to be debated are to be found there together with notices of degrees, lectures, *etc*.
5. The Oxford Magazine is not an official University publication but is a forum for the expression of opinion mainly by members of the university. Much of the views expressed are directed to exert pressure on the university's decision-making that is made in a democratic manner by

Convocation or Congregation. The Magazine also has articles on matters of general interest.
6. Each college publishes its own reports giving information about fellows and students, management structure and finances.
7. Full information about the Chemistry School at Oxford today together with details of the research interests of each member of the staff are available on the Oxford University Chemistry website www.chem.ox.ac.uk.
8. The new central organisation of the university is under Divisional Boards that were formed in 2000 in order to manage the faculties, including their financing and creation of posts. The divisions at present are: Humanities; Mathematical, Physical and Life Sciences, which includes physics, chemistry, chemical biology, zoology and plant sciences; Biochemistry, Biophysics and Medicine; and the Department of Continuing Education. The separation of subject matters is clearly somewhat arbitrary.
9. On the titles: Demonstrator, Lecturer, Professor.
 The title lecturer and reader have been used loosely in Chapter 3 as they are not clearly employed in documents. The intention of the university in creating the position of demonstrator, sometime after 1850 was, in so far as we can discover, to assist in the management and practical teaching in laboratories under a professor. The professors used them later as research colleagues together with *bona fide* assistants but they were not necessarily associated with a college. Later again colleges managed to couple many of them with college fellowships, so saving money. There was then the anomalous situation of some demonstrators in better financial and social standing than others. In 1964 the university and colleges agreed on a joint appointment scheme of fellow/lecturer, the title lecturer, replacing demonstrator. The university position of research officer created subsequently has again left a group of scientists outside the college system. Professorships were appointed by the university and were placed in charge of departmental laboratories, or for academic personal distinction (labelled Schedule A). From 1996 the university awarded Titles of Distinction, professorships, to lecturers on the basis of their international standing. Generally speaking these awards arose in order to make especially scientists in Oxford have equal title to those elsewhere in the world and especially in USA. As a consequence, over 50% of the lecturers in chemistry, and usually the older ones, have been "promoted" to professor but without any change in duty or salary. This formal change is not easily fitted into an historical account in which we wish to separate the role of the professors who are heads of departments, plus a very few elected to personal chairs, from those under their wings who were collectively called lecturers until 1996. We shall therefore keep this distinction in title as a matter of convenience and clarity, hoping to cause no offence. Reader is today also a university appointment at an intermediate level between lecturer (demonstrator) and professor.

10. In the period from 1945–1965 it was common conversational jostling that gave rise to jocular characterisation of University and other people, not always kindly.
11. The Syllabus for chemistry examinations, Part I, in 1950 reads; Candidates will be expected to show an acquaintance with I. Inorganic Chemistry II. Organic Chemistry III. General and Physical Chemistry. There will also be a practical examination in each of these subjects. The syllabus is much the same even today, while certain other subjects are more closely defined.
12. Note on The Biochemistry Department.[102]
    The Biochemistry Department at Oxford has remained very separate from Chemistry owing perhaps to its origin from physiology. Traditionally, undergraduates reading chemistry have not been directed to take an interest in biological sciences and courses in biochemistry became more directed to medicine than chemical problems. The selection of students for the biochemistry course is from those at school taking more descriptive science and with little emphasis on mathematics. Hence, the teaching of all branches of chemistry in biochemistry is of a lower mathematical content. Research has turned further away from chemistry toward genetics recently, which receives little notice in chemistry. Oxford chemistry graduates have frequently provided the more physical chemical aspects of biochemistry, note for example Radda, Campbell, S. Ferguson, M. Acheson. Because this separation is so marked we refer the reader for the history of biochemistry to the book by Stocken and Ord,[102] and we do not describe it here.

REFERENCES

1. B. Harrison, ed. *The History of the University of Oxford, Vol. VIII, The Twentieth Century*, Clarendon Press, Oxford1994.
2. Oxford University Calendars 1945–2008, Oxford University Press, Oxford, see Note 1.
3. Statutes, Decrees and Regulations of the Oxford University, Oxford University Press, Oxford, see Note 2 and University of Oxford Statutes.
4. University Financial Reports, Oxford University Gazette, Oxford University Press, Oxford, see Notes 3 and 4 and Appendix 2.
5. Appendix 1 describes the laboratories and their histories.
6. College independent statutes and annual statements are available from the Colleges.
7. Oxford University Magazine acts as an outlet for opinion. Published by Delegates of the University Press, Oxuniprint, Kidlington, Oxford, see especially issues of 2006 to 2008, see Note 5.
8. Section 7.4 will describe the changes in attitude to the teaching of chemistry over the 1945–2008 period.
9. This quote is attributed to Shirley Williams.
10. BBC report 29th January 1985.

11. Research Assessment Exercises are carried out by The Research Councils for The Higher Education Funding Council for England (HEFCE).
12. Financial Assessments are carried out by HEFCE.
13. Divisional Boards, see Notes 1 and 8.
14. J.C. Smith, *The Development of Organic Chemistry at Oxford*, 2 vols. unpublished Archives of The Dyson Perrins Laboratory (organic chemistry) Oxford, Radcliffe Science Library, Oxford, Ref. No. 1933 d. 302.
15. R.F. Barrow and C.J. Danby, *A History of the Physical Chemistry Laboratory, 1941–1991*, published by the laboratory 1991.
16. W.G. Richards, appointed 1996 for ten years. S.G. Davies appointed 2007.
17. Oxford University Research Assessment Exercise, 2008.
18. Prof. J.N.I. Connor of Manchester University kindly provided the information concerning this university from their 1933 to 1952 Departmental Reports.
19. Butler Education Act (1944) and its effect on Oxford, see note 1.
20. Information kindly supplied by the Chairman of The Chemistry School, 1996–2007, W.G. Richards.
21. The Syllabus from the University of Oxford Examination Statutes, 1950, is given in Note 11.
22. N.V. Sidgwick, *The Chemical Elements and their Compounds*, 2 vols Oxford University Press, Oxford, 1950.
23. C.N. Hinshelwood, *The Structure of Physical Chemistry*, Oxford University Press, Oxford, 1951.
24. R. Robinson, *The Structural Relations of Natural Products*, Oxford University Press, Oxford 1955.
25. E.J. Bowen, *The Chemical Aspects of Light*, Oxford University Press, Oxford, 1942.
26. H.W. Thompson, *A Course in Chemical Spectroscopy*, Oxford University Press, Oxford, 1938.
27. R.P. Bell, *Acid-Base Catalysis*, Oxford University Press, Oxford, 1941.
28. R.P. Bell, *The Proton in Chemistry*, Cornell University Press, Ithaca, New York, 1959.
29. J.H. Wolfenden, *Numerical Problems in Advanced Physical Chemistry*, Oxford University Press, Oxford 1938.
30. J.W. Linnett, *Wave Mechanics and Valence*, Methuen, London, 1960.
31. H.J. Emeleus and J.S. Anderson, *Modern Aspects of Inorganic Chemistry*, Routledge and Kegan Paul, 3rd edn. London, 1960.
32. R.O.C. Norman and J.M. Coxon, *Principles of Organic Synthesis*, Blackie Academic & Professional, London, 3rd edn. 1993 (1st edn. 1968 pub. Methuen).
33. C.S.G. Phillips and R. J. P. Williams, *Inorganic Chemistry*, 2 vols., Oxford University Press, Oxford, 1965.
34. P.W. Atkins, F. Armstrong, T. Overton, J. Rourke and M.T. Weller, *Inorganic Chemistry*, Oxford University Press, Oxford, 4th edn. 2005. (1st edn. as D.F. Shriver, P.W. Atkins and C.H. Langford (1990)).

35. P.W. Atkins, *Physical Chemistry* (8th edn.) Oxford University Press, Oxford, 2005.
36. P.W. Atkins, *The Second Law – Energy, Chaos and Form*, Scientific American Books, New York, 1994.
37. P.W. Atkins, *The Periodic Kingdom*, Phoenix Press, London, 1996.
38. P.W. Atkins and R. S. Friedman, *Molecular Quantum Mechanics*, Oxford University Press, Oxford, 1999 (1st edn. 1970).
39. E.B. Smith, *Basic Chemical Thermodynamics*, Oxford University Press, Oxford 1973. E. Brian Smith, born 1933, Master St Catherine's College 1988–1993, Vice-Chancellor University of Cardiff 1993–2001.
40. Oxford Chemistry Primers. This is a series of 50 short undergraduate books covering all branches of chemistry to the advanced level examination papers at Oxford. Many are written by Oxford fellow/lecturers. They were published by Oxford University Press, Oxford during the last 20 years.
41. C.N. Hinshelwood, *The Chemical Kinetics of the Bacterial Cell*, Clarendon Press, Oxford, 1946.
42. C.N. Hinshelwood, *The Kinetics of Chemical Change*, Oxford University Press, Oxford, 1940.
43. C.N. Hinshelwood, The Nobel Prize for Chemistry was awarded with N.N. Semenov "for research into the mechanisms of chemical reactions".
44. W. Hume-Rothery and G.V. Raynor, *The Structure of Metals and Alloys*, The Institute of Metals, London, 1956.
45. W. Hume-Rothery, *Atomic Theory for Students of Metallurgy*, 1st edn., Institute of Metals, London 1944.
46. H. Eyring, D. Henderson and W. Jost, *Physical Chemistry an Advanced Treatise*, Academic Press, New York, 1971.
47. J.S. Anderson, *Current Problems in Non-Stoichiometry*, Advances in Chemistry, Publ. 39, American Chemical Society, 1963, see ref. 61.
48. Note in particular the influence of Professor F.A. Lindemann (later Lord Cherwell) in physics especially from 1925–1940. See R. Fox and G. Gooday. *Physics in Oxford 1839–1939*, Oxford University Press, Oxford, 2005.
49. A.R. Todd. The Nobel Prize for Chemistry in 1957 was awarded "for his work on nucleotides and nucleotide coenzymes".
50. J.W. Cornforth. The Nobel Prize for Chemistry in 1975 was awarded with V. Prelog for their work on "the stereochemistry of reactions".
51. D.H.R. Barton. The Nobel Prize for Chemistry in 1969 was awarded with O. Hassel "for their contribution to the development of the concept of conformation".
52. Ewart R.H. Jones (1911–2002), Professor of organic chemistry Manchester 1947–1955. Member of many science committees, knighted 1963, see J. Jones, *Biog. Mem. Fell. Roy. Soc.*, **49**, 263, (2003).
53. F.J.C. Rossotti and H. S. Rossotti, *The Determination of Stability Constants*, McGraw-Hill, New York, 1961.
54. W.A. Waters, *The Chemistry of Free Radicals*, Clarendon Press, Oxford 1947.

55. B.R. Brown, *The Organic Chemistry of Aliphatic Nitrogen Compounds*, International Series of Monographs on Chemistry, Clarendon Press, Oxford, 1994.
56. see reference 33, Vol.2.
57. C.A. Coulson, *Valence*, Oxford University Press, Oxford, 1953.
58. L. Pauling, *The Nature of the Chemical Bond*, Cornell University Press, 1st edn. 1940. Pauling was awarded the Nobel Prize for Chemistry in 1954 "for his research into the nature of the chemical bond". He was also awarded the Nobel Peace Prize.
59. L.E. Orgel, *An Introduction to Transition-Metal Chemistry: Ligand Field Theory*, Methuen, London, 1960.
60. M.J.S. Dewar, *An Introduction to Modern Organic Chemistry*, Oxford University Press, Oxford, 1965.
61. J.S. Anderson, (1908–1992), Professor of Chemistry Melbourne University, Australia 1954–55, National Chemistry Laboratory 1959–63, see B.G. Hyde and P. Day, *Biog. Mem. Fell. Roy. Soc.*, **38**, 1, (1992).
62. H.A.O. Hill and P. Day (eds.), *Physical Methods in Advanced Inorganic Chemistry*, Interscience, New York, 1968.
63. A.K. Cheetham and P. Day, *Solid State Chemistry Techniques*, Clarendon Press, Oxford, 1988. Cheetham was later professor in USA and returned as a professor of Material Science to Cambridge in 2007.
64. D.M.P. Mingos, *Essential Trends in Inorganic Chemistry*, Oxford University Press, Oxford, 1998. Mingos was Professor of Chemistry, Imperial College, London and then Principal, St Edmund Hall, Oxford.
65. J.B. Goodenough, *Magnetism and the Chemical Bond*, Interscience, New York, 1963. US Citizen, born 1922, Senior Scientist and Professor Massachusetts Institute of Technology, 1952–1976, Professor in Texas, USA, 1986–.
66. F.A. Cotton and G. Wilkinson, *Advanced Inorganic Chemistry*, Interscience, New York, 1966.
67. C.M. Dobson, Head of Oxford Centre for Molecular Studies 1998–2002 then Professor in Cambridge 2002– and Master, St John's College, Cambridge. 2007–.
68. P.A. Cox, *Transition Metal Oxides*, Clarendon Press, Oxford, 1995 and several other inorganic chemistry books at an advanced level.
69. G. Wilkinson. The Nobel Prize for Chemistry in 1973 was awarded with E.O. Fischer "for their work on the chemistry of organo-metallic compounds".
70. G.E. Coates, M.L.H. Green and K. Wade (eds.), Vol. II, *The Transition Elements*, by M.L.H. Green, Methuen, London, 1968.
71. G.E. Coates, M.L.H. Green, P. Powell and K. Wade, *Principles of Organometallic Chemistry*, Methuen, London, 1968.
72. J.H. Jones, *The Chemical Synthesis of Peptides*, International Series of Monographs on Chemistry, Oxford University Press, Oxford, 1991.
73. R.E. Richards, born 1922, was educated in Oxford, Dr Lee's Professor of Chemistry 1964–70, knighted, 1977. He was chairman of The Enzyme

Group 1970–1985 Warden of Merton College 1969–84 and Vice-Chancellor of Oxford University 1977–1981.
74. R.A. Dwek, I. D. Campbell, R. E. Richards and R.J.P. Williams (eds), *Nuclear Magnetic Resonance in Biology*, Academic Press, London, 1977.
75. Frederick S. Dainton, (1914–1997), Professor of Physical Chemistry Leeds 1950–1965, Oxford 1970–1973, Vice-Chancellor Nottingham 1965–1970, member of several government committees, knighted 1971, baron 1986, see P. Gray and K. Irvine, *Biog. Mem. Fell. Roy. Soc.*, **46**, 86, (2000).
76. F.S. Dainton, *Chain Reactions*, Methuen, London, 1956.
77. K.A. McLauchlan, *Magnetic Resonance*, Oxford University Press, Oxford 1972.
78. R.P. Wayne, *The Chemistry of Atmospheres*, Clarendon Press, Oxford, 1993 and *Principles and Applications of Photochemistry*, Oxford University Press, Oxford, 1988.
79. W.G. Richards and P.R. Scott, *Structure and Spectra of Atoms*, J. Wiley and Sons, London, 1976. Richards was the Chairman of Chemistry, Oxford, 1996–2007.
80. E.A. Vincent, *Geology and Mineralogy at Oxford*, 1860–1986. Published by the author, Earth Sciences Department, Oxford.
81. G. Ferry, *Dorothy Hodgkin: a Life*, Granta Publications, London, 1998. Nobel Prize for Chemistry in 1964 was awarded to Dorothy Hodgkin for "her determination of the structures of important biochemical molecules", see G. Dodson, *Biog. Mem. Fell. Roy. Soc.*, **48**, 179, (1994).
82. David C. Phillips, (1924–1999), Professor of Biophysics Oxford 1966, knighted 1979, Lord Phillips of Ellesmere, see L. Johnson *Biog. Mem. Fell. Roy. Soc*, **46**, 377, (2000).
83. S. Mann, J. Webb and R.J.P. Williams (eds.), *Biomineralisation*, VCH, Weinheim, Germany, 1991.
84. The Enzyme Group. There is an account of it in the Bodleian Library Special Collections, Shelfmark MSS. Eng. 2662–81. d. 2251.
85. G.K. Radda (1960), Professor of Biochemistry, Oxford (1980), Chairman Medical Research Council, UK.
86. John S. Rowlinson, born 1926, Professor of Chemical Technology (London) 1961–1973, Physical Secretary, The Royal Society 1994–1999, knighted 2000.
87. J.S. Rowlinson and B. Widom, *The Molecular Theory of Capillarity*, International Series of Monographs on Chemistry, Oxford University Press, USA 1982.
88. J.S. Rowlinson, *Cohesion: A Scientific History of Intermolecular Forces*, Cambridge University Press, Cambridge, 2002.
89. J. Simons, born 1934, Professor of Physical Chemistry, Nottingham 1981–1993.
90. J. Klein, born 1949, Senior Scientist Weizmann Institute, Israel 1988–1993.

91. Jack E. Baldwin, born 1938, Professor of Organic Chemistry, King's College London, 1972, Massachusetts Institute of Technology, 1972–1978.
92. M.L.H. Green, born 1936, Professor of Inorganic Chemistry Oxford, 1986–2000.
93. R.J.P. Williams, born 1926, Napier Royal Society Research Professor 1975–1991
94. J.J.R. Fraústo da Silva and R.J.P. Williams, *The Biological Chemistry of The Elements*, Clarendon Press, Oxford 2nd edn 2001, (1st edn 1991).
95. R.J.P. Williams and J.J.R. Fraústo da Silva, *The Natural Selection of The Chemical Elements*, Oxford University Press, Oxford, 1996.
96. R.J.P. Williams and J.J.R. Fraústo da Silva, *Bringing Chemistry to Life*, Oxford University Press, Oxford, 1999.
97. R.J.P. Williams and J.J.R. Fraústo da Silva, *The Chemistry of Evolution*, Elsevier, Amsterdam, 2006.
98. P.P. Edwards and C.N.R. Rao (eds), *The Metallic and Non-metallic States of Matter*, Taylor and Francis, London 1985.
99. P.P. Edwards and C.N. Rao (eds), *Metal-Insulator Transitions Revisited*, Taylor and Francis, London, 1995.
100. P.P. Edwards, born 1949, Professor of Chemistry, Birmingham, 1991–1999.
101. Oxford University website is www.ox.ac.uk and that for chemistry is www.chem.ox.ac.uk
102. M.G. Ord and L.A. Stocken, *The Oxford Biochemistry Department: Its History and Activities 1920–1985*, Joshua Associates Ltd., Oxford, 1990.
103. R. Curtis, C. Leith, J. Nall and J. Jones, *The Dyson Perrins Laboratory and Oxford Organic Chemistry 1916–2004,* Published by John Jones, Balliol College, Oxford, 2008.

Index of Names

Abraham, D.B., 258, 266
Abraham, E.P., 142, 188–9, 254
Acheson, R.M., 286
Achmatowicz, O., 158
Acland, H.W., 89–96
Adams, W., 65–6
Adlington, R.M., 267, 269
Airy, G.B., 104
Akers, W.A., 175, 177
Albery, W.J., 253, 256
Alcock, J., 63
Alcock, N., 57–9, 63–4, 68
Aldrich, G.O., 79–80
Aldrich, H., 62
Allen, P., 175–8
Ampère, A.-M., 86, 114
Anderson, H.L., 268–9
Anderson, J.S., 201, 207, 220–2, 230, 240, 244–56, 289
Anderson, T., 103
Angel, A., 138
Appel, H., 163
Applebey, M.P., 138, 142, 147, 152, 164, 168, 176–7
Arden, J., 56
Aristotle, 4, 6, 19–22, 29–30, 33–7
Armstrong, F.A., 265, 268, 272
Armstrong, H.E., 108, 112, 133
Arndt, F., 162
Arrhenius, S.A., 114, 116
Ashmole, E., 17–8, 40–2, 46
Atkins, P.W., 221, 255
Atterbury, L., 62

Aubrey, J., 23
Austin, W., 57, 66, 68, 71
Avogadro, A., 85, 114

Babbage, C., 87, 90
Bacon, F., Lord Verulam, 22
Bacon, R., 17
Badger, R.M., 154
Baeyer, A. von, 112, 133, 148
Bailey, A.S., 200, 242
Bain, C.D., 266
Baker, H.B., 112, 117–8, 120, 133, 136
Baker, W., 132, 142, 158, 161, 169, 172, 188–9
Baldwin, J.E., 203, 207, 250, 260, 267–9, 291
Bamberger, E., 148
Barker, T.V., 122, 138
Barltrop, J.A., 188, 242
Barratt, S., 175
Barrow, R.F., 231, 243
Bartholin, C., 39
Bartholin, T., 39
Barton, D.H.R., 235
Bathhurst, R., 23, 26
Battle, P.D., 271–2
Baumé, A., 60
Bayley, C., 59
Bayley, F., 59, 64, 70
Bayley, J.H.P., 207, 266, 272, 284
Baynes, R.E., 119
Beattie, I.R., 244
Beddoes, T., 57, 65–71, 79

Index of Names 293

Beer, P.D., 272
Bel, *see* Le Bel
Bell, R.P., 132, 141, 150–5, 191, 200, 219, 227–8, 231, 243–5, 253–4
Berkeley, 8th Earl of, 122, 168
Bernal, J.D., 167, 174–5, 188, 256
Berthelot, P.E.M., 115
Berthollet, C.L., 114
Berzelius, J.J., 7, 99
Binnie, M., 227
Biot, J.B., 99
Birch, A.J., 161, 233
Black, J., 54–5, 60, 65, 68, 71
Blackett, P.M.S., *later* Lord, 167
Blackstone, W., 63
Boerhaave, H., 54, 60, 63, 66, 82, 114
Boltzmann, L., 119
Boole, G., 103
Boscovich, R.J., 86
Boulton, M., 69
Bourdillon, R.B., 139
Bourne, R., 57–8, 70–1, 80–1
Bowell, J., 28
Bowen, E.J., 132, 139, 146–53, 163, 191, 200, 219, 227, 231, 243
Bowman, H.L., 173–4
Bowra, C.M., 229
Boyle, R., 6–7, 22–44
Bradley, J., 45, 57–8
Bragg, W.H., 174–5
Brande, W., 82–4, 97
Braterman, P.D., 248
Bredig, G., 148
Bredt, J., 110
Brewer, F.M., 168, 174, 191, 200, 228, 239, 279
Brewster, D., 87–8
Bridgewater, 8th Earl of, 83
Brockway, L.O., 155
Brodie, B.C., jr, 89, 96–107, 113, 115, 118, 269
Brodsky, A., 159
Brønsted, J.N., 153–4
Brongniart, A., 81
Brouard, M., 266
Brown, B.R., 242, 262

Brown, J. Michael, 236, 252, 254, 268–9
Brown, J. Milton, 266
Brown, R., 88
Buckingham, A.D., 244, 246
Buckland, W., 82, 84–94, 98
Bugge, T., 58
Bulley, F., 104
Bunn, C.W., 177–8, 189
Bunsen, R.W., 92, 116
Burn, P.L., 268–9, 284
Burrows, G., 104
Busby, R., 61–2

Caldin, E.F.H., 142
Campbell, I.D., 254, 259–62, 265, 271, 286
Camper, P., 63
Cannizzaro, S., 101
Caroline, Queen, 62
Carpenter, H.C.H., 169
Carrington, A., 207, 266
'Carroll, L.', *see* Dodgson, C.L.
Chain, E.B., 188–9
Chambers, E., 54
Chapman, D.L., 120, 132, 138, 144, 149, 156, 167, 192
Chapman, M., 120, 149
Chaptal, J.A.C., 60
Charles I, King, 23–4
Chattaway, F.D., 111, 119, 137, 143, 148, 157–8, 162, 188, 200
Chaucer, G., 18, 20
Cheetham, A.K., 257, 289
Cherwell, Viscount, *see* Lindemann, F.A.
Child, M.S., 207, 258, 266–7
Christie, M.I., 213, 244
Clarke, S.J., 271–2
Clary, D.C., 266
Clemo, G.R., 138, 158
Clerk, J. 26–30, 42
Clifton, R.B., 133
Clusius, K., 152, 154, 192
Cocksedge, H.E., 176–7
Coles, B.A., 242, 266

Compton, R.G., 266, 284
Comte, A., 92, 101
Conant, J.B., 172
Conroy, J., 118
Conyers, [W?], 40
Cooper, S.R., 251, 272
Cornforth, J.W., 161, 188, 233–4, 254
Cornforth, R., 188
Cotton, F.A., 250
Cox, P.A., 251, 272
Coxon, J.M., 253
Cronshaw, G.B., 119
Crookes, W., 100
Crosse, J., 25, 28–32
Crowfoot, *see* Hodgkin, D.M.C.
Crowfoot, J.W., 174
Crowther, J.G., 167
Cullen, W., 54–5, 59, 65
Curzon, Lord, 133, 140

Dainton, F.S., *later* Lord, 143, 168, 171, 207, 255, 263, 290
Dallowe, T., 54
Dalton, J., 44, 66, 80, 84–8, 99, 109
Danby, C.J., 191, 227–8, 242, 254
Daniell, J.F., 114
Darwin, C.R., 90, 96
Daubeny, C.G.B., 44, 83–98, 113–4, 117, 121
Davies, S.G., 207, 252, 268–9, 284
Davies, W., 158
Davis, B.G., 268–9, 284
Davis, J.J., 26, 268, 272, 284
Davoud, J.G., 191
Davy, H., 55, 69, 82, 84, 87
Day, A.C., 254
Day, P., 238–51, 270
Day, W., 28
Dean, A.C.R., 242, 256, 266
Deane, T., 95
Debye, P.J.W., 149, 155, 172, 192
Democritus, 23
Demuth, W.H.H., 176
Denning, R.G., 250, 272
Dennis, L.M., 168, 171
Derby, 14th Earl of, 96

Devonshire, 7th Duke of, 107
Dewar, M.J.S., 171, 188, 233, 245
Dickens, P.D., 244
Dilworth, J.R., 268, 272
Dixon, H.B., 109, 116–20, 133–4
d'Leny, W., 177
Dobson, C.M., 251, 259–61, 271–2, 284, 289
Dobson, G., 257
Dodgson, C.L., 115
Donkin, W. Fishburn, 102, 110
Donkin, W. Frederick, 107, 110
Donnan, F.G., 178
Donohoe, T.J., 267, 269
Downs, A.J., 252, 270, 272
Doye, J.P.K., 266–7
Dulong, P.L., 85, 114
Dumas, J.B., 97, 114
Duncan, P., 89
Dundas, H., *later* 1st Viscount Melville, 69–70
Dwek, R.A., 254, 259–61, 265, 268
Dwight, J., 49
Dyson Perrins, C.W., 135

Edwards, P.P., 271–2, 291
Egdell, R.G., 265, 271–2
Egerton, A.C.G., 170
Egleton, F. & W., 28
Einstein, A., 172
Eland, J.H.D., 254, 266
Elford, P., 114
Eliot, T., 71
Emeleus, H.J., 220, 248
Esson, E., 110
Esson, W., 85, 115–6
Eucken, A., 154
Everett, D.H., 191
Eyring, H.J., 228

Fairbanks, H.J., 269, 284
Faraday, M., 83, 88, 101, 104, 114
Farnell, L.R., 165–6
Fawcett, E.W., 177
Federov, E.F., 122
Fell, M., 25

Index of Names

Fell, S., 24
Fender, B.E.F., 248, 251
Ferguson, F., 286
Fieser, L.F., 159
Fischer, E.O., 252, 289
Fischer, H.E., 112
Fisher, W.W., 104, 107, 110, 114
Fleet, G.W.J., 268–9
Fleming, A., 188
Florey, H.W., *later* Lord, 188
Foord, J.S., 266
Forbes, J.D., 88
Fordyce, G., 56
Fourcroy, A.F., 60, 71
Fowler, A., 154
Fox, J., 68
Frankland, E., 99, 112
Fraústo da Silva, J.J.R., 271
Frederick, Prince of Wales, 64
Freeman, R., 254, 256, 261, 266
Freeth, F.A., 175–7
Freind, J., 57, 62, 64, 66, 266
Frewin, R., 57–64, 72
Froude, W., 85
Fry, S.M., 174–5
Fulkes, J., 28

Galilei, G., 21, 30
Garden, A., 83
Gardner, J., 110
Garrod, A., 147
Gasser, R.P.H., 255
Gatty, O., 151, 154
Gaut, G., 175
Gay-Lussac, J.L., 84–5, 114
George, H.J., 138, 149
Gerhardt, C.F., 97, 99
Gibbon, E., 60, 143
Gibbs, J.W., 115
Gibson, R.O., 177
Gilbert, D., 65
Gilbert, J.H., 87
Goldsworthy, L.J., 242
Goodenough, J.B., 207, 222, 248, 250–1, 289
Gordon, K., 176–7

Gordon, R.G., 207
Gouverneur, V., 268–9
Gouy, G., 120
Graham, G., 45
Graham, T., 85
Greatorex, R., 31, 49
Green, J.C., 255, 265, 270, 272
Green, M.L.H., 203, 207, 222, 244, 249–52, 268, 270–2, 284, 291
Gregory, D., 62
Griffiths, J.H.E., 257
Grignard, V., 160
Grimm, J., 58
Groth, P.H.R. von, 121, 151
Guericke, O. von, 30
Guldberg, C.M., 116
Gulland, J.M., 158
Guyton de Morveau, L.B., 68

Haber, F., 154
Hall, F.C., 137, 162
Halley, E., 44–5
Halsall, T.G., 242
Hammick, D. Ll., 132, 137, 142–3, 155, 157, 161–2, 200, 219, 231, 242, 258
Hamnett, A., 251, 265
Hancock, G., 255, 265–6, 284
Hannes, E., 43, 57, 61–2
Harcourt, A.G.V., 104, 107–20, 152
Harcourt, R.M., 110
Harcourt, W.V., 87–8
Harris, J., 117
Hartley, E.G.J., 121–2, 137, 139, 168
Hartley, H.B., 118–9, 122, 139, 142–7, 150–5, 176, 200
Hartlib, S., 24
Harvey, W., 21, 34
'Hasolle, J.', *see* Ashmole
Hassel, O., 288
Hatchett, C., 70
Haüy, R.J., 81
Hauksbee, F., 60
Haworth, R.D., 158
Helmont, J.B. van, 19–22, 33–8, 42, 66
Henry, T., 56
Heyes, S.J., 272

Higgins, B., 56, 66–8
Higgins, T., 66
Higgins, W., 57, 65–7
Hill, H.A.O., 207, 249–50, 265, 268, 271–2, 284
Hinds, S., 90
Hinshelwood, C.N., 9–10, 132, 138, 143–56, 164, 167, 173–9, 187–92, 199–207, 218–33, 237–40, 245, 247, 279
Hirsch, P.B., 248
Hitchins, A.F.R., 166
Hittorf, J.W., 114
Hodgkin, D.M.C., 132, 141, 163–4, 173–9, 187–92, 199–201, 206–7, 218–33, 237–40, 245, 247, 279
Hodgkin, T., 175, 245
Hodgson, D.M., 268–9
Hoff, J.H. van 't, 105, 110, 114
Hofmann, A.W., 98, 100
Holroyd, G.W.F., 112
Hooke, R., 23–6, 29–40, 43–4, 61
Hope, E., 138, 142, 157–8, 161, 200, 219, 241
Hope, T., 84
Hore, P.J., 265–6
Hornsby, T., 57–8, 70
Howard, B.J., 266
Hughes, T., 57, 63
Hume-Rothery, W., 132, 164, 168–70, 173, 178, 200, 228, 240, 245, 248
Hutchison, W.K., 142, 175
Hutton, J., 65
Huxley, T.H., 96, 104

Imori, S., 166
Ing, H.R., 158
Ingold, C.K., 136, 160–3, 172, 178, 232
Irving, H.N.M.H., 141, 148, 200, 220, 228, 243, 248

Jackson, C., 65, 70
James, E.J.F., *later* Lord, 148
Jameson, R., 84
Jenkins, L., 120
Jervis-Smith, F.J., 117

Jeune, F., 89–90, 94
Johnson, S., 65
Jones, E.R.H., 199, 201, 207, 229, 235–6, 242, 247, 253–4, 258, 260, 267, 288
Jones, J.H., 253, 269
Jouguet, É., 120
Jowett, B., 90, 109

Kearton, C.F., *later* Lord, 168, 175–6
Keill, James, 62
Keill, John, 62
Keir, J., 60, 67
Kekulé, A., 101, 105, 110
Kelvin, Lord, *see* Thomson, W.
Kenyon, J., 138
Kermack, W.O., 138
Kerr, R., 60, 71
Kidd, J., 57, 67, 80–90
King, F.E., 132, 141, 161–2, 188, 200, 233, 242
Kirwan, R., 66, 81
Kistiakowsky, G.B., 168
Klein, J., 207, 264, 266, 290
Knowles, J.R., 221, 253, 258–9
Koepfli, J.B., 159
Krebs, H.A., 274

Ladenburg, A., 112
Laidler, K.J., 143
Lambert, B., 9, 105, 110–3, 122, 139, 147, 164, 168–70, 191, 200, 228, 231, 237–8
Lambert, J.D., 152–5, 187, 191, 200, 231, 255
Lane brothers, 43
Lankester, E.R., 104
Lapworth, A., 158
Latimer, W.M., 111
Laurent, A., 83, 99
Lavater, J.R., 57, 62
Lavoisier, A.L., 7, 34, 45, 53, 55, 60, 71, 80–1, 84
Lawes, J.B., 87
Lawrence, T., 63
Lawson, D.R., 177

Index of Names

Leacock, S., 142
Le Bel, J.A., 105
Lee, M., 64, 81, 113
Leech, M., 28
Lemery, N., 60, 62
Leny, *see* d'Leny
Lewis, G.N., 171
Lewis, W., 56
Liddell, H.G., 85, 90, 104
Liebig, J., 85, 87, 92, 97
Lindemann, F.A., *later* Viscount Cherwell, 169–70, 172–4, 288
Linnett, J.W., 168–9, 189–92, 200, 219, 228, 239–40, 243–4, 248
Liveing, G.D., 134
Locke, J., 24, 26, 61
Logan, D.E., 207, 266–7
Longuet-Higgins, H.C., 221, 244–5
Louis XVI, King, 69
Lowe, G., 207, 221, 253, 259, 269, 284
Lower, R., 35, 39
Lowth, R., 63
Lucretius, 23
Lunardi, V., 68
Lutyens, W., 177
Lydell, J., 23
Lyell, C., 90

Macclesfield, 2nd Earl of, 57
McLauchlan, K.A., 255, 265–6
Macquer, P.J., 60
Madan, H.G., 104
Madden, P.A., 266–7
Manley, J.J., 117–8
Mann, S., 270
Manolopoulos, D.E., 266
March, N.H., 207, 258, 266
Marcus, R.A., 207
Marsh, J.E., 105, 108–10, 114, 137
Marsh, J.K., 166
Maskelyne, *see* Story Maskelyne
Mason, F.A., 138
Maxwell, J.C., 103
Mayow, J., 34–5, 40, 42–4
Meakins, G.D., 242
Mendeleev, D.I., 103, 105

Meyer, V., 112
Michael, A., 160
Michels, A.A.M.F., 177
Miers, H.A., 121–2, 134
Miller, W.A., 107
Mingos, D.M.P., 250, 289
Misciatelli, P., 166
Mitscherlich, E.A., 114
Moelwyn-Hughes, E.A., 152
Moffitt, W.E., 244
Moloney, M.G., 267, 269, 284
Moore, T.S., 110–1, 113
Moseley, H.G.J., 118–9, 187
Mountbatten, 1st Earl, 187
Mountford, P., 270, 272
Moyes, H., 56
Murchison, R.I., 88

Nagel, D.H., 114, 118, 151
Nernst, H.W., 120
Neumann, C., 60
Newman, J.H., 87
Newton, I., 54
Nicholls, F., 63
Nicholson, W., 60
Norman, R.O.C., 221, 242, 253
Norrish, R.G.W., 153
North, F., *styled* Lord, 69–70
Noyes, W.A., 191
Nuffield, Viscount, 157, 190

Odling, W., 8, 99–113, 117–8, 123, 133–4.
Ogston, A.G., 171
O'Hare, D.M., 270–2
Orchard, A.F., 250, 255, 265, 272
Orgel, L.E., 221, 245, 247
Orr-Ewing, J., 175
Ostwald, F.W., 101, 114, 119, 171

Paracelsus, 20–2, 35–7
Parkes, G.D., 141, 148, 179, 200
Parsons, J., 57–8, 65–6
Parsons, T.R., 174
Pasteur, L.J., 37
Paton, W.B., 256

Pattison, M., 90
Pauling, L.C., 155, 163–4, 245
Peach, J.M., 269
Pechmann, H. von, 171
Pegge, C., 83
Penfold, J., 265
Percival, J., 109
Perkin, W.H., jr, 8–11, 112, 123, 132–48, 157–66, 169–71, 176, 178, 200, 207, 210, 213, 232–5, 247, 253
Perkin, W.H., sen., 112
Perrin, M., 175, 177
Perrins, *see* Dyson Perrins
Perry, J., 112
Perutz, M.F., 175
Peterborough, 3rd Earl of, 62
Petit, A.T., 85, 114
Petty, W., 23–4
Pfaundler, L., 115
Philip, J.C., 136
Phillips, C.S.G., 200, 221, 239, 243–4
Phillips, D.C., *later* Lord, 257, 259–60, 290
Phillips, J., 87, 96
Pilling, M.J., 266
Planck, M.K.E.L., 154
Plant, S.G.P., 158, 161, 200, 241–2
Plot, R., 17, 40, 42, 57
Plumptre, F.C., 90
Polgar, N., 200, 242
Pope, W.J., 123, 132, 134, 136, 178
Porter, M.W., 122
Porter, R.R., 260
Portland, 3rd Duke of, 79
Pott, P., 66
Potter, W., 28
Poulton, E.B., 133, 141
Powell, B., 85, 89–90, 94, 96, 98
Powell, H.M., 132, 170, 173–5, 178, 191, 200, 207, 219–20, 231, 237, 244, 256–8
Prelog, V., 288
Price, J., 69
Price, W.C., 265
Prichard, C.R., 177
Priestley, J., 34, 45, 54–6, 60, 68

Pringle, J.W.S., 257
Prout, C.K., 256–7, 272
Prout, W., 83
Pusey, E.B., 85, 94

Radda, G.K., 253, 259, 286, 290
Raikes, H.R., 139, 146, 150–2, 200
Ramsay, W., 108, 133
Raynor, G.V., 170
Richards, R.E., 189, 200–1, 207, 231, 240, 243, 247, 254–65, 289–90
Richards, T.W., 152–3
Richards, W.G., 266–7, 282, 284
Rideal, E.K., 172
Rigaud, S.P., 71, 87, 89
Riley, D.P., 175
Ripley, G., 18, 20, 35
Robertson, J., 267, 269
Robinson, C.V., 269
Robinson, G.M., 233
Robinson, M.J., 160, 254
Robinson, R., 8–11, 155–67, 173–8, 187–91, 199–207, 218–9, 225–6, 229–36, 247, 257
Rodebush, W.H., 111
Rollett, J.S., 267
Roscoe, H.E., 107, 117
Rose, J.D., 175
Rosseinsky, M.J., 272
Rossotti, F.J.C., 240, 244, 249, 272
Rouelle, G.F., 66
Rowlinson, J.S., 207, 221, 263–4, 266, 290
Ruskin, J., 95–6
Russell, A.S., 146, 149, 165
Russell, J., 90
Rutherford, E., *later* Lord, 118, 145
Rydon, H.N., 161

Sadler, J., 57, 68–9, 71
Sanger, F., 234
Savile, H., 40
Scheele, C.W., 7
Schiebold, E., 173
Schofield, C.J., 268–9, 284
Schweppe, J.J., 69

Index of Names

Scott, W., *later* Baron Stowell, 65
Selbourne, 1st Earl of, 108
Semenov, N.N., 288
Seward, M., 109
Shaw, P., 54, 56, 60
Shuchirch, W., 18
Sibthorp, J., 68–9
Sidgwick, N.V., 9, 109–13, 132, 137,
    142–6, 152, 155–7, 162–73, 192,
    199–228, 233, 236–8, 247
Siegbahn, K.M.G., 265
Simons, J.P., 207, 264, 266, 290
Simpson, C.J.S.M., 265–6
Smith, E.B., 222, 255, 266, 288
Smith, F.J., *see* Jervis-Smith
Smith, G., 90
Smith, H.J.S., 98–9, 102, 104, 107, 118
Smith, J., 57–8, 64, 66
Smith, J.C., 112, 161, 188, 191, 200, 242
Smith, L.J., 272
Smithson, J., 65
Soddy, F., 108–12, 145–9, 154–7,
    164–73, 178, 189, 200, 207, 226,
    228, 238, 279
Softley, T.P, 266
Springall, H.D., 161
Stacy, H.P., 70–1
Stahl, G.E., 54, 66
Stahl, P., 25–6, 28, 42
Stamp, J.C., *later* Lord, 151
Stanley, A.P., 90, 94
Staveley, L.A.K., 152–5, 189–92,
    200, 219, 228, 243
Steel, J.L.S., 176–7
Stevens, T.S., 158
Stock, J., 68
Stokes, G.G., 103
Story Maskelyne, M.H.N., 85–6, 92,
    94, 98, 113, 121
Stowell, Baron, *see* Scott
Strauss, F.B., 163, 234, 242
Sutherland, G.B.B.M., 189
Sutton, L.E., 132, 141, 143, 152, 155,
    162–3, 172, 189–91, 200, 219,
    227, 231, 242, 245, 255
Swallow, J.C., 177

Tait, A.C., 85, 90
Taylor, T.W.J., 152, 161–2, 172, 179,
    187–8, 200, 219, 242
Taylor, W., 28
Thaller, V., 242
Thatcher, M.H., *later* Lady, 202
Thomas, J., 63
Thomas, R.K., 255, 265–6
Thompson, H.W., 132, 143–4, 152–5,
    164, 167–8, 176, 189–91, 200, 207,
    219, 228–9, 239–42, 255, 262
Thomson, T., 85
Thomson, W., *later* Lord Kelvin, 65, 68
Thomson, W., Archbishop, 85
Thorpe, J.F., 136
Tilden, W.A., 108
Tillyard, A., 28–9
Timmel, C.R., 265, 270
Tizard, H.T., 109, 111–2, 170
Todd, A.R., *later* Lord, 161, 164,
    233–4, 254
Tomlinson, M.L., 213, 242
Toone, S., 28, 30
Topley, B., 175
Torricelli, E., 30
Townsend, J.S.E., 133–4
Trikojus, V.M., 159
Turner, D.W., 207, 254, 265–6
Turner, E., 88
Tyndall, J., 97, 116

Uffenbach, Z.C. von, 41, 44–5, 60–2
Usherwood, H., 136

Van 't Hoff, *see* Hoff
Veley, V.H., 108–9, 114, 118
Venanzi, J.M., 244, 250, 252
Vernon Harcourt, *see* Harcourt, A.G.V.
Verulam, Lord, *see* Bacon, F.
Vesalius, A., 39
Vigani, J.F., 62
Vincent, W., 80

Waage, P., 116
Walden, A.F., 109, 111, 113, 119, 152,
    168

Walker, A., 56
Walker, Richard, 70
Walker, Robert, 82, 89
Wall, J., 65
Wall, M., 57–8, 62, 65–8, 71–2
Wallace, M.I., 266
Walls, E., 175
Waltire, J., 56
Ward, J., 30, 38–40, 42
Warrell, J.E., 150
Warren, H., 134–5
Waterhouse, P., 134
Waters, W.A., 207, 221, 242, 253
Watkin, D.J., 257, 272
Watson, R., 60, 72
Watt, J., 69
Watts, J., 108, 110–1, 113–4
Wayne, R.P., 86, 255, 266
Wayneflete, William of, 103
Wedgwood, J., 60
Weiler, G., 163, 234
Weissberger, A., 162
Weizmann, C., 136
Wellington, 1st Duke of, 90
Wells, A.F., 174
Wheatstone, C., 90
White, C., jr, 51, 61
White, C., sen., 48, 61
White, J.W., 255, 265–6
Whiteside, J., 44, 57
Whitham, G.H., 236, 254
Whiting, M.C., 242
Whyte, C., see White, C.
Widom, B., 207, 264
Wilberforce, S., 96
Wilden, W., 28, 42

Wilhelmy, L., 115
Wilkins, J., 6–7, 24, 34
Wilkinson, G., 242, 250, 252, 270
Williams, D., 94
Williams, G., 81
Williams, R.J.P., 200, 221–2, 239–44, 248–9, 258–9, 268, 270–2, 291
Williams, S.V.B., 286
Williamson, A.T., 152
Williamson, A.W., 99–101, 103
Willis, T., 23–9, 35–44
Willstätter, R., 112, 151, 163
Wilson, D.R., 118
Wilson, E.B., 168
Wilson, G., 59
Winmill, T.F., 111
Wolfenden, J.F., 142, 149–51, 191, 200, 227, 231
Wollaston, W.H., 60, 83
Wong, L.L., 268, 270, 272, 284
Wood, A., 25–6
Woodforde, W., 63
Woodward, B., 95
Woodward, L.A., 149, 152, 155, 192, 200, 219, 228, 243
Woodward, R.B., 164, 232, 234, 242, 253
Wren, C., 23
Wüthrich, K., 260
Wurtz, A.C., 99
Wyndham, T.H.G., 104, 107, 110
Wynne-Jones, W.F.K., *later* Lord, 153

Xiao, T., 284

Young, G.T., 242, 253
Yule, C.J.F., 117

Index of Subjects

Abbot's Kitchen, 7, 12, 95–9, 105–6, 110, 164–5, 275–8
acetone, 137
acetylene, 235, 253
acid-base catalysis, 152–3, 219
acoustics, 91
agriculture, 64, 82, 86–7, 96
Air Board, 137
air pump, 22, 29–38, 44
Air Raid Precautions (ARP), 191
Albright and Wilson, 175
alchemy, 17–20
Alembic Club, 112–3, 118, 172–3, 175, 221
alkaloids, 134–5, 158, 160, 232
alloys, 67, 169–70, 243, 245, 271
anaesthesia, 116, 256
anatomy, 7, 41, 52, 61, 64, 80
Angel Inn, 25, 28
anthocyanins, 232
antibiotics, *see also* drugs, 188–9, 254, 267
apothecaries, 21, 26–30, 69–70, 72, 82
architects, 23, 94–5, 134–5, 166–7, 278–80
Armourers' and Braziers' Co., 169
Ashmolean Museum, *see* Museum
Ashmolean Society, 87, 89
astronomy, 40, 42, 57, 91
  Savilian Professor of, 40, 57, 78, 110
atomic bomb, 149, 192
  number, 118
  symbols, 86, 88
  theory, 22–3, 66, 80, 84–9, 99–101, 108

Auvergne, 84
aviation fuel, 191
Avogadro's hypothesis, 85, 101, 114

Bachelor of Arts (BA), 52, 56, 107
  of Natural Science (BNS), 107
  of Science (BSc), 99, 109, 122, 141
bacterial kinetics, 152, 192, 225, 228–9, 256
balloon, hot air, 68–9
Barclay's bank, 135
Barker index of crystals, 122
Beam Hall, 25
Bedford lecturer, 116, 154
biochemistry, *see also* chemistry, biological, 34–5, 123, 174
Biochemistry department, 12, 253, 258–61, 265, 272, 286
biophysics, 256–8
bio-sensors, 268, 272
Birch reduction, 161
blood, 21, 32–5, 37–8, 44, 67
Boake Roberts and Co., 138
Boards of Faculties, 106
Bodleian library, *see also* Radcliffe Science Library, 6, 18, 69
Bostar Hall, 25, 28
Botanical Gardens, 25, 40, 113
botany, 58, 69, 86, 91, 96, 141
  Sheradian Professor of, 69
Boyle's law, 35
brewing, 36, 53
Bridgewater treatises, 83

Bristol, 84, 88
British Anti-Lewisite (BAL), 191
British Association for the Advancement of Science (BAAS), 84–9, 94, 96, 102–3, 107, 115, 117, 119
British Dyes, 138, 157–8
British Dyestuffs Corporation, 135, 157, 162, 176
British Museum, 121
Bruker, 259
Brunner Mond, 138, 175–6
Bush, W.J., and Co., 138
'Butler' education act (1944), 213–6

caloric, *see* heat
Cambridge Philosophical Society, 89
 University, 2, 58–9, 72, 86, 92–3, 112, 123, 132, 144, 178, 214, 234
capillarity, 264
carbon atom, 102, 105
catalysis, 244, 252, 254, 270, 272, 284
cephalosporin, 254
ceramics, 67
Chairman of Chemistry, 203, 212, 269, 273
Chancellor, 4–5, 79, 90, 96, 104, 133
chaperones, 109
Chapman–Jouguet layer, 120
charcoal, 33, 190–1
Chatham House, 191
chemical biology, *see* chemistry, biological
chemical calculus, Brodie's, 101–3
Chemical Club, 112
Chemical News, 100, 102–3
Chemical Society, *see also* Royal Society of Chemistry, 89, 96, 101–5, 112, 132, 137, 172–3, 235
Chemical Warfare, Brigade and Committee, 137
chemistry, inorganic, organic, physical, *passim; see also* biochemistry, electrochemistry, photochemistry *and* radiochemistry
 analytical, 96, 117, 220
 atmospheric, 255

 bio-inorganic, 243, 271–2
 biological, 11, 215, 222, 228–9, 234, 240, 246–9, 257, 266–9, 272, 278
 industrial, 137–8, 156–7, 162, 166, 252–3
 medicinal, 21, 37–42, 53, 64, 80
 organo-metallic, 243–4, 250–4, 268–72
 pharmaceutical, 252
 philosophic, 55, 80
 pneumatic, 45, 55, 58
 soft matter, 264
 solid state, 230, 240, 244, 248–51, 266, 271–2
 theoretical, 230, 244–6, 258, 264–7
chloroform inhaler, 116
cholesterol, 220, 234, 257
Christ Church, 24–5, 61–5, 73, 80, 93
chromatography, 239, 243
chronograph, tram, 117
Clarendon building, 6, 93
Classical Association, 227
classics, *see* Literae Humaniores
clathrates, 256
coal tar, 83
colleges, *see also* laboratories, college, 2–5, 13, 141–2, 196, 238–9
combustion, 22, 32–5, 44, 53–5, 80, 192
commissions, parliamentary and royal
 Oxford and Cambridge (1850), 7, 15, 90–4
 'Cleveland' (1871), 108
 'Devonshire' (1873), 7, 107–8
 'Selbourne' (1877), 108
 'Asquith' (1920), 15, 145, 148
Common University Fund (CUF), 10, 108–9
Commonwealth Fund, 168
computer modelling and simulation, 267
Congregation, House of, 3, 5, 80, 93, 96, 107, 148
conscription, 192
constitution of the university, *see* government
Convocation, House of, 80, 84, 89–94, 98–9, 104
cooling mixtures, 70

Index of Subjects 303

cordite, 137
Courtaulds, 175
crucibles, 41, 67–8
crystallography, *see also* mineralogy, 9, 121–2, 134, 173, 219–20, 256–8
  x-ray, 122, 163–4, 170, 173–8, 188–9, 228, 230, 234, 244, 256, 272

Daubeny laboratory, *see* laboratories, college
Deep Hall, 20, 25–6, 29–32
degrees, *see also*, Bachelor, Doctor, *and* Master, 56–8, 88
demonstrators, 7–10, 93, 110–1, 114, 132, 158, 161, 196–7, 208, 226, 285
  Aldrichian, 104, 107, 110, 114, 147
departments, *see* laboratories, university
Department of Explosive Supplies, 137–8
Department of Scientific and Industrial Research (DSIR), 138, 155, 172
dipole moments, 155, 163, 231
Divinity School, 6
Doctor of Civil Law (DCL), 42, 82, 88
  of Philosophy (DPhil), 140–1, 211, 224
  of Science (DSc), 88, 116, 118, 141
drugs, *see also* antibiotics, 21, 136–7, 188–9, 252, 254, 284
dry reactions, 116–7
dyestuffs, 26, 136–7
Dyson Perrins trust, *see* laboratories, university

Edinburgh University, 59, 65, 84
electoral boards, 104, 134, 206–7
Electrical Laboratory, 118
electrochemistry, 119, 151, 219, 256, 265–6, 284
electrolysis, 114
electron diffraction, 155, 163, 231, 249
  microscopy, 248, 257–8
  transfer, 248
electronic theory of organic reactions, 160–1, 173
elements, 19–22, 35–7, 53, 80, 82, 100–3, 172

Encyclopaedia Britannica, 82
energy transfer, 154, 231
Engineering Science, 169
enzymes, 250, 253, 269
Enzyme group, 249, 253–4, 258–61, 268
esters, 115
examinations, 89, 91–3, 100, 106–8, 144, 213–8, 222
  entrance, 215
  examiners, appointment of, 92, 106
  external, 108
Experimental Philosophy, Professor of, 81, 84, 93
  Reader in, 82
explosions and explosives, 111, 116–7, 120, 136–8, 152, 177

Faraday Society, *see also* Royal Society of Chemistry, 132, 172, 229, 235
  Discussions, 132, 154, 191
fellowships, 4–5, 10, 93, 132, 141–2, 146, 152, 157, 161–2, 196–8, 208–11, 226, 261–2, 285
fermentation, 36–7, 43, 63, 188–9
ferrocene, 242
finance, 11, 85, 94, 105, 133–6, 145–7, 149–50, 157, 161, 167, 174, 190–1, 197, 201–3, 211–2, 241, 274, 282–3
Foxcombe, 122, 168
Franks Commission, 15

Gallipolli, 118, 138, 187
gas chromatography *see* chromatography,
  diffusion, 192
  kinetic theory of, 101, 103, 119
  referee, 116
  warfare, 136–7, 139, 190–1
Gas, Light and Coke Co., 151
geology, 65, 81–2, 84, 91, 121
Geometry, Savilian Professor of, 40, 44, 64, 66, 78, 88, 98, 107
German chemistry, 54, 92, 132, 137, 140, 154, 159
  refugee scientists, 162–3, 188

glassblowing, 111, 151, 238
Glastonbury, *see* Abbot's Kitchen
glyco-biology *and* -chemistry, 260, 267–8
Glycoform, 284
glyptic formulae, 102–3
Goldsmiths' Company, 154
Goldwin Smith professorship, 110
Gouy-Chapman layer, 120
government of the chemistry school, 13–4, 203
  of the university, 2–6, 285
'Greats', *see* Literae Humaniores
Greek, *see also* Literae Humaniores, 106–7, 109
Gresham College, 24, 42
gunpowder, 26, 32–3, 44
Guy's hospital, 80, 83, 104–5

haemoglobin, 175
Hallé orchestra, 159
Harcourt scholarship, 174
heat, 22, 33, 53–5, 71,
  specific, 85, 114
Hebdomadal Board, *later* Council, 85, 93, 110, 133, 136, 147–8
Henry fellowship, 168
high pressure reactions, 177–8
Higher Education Funding Council (HEFCE), 203, 263, 283
Hinshelwood lectures, 224
history, *see also* Museum of the History of Science,
  Camden Professor of, 40, 62, 72
History and Law school(s), 109, 215
Hulme Hall, Manchester, 59
hydrogen bond, 111, 245

Imperial Chemical Industries (ICI), 113, 157, 162, 168, 174–6, 188
Imperial and Foreign Corp., 165–6
industrial careers, 106, 175–8
  research, 137–8, 162, 166
Inhibox, 284
Institute of Chemistry, *later* Royal, 105–6, 110, 173, 235

Institute of Metals, 170
Institut von Laue Langevin, 249
insulin, 174, 257
intermolecular forces, 255–6, 264
Irving-Williams series, 243
isomers, 85–6, 89, 100–1, 103
isomorphism, 85–6, 114

Jacobite, *see* Tory
Japan Prize, 251
JEOL, 249
Junior Scientific Club, 112

Karlsruhe congress, 101
Keeper of the Museum, 96
kinetics of reactions, *see* reactions

laboratories, college, 113–23, 145–8, 197
  Balliol, 92, 98, 112–3, 116–7, 122, 275
  Balliol-Trinity, 7–9, 109, 113, 117, 119, 133–4, 145, 147–57, 169, 189, 275
  Christ Church, 6–7, 64, 70, 104, 108, 113–6, 145, 148, 275
  Jesus, 119–20, 138, 145, 149, 189, 225, 275
  Magdalen (Daubeny), 7, 91–2, 96, 98, 113, 117, 145, 148–9, 275–6, 278
  Queen's, 118–9, 145, 148, 275
laboratories, university, *see also*
  Abbot's Kitchen *and* Old Chemistry Department
  Central, 276
  Crystallography, 9, 276
  Inorganic (ICL), 9–10, 189, 199, 204, 220, 225, 228, 237–43, 275–6, 279–80
  Organic *or* Dyson Perrins (DP), 9–10, 12, 135–8, 147–8, 158, 161–4, 171, 187, 189, 225, 275–80
  New (Pharmacology), 12, 275
  Physical (PCL) *now* Physical and Theoretical (PTCL), 9–10, 12, 19, 148, 189–90, 225–7, 231, 275–7, 279
  Research, 269, 275–81

Index of Subjects

Laboratory, The, 102–3
lasers, 255, 265, 284
Latin, *see also* Literae Humaniores, 107, 109
law, *see* history
lectures, 57, 59, 63, 81–4, 89, 98, 107–9, 134, 218–20
 public, 56, 62, 65–7
lecturers, *see also* demonstrators, 10, 57, 93, 196–8, 208–11, 285
Leiden University, 58, 63
Lever Brothers, 175
liquids, 243, 264, 266–7, 270
Literae Humaniores ('Greats'), 6, 81, 84, 91–2, 111, 131, 215
London, Midland and Scottish Rly, 151, 176
London University, King's College, 85
 University College, 85, 178
Lord Mayor, 168, 220
lysozyme, 188, 257, 259

Magdalen College, 103–4, 134
magnetic fields, 265
Manchester University, 8, 117, 133–5, 157–8, 232, 235–6
mass action, law of, 116
mass spectrometer, 238, 268–9
Master of Arts (MA), 57, 59, 63, 65, 80, 93, 99, 107
 of Natural Science (MNS), 107
material science, *see also* metallurgy, 11, 243, 269
mathematics, 82–3, 92, 98, 109, 120, 123, 131, 144, 216–7
 Rouse Ball Professor of, 244, 258
Mathematics and Physics school, 81, 89
mechanical philosophy, *see* physics
Medical Research Council (MRC), 138, 235
medicine, 37–40, 52–5, 58, 69, 82, 98
 Aldrichian Professor of, 71, 79–81
 Lichfield Professor of Clinical, 65–6, 71
 Regius Professor of, 63, 80, 83, 86, 147
Medisense, 284

metallurgy, *see also* material science, 7, 53, 164, 168–70, 178, 228, 243, 277
Millard lecturer, 116
mineralogy, *see also* crystallography, 65, 81–4, 107, 113, 121–3, 170, 178, 277
 Professor of, 121, 134
Ministry of Munitions, 136, 138–9
 of Supply, 189, 191
molecular biology, *see also* biophysics *and* chemistry, biological, 229, 234, 257
Monsanto, 268
Morley laboratories, Manchester, 134–5
mountaineering, 110, 233, 262
Museum, Ashmolean, 6–7, 9, 17, 40–5, 60–3, 67, 83, 86, 89, 91, 93, 98, 275
 University, 7, 9, 12, 89, 91, 93–108, 118, 113, 120
Museum of the History of Science, 40–1, 67
Museum House, 95, 275, 277
music, 159, 233
mustard gas, 136–7

nanochemistry, 270
naphthalene, 83
National Chemical Laboratory, 248
National Union of Scientific Workers, 167
Natural History Society, 89
natural philosophy, *see* physics
 Sedleian Professor of, 40, 57
Natural Sciences Dining Club, 229
Natural Sciences Faculty Board, 121, 145, 147, 159
Natural Sciences school, 86, 91–2, 98, 109
Nature, 119, 221
neutron scattering, 255, 265
nitrogen, 111, 171–2, 236
Nobel Industries, 176
Nobel prizes, 132–3, 139, 146, 149, 152, 158–64, 188–9, 207, 218, 227, 238, 245, 252, 257, 260, 265, 274
North Report, 15

Officina Chimica, *see* Museum, Ashmolean,
Old Chemistry Department, 120–2, 137, 164–71, 178, 189
Opsys, 284
optics and optical properties, 91, 244, 246
osmotic pressure, 122, 168
Oxfam, 245
Oxford Advanced Surfaces, Asymmetry, Biosensors, Catalysts, Molecular, Nanolabs, 284
Oxford Centre for Molecular Sciences (OCMS), 260, 271
Oxford Chemistry Primers, 222
Oxford Instruments, 25
Oxford Medical Diagnostics, 265, 284
Oxford University Press (OUP), 94, 153, 172
oxidation, 267–8
Oxtox, 284
ozone, 103, 120

palaeontology, 82
papain, 253
parliament, *see* commissions
Pathology department, *see* Sir William Dunn School
Pegasus football team, 262
penicillin, 164, 188–9, 220, 234, 257, 267
peptides, 253
periodic classification, 101, 103, 105, 219, 237
pharmacology, *see* also drugs, 12, 82, 158, 217, 256
Pharminox, 284
Philosopher's stone, 37
Philosophical Club, 24
philosophy, *see* Literae Humaniores
Philosophy, Politics and Economics (PPE), 131, 215
phlogiston, 54–5, 66, 71, 80–1
phosgene, 136
photochemistry, 86, 120, 149–53, 219, 264
photography, 98
photometer, 116

Physical Sciences Board, 142, 147, 155, 203, 212
physics, 12, 57, 81, 91–2, 114, 118–9, 122–3, 131, 144, 192, 215, 235
   Wykeham Professor of, 133
physiology, 12, 91
Plessey, 175
Pneumatic Institute, Clifton, 68–9
polymers, 122, 268
polythene, 177–8
porcelain, *see also* ceramics, 65
positivism, 88, 92, 100–1, 109
practical courses and examinations, 60, 85, 92, 98–9, 105, 108–9, 121, 142, 220, 223
Praelector, Aldrichian, 80, 104
Privy Council, 104
proctors, 80, 96, 106
professors of chemistry, 17–18th centuries, 40, 43, 57, 60–3
   Aldrichian, 67, 79–84, 93, 96, 103, 105, 121
   Chemical Biology, 206–7, 266
   Crystallography, 174
   Dr Lee's, 9, 133, 145, 153–6, 164, 206–8, 225, 238, 263–6
   Eastman, 245
   IBM, 207
   Inorganic, 11, 206–7, 230, 248
   personal (ad hominem), 173, 206–7, 285
   Royal Society, 207, 222, 266, 271
   Theoretical (*later* Coulson), 206–8, 258, 266–7
   titular, 207, 285
   Waynflete, 103, 105, 110, 121, 132–5, 206–8, 269
proteins, 161, 174, 249, 256, 259, 267
Psychology laboratory, 12, 277
Public Analyst, 110

quantum mechanics, 169, 217, 230, 244–6
Queensferry Ordnance Factory, 138

radar, 192
Radcliffe Camera, 6

Index of Subjects 307

Radcliffe Infirmary, 58, 65–7, 70–1, 81
Radcliffe Observatory, 58
Radcliffe Science Library, 96–7, 105–6, 165, 220, 222
radiochemistry and radium, 145–6, 149, 164–6
Ramsay fellowship, 166
Randolph Hotel, 113
rare earths, 118
reaction dynamics, kinetics and mechanism, 104, 115–6, 119–20, 152, 154, 163, 192, 218, 228, 240, 243, 253–5, 266–7
readers, 57, 68, 71–3, 80, 285
　Aldrichian, 80
　in Experimental Philosophy, 89
　(Dr) Lee's, anatomy, 57–8, 64–5, 81, 86, 89
　(Dr) Lee's, chemistry, 86, 93, 104, 113, 117, 165
　(Dr) Lee's, physics, 93
　in Mineralogy, 92
Reform act (1832), 87
Registrar, 145, 147
religious tests, 71, 88, 93, 98–9
ReOx, 284
Research Assessment Exercise (RAE), 203, 283
Research Councils, 203, 210, 212, 236, 258, 263
Rex Richards Laboratory, 261, 277
Rhodes scholars, 152, 166, 191, 227
Robbins Report, 15
Robinson lectures, 224
Rockefeller Foundation, 155, 161–2, 188, 233, 257
Rothampsted, 87
Royal College of Chemistry, 85, 98
Royal College of Physicians, 87, 104–5
Royal College of Science, 122
Royal Flying Corps, 139
Royal Institution, 69, 82, 84, 92, 97, 101, 104, 249, 257
Royal Ordnance Factory, *see* Queensferry
Royal School of Mines, 169

Royal Society, 6, 25–6, 34, 42, 45, 61, 87, 102–4, 132, 167, 172, 227–8, 233, 264
　Bakerian lecture, 115, 117
　conversazione, 117
　Research fellows, 198, 211
　Research professors, *see* professors
　Warren fellowship, 170
Royal Society of Chemistry, *see also* Chemical Society, Faraday Society, Institute of Chemistry, *and* Society of Analytical Chemistry, 235
Rural Economy, Professor of, 86, 96
Rutherford Laboratory, 265

safety, 217, 220, 223, 269, 280
St Bartholomew's hospital, 66, 71, 111, 119
St George's hospital, 96, 110–1
salaries, 10, 69, 81, 85, 94, 121, 134, 150, 191, 198
schools, Dulwich College, 117
　grammar, 214
　Westminster, 43, 61, 65
school examinations, 215–6
　teachers, 117, 151, 224
Science Masters' Association, 151
Scottish universities, 58, 112
Sheldonian Theatre, 6
Shell, 235
Sir William Dunn School of Pathology, 188, 254
Social Credit, 164
Society of Analytical Chemistry, 235
Society of Chemical Industry, 233
solutions, 114, 118–9, 122, 152, 168, 243, 253
Special Operations Executive (SOE), 191

spectroscopy, electron spin resonance (ESR), 242, 249, 255, 265–6, 271–2
　infra-red, 153, 163–4, 168, 189, 191, 219, 240, 266
　microwave, 266

nuclear magnetic resonance (NMR), 164, 231, 249, 254, 256–61, 265–6, 272
photo-electron, 255, 265–6, 272
Raman, 149, 155, 192, 243, 250, 272
ultra-violet and visible, 154, 163, 231, 240, 266
spin-out companies, 212, 283–4
statistical mechanics, 119, 246
stereochemistry, 110, 252
steroids, sterols, *see also* cholesterol, 160–1, 174, 187–8, 235
sugars, *see also* glycobiology, 115, 268–9
superphosphate, 87
surfaces, 244, 266, 284
Sutcliffe Speakman, 191
syllabuses, 91, 286

terpenes, 110, 235
thermodynamics, 114–5, 119, 154, 170, 192, 218–9, 243
Thoresby colliery, 79
Tory, 56, 62, 64
Tractarian movement, 87, 95
Tradescant family, 41–2
'Tube Alloys', *see* atomic bomb
tutors and tutorials, 13, 98, 114, 141–5, 175, 201, 218, 223
types, theory of, 99

ultrasonic dispersion, 154, 231

undergraduate course, numbers, 87, 92, 108, 131, 199, 201, 213–6
Part 1, 131, 140, 216–23
Part 2, 131, 140, 142–5, 152, 157, 168, 179, 210–1, 216, 223–4
United Alkali Co., 176
uranium, 192

valency, 101, 103, 171, 219, 236–7
vanadium, 107
Vastox, 284
Vice-Chancellor, 5, 80–1, 90, 96, 104, 106, 147, 165, 196, 263
vitamin B_{12}, 164, 249, 257, 272
volcanoes, 84, 86
von Laue Langevin Institut, see Institut

War Inventions Board, 139
water analysis, 42–3, 53, 59, 63, 69, 86
Welcome Foundation and Trust, 175
Whig, 57, 64, 90
women, admission of, 13, 109–10, 113, 131, 213–6
fellows, 174–5, 213, 244
World War, first, 107, 111, 118, 136–40, 167
second, 107, 187–94

York, 87–8

Zeitschrift für physikalische Chemie, 114
Zoology department, 12, 222, 257
Zyentia, 284